T0341108

Techniques for Designing and Analyzing Algorithms

Chapman & Hall/CRC Cryptography and Network Security Series
Series Editors: Douglas R. Stinson and Jonathan Katz

Secret History: The Story of Cryptology, Second Edition
Craig P. Bauer

Data Science for Mathematicians
Nathan Carter

Discrete Explorations
Craig P. Bauer

Cryptography: Theory and Practice, Fourth Edition
Douglas R. Stinson and Mary P. Paterson

Cryptology: Classical and Modern, Second Edition
Richard Klima and Neil Sigmon

Group Theoretic Cryptography
Maria Isabel Gonzalez Vasco and Rainer Steinwandt

Advances of DNA Computing in Cryptography
Suyel Namasudra and Ganesh Chandra Deka

Mathematical Foundations of Public Key Cryptography
Xiaoyun Wang, Guangwu Xu, Minggiang Wang, Xianmeng Meng

Guide to Pairing-Based Cryptography
Nadia El Mrabet and Marc Joye

Techniques for Designing and Analyzing Algorithms
Douglas R. Stinson

https://www.crcpress.com/Chapman--HallCRC-Cryptography-and-Network-Security-Series/book-series/CHCRYNETSEC

Techniques for Designing and Analyzing Algorithms

Douglas R. Stinson

CRC Press is an imprint of the
Taylor & Francis Group, an **informa** business

First edition published 2022
by CRC Press
6000 Broken Sound Parkway NW, Suite 300, Boca Raton, FL 33487-2742

and by CRC Press
2 Park Square, Milton Park, Abingdon, Oxon, OX14 4RN

CRC Press is an imprint of Taylor & Francis Group, LLC

Library of Congress Cataloging-in-Publication Data

Names: Stinson, Douglas R. (Douglas Robert), 1956- author.
Title: Techniques for designing and analyzing algorithms / Douglas R.
Stinson.
Description: First edition. | Boca Raton : C&H\CRC Press, 2021. | Includes
bibliographical references and index.
Identifiers: LCCN 2021000962 (print) | LCCN 2021000963 (ebook) | ISBN
9780367228897 (hardback) | ISBN 9780429277412 (ebook)
Subjects: LCSH: Algorithms--Textbooks.
Classification: LCC QA9.58 .S76 2021 (print) | LCC QA9.58 (ebook) | DDC
518/.1--dc23
LC record available at https://lccn.loc.gov/2021000962
LC ebook record available at https://lccn.loc.gov/2021000963

ISBN: 9780367228897 (hbk)
ISBN: 9781032024103 (pbk)
ISBN: 9780429277412 (ebk)

DOI: 10.1201/9780429277412

Publisher's note: This book has been prepared from camera-ready copy provided by the authors.

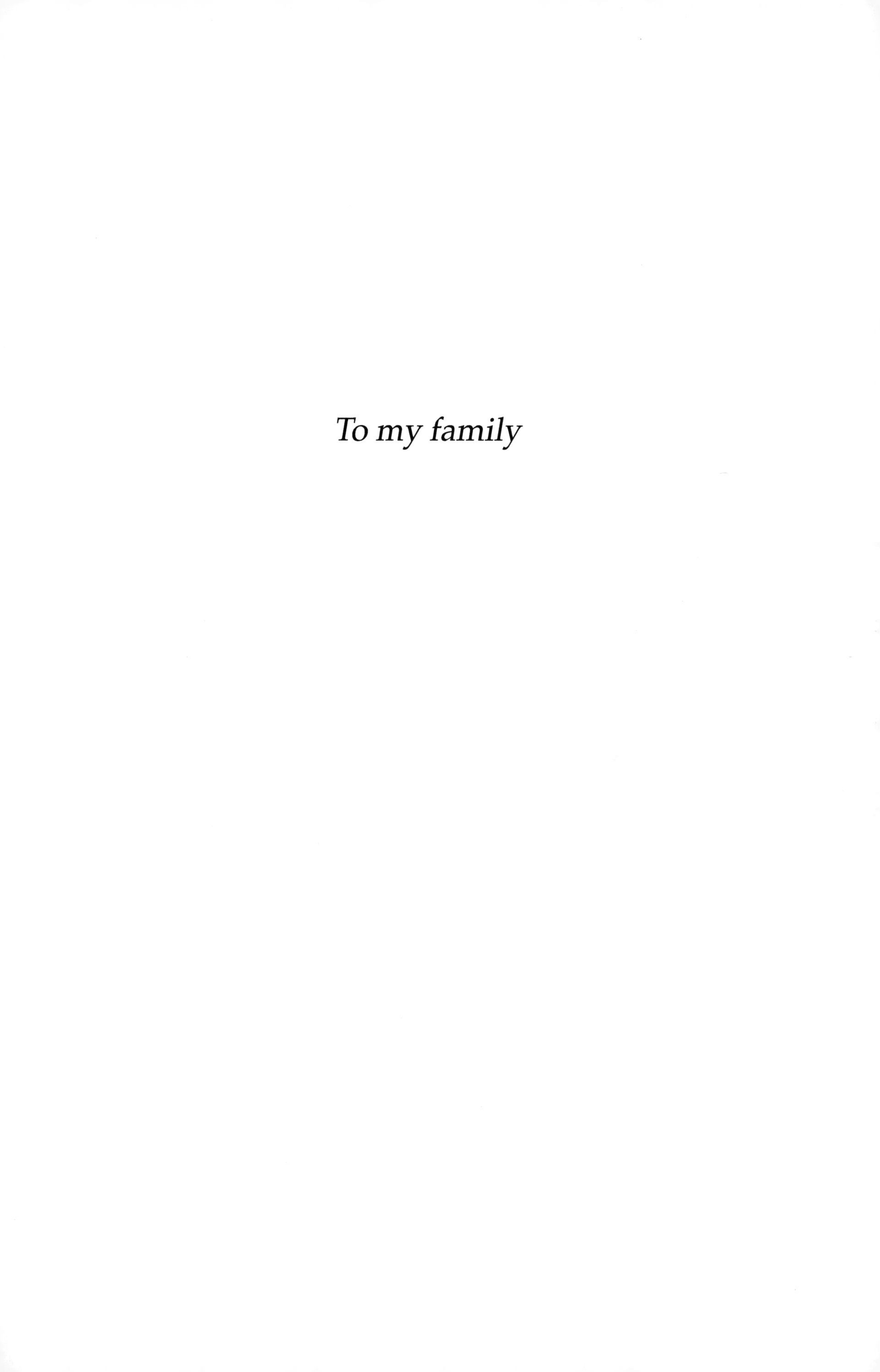

To my family

Contents

Preface

I started teaching courses on the design and analysis of algorithms in the early 1980s. At that time, there was only one available textbook, entitled *Fundamentals of Computer Algorithms*, by Horowitz and Sahni. I typeset my course notes and published a short book entitled *An Introduction to the Design and Analysis of Algorithms* in 1985. The publisher was the Charles Babbage Research Centre, a small in-house publisher at the University of Manitoba that was headed up by Ralph Stanton. A number of people at various universities used this book in their courses over the years, but it eventually went out of print.

I suppose my book was one of the earlier examples of desktop publishing. I prepared the book on my new (at the time) Macintosh 512K computer. It was popularly known as the "Fat Mac" because the amount of RAM had been increased from 128K (in the original version) to 512K. Typesetting software was quite primitive at the time, and I recall preparing mathematical equations as graphic objects and pasting them into the document in appropriate places.

After teaching this material for many years and developing various versions of course slides and notes, I decided to write an updated textbook on the subject. Of course there have been many developments in algorithms during the last 35 years, but much of the basic theory in modern courses is very similar to material covered in courses of that era. However, my goal has been to write an up-to-date textbook incorporating various new developments that are appropriate for an introductory textbook.

This book is based on the curriculum of the third-year algorithms course that is taught in the David R. Cheriton School of Computer Science at the University of Waterloo. At Waterloo, computer science students take a second-year course in data structures and data management. In the second-year course, students are introduced to order notation and complexity analysis as well as a considerable amount of material on basic and advanced data structures, including stacks, queues, priority queues, search trees, etc. Various sorting algorithms are also studied and analyzed.

I am hoping that most students using this book will be familiar with much of the background material mentioned above. However, since various course sequences in algorithms exist, the first three chapters of this book provide a treatment of mathematical background, basic algorithm analysis and data structures. This will be review for many students, but it should still serve as a useful reference. For students who have not previously covered this material, a study of these chapters will provide an introduction that is suitable for the rest of the book.

Here is a summary of the material covered in the book.

- Chapter 1 presents various mathematical topics, such as order notation, relevant mathematical formulae, probability theory and random variables.
- Chapter 2 examines various algorithm analysis techniques, including loop analysis. We also introduce reductions, which are revisited again in Chapter 9. Finally, we provide some examples of exact analysis in this chapter.
- Chapter 3 discusses various basic concepts in data structures, including stacks, queues, priority queues, binary search trees and hash tables.
- Chapter 4 treats divide-and-conquer algorithms. This chapter also contains a fairly detailed discussion of recurrence relations.
- Chapter 5 is a presentation of various greedy algorithms.
- Chapter 6 covers dynamic programming and memoization.
- Chapter 7 describes various graph algorithms, including breadth-first and depth-first search, minimum spanning trees and shortest path algorithms.
- The topic of Chapter 8 is backtracking algorithms, including pruning techniques and branch-and-bound. This is a topic that is not often found in textbooks on the design and analysis of algorithms.
- Chapter 9 presents the basic theory of NP-completeness, including the complexity classes **P** and **NP**, and NP-hard problems. Approximation algorithms are introduced and we also give a brief treatment of undecidability.

There are many textbooks on algorithms that contain more than 1000 pages of material. I have deliberately written a shorter book in which I have attempted to treat the essential concepts of the subject, by choosing a representative sample of useful algorithms to illustrate the most important design techniques. I have been careful to provide detailed pseudocode descriptions of the algorithms along with illustrative algorithms. Proofs of correctness of algorithms are also included when it is appropriate to do so. Finally, I have tried to present all the material in the book with a suitable amount of mathematical rigor.

Design and analysis of algorithms can be a difficult subject for students due to its sometimes abstract nature and its use of a wide variety of mathematical tools. My philosophy is always to try to make the subject as straightforward as possible. Ideally, after reading and understanding the material in this book, students will be able to apply the basic design principles to various real-world problems that they may encounter in their future professional careers.

I have received useful feedback regarding the content of this book from various people. In particular, I would like to thank Michael Dinitz, K. Gopalakrishnan, and Don Kreher for their comments. I have also gained much useful advice over the years from various colleagues with whom I have co-taught this material.

Douglas R. Stinson
December 2020

Chapter 1

Introduction and Mathematical Background

In this chapter, we discuss mathematical background, including terminology and basic concepts relating to algorithms, order notation (to specify the complexity of algorithms) and various relevant mathematical formulae. Finally, we provide a brief introduction to probability theory and the use of random variables. This chapter can be considered as a review or reference for readers familiar with this background material.

1.1 Algorithms and Programs

In this book, we are studying algorithms, concentrating on their design and analysis. In this section, we explain exactly what we mean by these terms, and we give some motivation for studying these aspects of computer science.

Our basic premise is that we have a problem that we wish to solve by writing a computer program. A computer program is simply a series of instructions that that will be performed by the computer, stated in a precise fashion in a particular programming language, e.g., C++. Yet the subject of this book is algorithms, which the dictionary defines as "a set of rules for solving a problem in a finite number of steps." Expanding on this slightly, we will define an ***algorithm*** as a step-by-step process for carrying out a series of computations to achieve a desired result, given some appropriate input. We want to make precise the distinction between studying algorithms, as opposed to studying computer programs or programming.

In view of the definition above, we can say that an computer program is an *implementation* of an algorithm, using a specified computer language, such as C++, Java, Python, etc.. The method being used to solve the problem is embodied in the algorithm. The algorithm is an abstraction of the computer program. As such, it can be studied without referring to any particular computer, computing environment, programming language, compiler or interpreter, etc. It describes the essential method that is being used, without getting bogged down in small details concerning the implementation. This is one main advantage of studying algorithms as opposed to computer programs.

In this book, we will present all algorithms using ***structured pseudocode***. The flow of the algorithm is controlled using various loop structures, such as **for** loops and **while** loops. Conditions are tested using the familiar "if-then-else" constructs.

DOI: 10.1201/9780429277412-1

We collect groups of statements together using a brace bracket rather than a "begin-end" grouping or something similar. This allows for a nice visual presentation of the algorithms in the book. Another convention we adopt is the use of a left arrow ($\leftarrow$) to denote an assignment of a value to a variable.

We will say that an algorithm ALG ***solves*** a problem Π if, for every instance I of Π, ALG finds a valid solution for the instance I in finite time. A key requirement in this definition is the requirement that an algorithm must find a correct answer *for every instance I* of the problem Π.

1.1.1 Analysis of Algorithms

In this book, we study the ***design*** and ***analysis*** of algorithms. "Analysis" refers to mathematical techniques for establishing both the ***correctness*** and ***efficiency*** of algorithms. An algorithm is *correct* if it always gives a correct answer in a finite amount of time. A ***proof of correctness*** can be formal or informal. The level of detail required in a proof of correctness depends greatly on the type and/or difficulty of the algorithm. If we desire a formal proof of correctness of an algorithm we design, this can sometimes be accomplished through the use of techniques such as ***loop invariants*** and mathematical induction.

Given an algorithm ALG, we want also to determine how efficient it is. In other words, how much resources (time and space) will an implementation consume? Mathematical methods can be used to effectively predict the time and space required by an algorithm, without actually implementing it in a computer program. This is useful for several reasons. The most important is that we can save work by not having to actually implement different algorithms in order to test their suitability. Analysis that is done ahead of time can compare several proposed algorithms, or compare old algorithms to new algorithms. Then we can implement only the best alternative, knowing that the others will be inferior.

Analysis of algorithms can include several possible criteria:

- What is the ***asymptotic complexity*** of algorithm ALG? Specifying asymptotic complexity involves the use of order notation, such as O-, Ω-, Θ-, o- and ω-notation, which is defined formally in Section 1.3. For now, we just mention that the most frequently used notation is O-notation, which specifies an upper bound on growth rate of an algorithm. For example, an algorithm that requires quadratic time (or less) is an $O(n^2)$ algorithm. Analyzing the complexity of an algorithm can require a variety of techniques, of which loop analysis is one of the most basic methods. Loop analysis is discussed in Section 2.1.

- What is the ***exact number*** of specified computations done by ALG? Analysis of asymptotic complexity may allow us to deduce that an algorithm has quadratic complexity, which would be denoted as a $\Theta(n^2)$ algorithm. But this does not distinguish between situations where the number of operations performed is n^2 or $1000n^2$, for example. Perhaps an exact analysis could be done to determine that the number of operations is (exactly) $100n^2$. Some examples of exact analysis are provided in Section 2.4.

- How does the ***average-case complexity*** of ALG compare to the ***worst-case complexity***? Most of the time, we consider worst-case complexity because it provides a guaranteed bound that holds for all possible inputs. But sometimes average-case complexity yields useful information. For example, the famous QUICKSORT algorithm has average-case complexity $\Theta(n \log n)$, but the worst-case complexity is $\Theta(n^2)$. So this algorithm is widely used in practice despite the fact that its worst-case complexity is worse than several other sorting algorithms. Average-case and worst-case complexity are defined in Section 1.2.2 and some examples are studied in Section 2.4.

- Is ALG the most efficient algorithm to solve the given problem? That is, can we find a *lower bound* on the complexity of *any* algorithm to solve the given problem? We note that proving lower bounds is often very difficult and we do not address this topic very much in this book. However, a simple example of a lower bound using exact analysis is provided in Section 2.4.

- Are there problems that cannot be solved efficiently? Here and elsewhere, "efficiently" means in polynomial time, i.e., in time $O(n^c)$ for some positive integer c. This question is addressed using the theory of ***NP-completeness,*** which shows that a large class of problems (the ***NP-complete*** problems) are "equivalent" in terms of their difficulty (if one NP-complete problem can be solved in polynomial time, then they all can be solved in polynomial time). Although there is no proof, the NP-complete problems are commonly believed to be intractable. See Chapter 9 for a treatment of this theory.

- Another resource that can be important in the context of an algorithm is *memory* (also known as ***space complexity***).

- There are problems that cannot be solved by *any* algorithm. Such problems are termed ***undecidable***. See Chapter 9 for further discussion of this concept.

1.1.2 Design of Algorithms

"Design" refers to *general strategies* for creating new algorithms. If we have good design strategies, then it will be easier to end up with correct and efficient algorithms. Also, we want to avoid using *ad hoc* algorithms that are hard to analyze and understand.

Although there are tens of thousands of algorithms that have been proposed to solve various problems, there are only a handful of underlying design principles or techniques. These describe the general manner in which the algorithm operates and they serve to unify many disparate algorithms by giving them a logical organization. A knowledge and understanding of these design techniques is valuable, both in designing new algorithms as well as in understanding how given algorithms work.

Here are some examples of useful design strategies, many of which we will study in this book:

reductions

This refers to using an algorithm for one problem as a subroutine to solve a different problem. See Section 2.3 for an introduction to this topic.

divide-and-conquer

Given a problem instance I, a divide-and-conquer algorithm solves smaller instances of the same problem recursively and then uses these solutions to smaller instances to solve the original instance I. Divide-and-conquer algorithms are studied in depth in Chapter 4, where many examples are given.

greedy

These are algorithms for optimization problems that build up a solution step-by-step by making a "greedy" choice at each step. We note that proving correctness of greedy algorithms can be challenging! Greedy algorithms are the subject of Chapter 5.

dynamic programming

Dynamic programming, like divide-and-conquer, solves smaller problem instances in order to solve an instance I, but the solutions to smaller problem instances are used to fill in a table in a non-recursive manner. Dynamic programming is presented in Chapter 6. In that chapter, we also discuss memoization, which is a recursive variant of dynamic programming.

depth-first and breadth-first search

These are algorithms to visit all the vertices and edges of a graph in a systematic manner. Graphs are often used to model networks, e.g., computer networks, communication networks, social networks. There are many important practical algorithms for graphs that are presented in Chapter 7.

backtracking

Backtracking is a method of exhaustive search (similar to a depth-first search) that can be used to enumerate all the possible solutions to a given problem instance, and/or find the optimal solution. Backtracking is often used to solve relatively large instances of difficult (e.g., NP-hard) problems in practice. Backtracking algorithms are studied in Chapter 8.

Here are a few additional comments:

- Algorithms can also be categorized as ***serial*** (one processor) or ***parallel*** (multiple processors). In this book, we are only studying serial algorithms.
- Further, algorithms can be ***deterministic*** or ***randomized***. In this book, we are mainly studying deterministic algorithms, though we will also consider a few randomized algorithms (see Section 4.12, for example).

- Finally, we also study the concept of ***nondeterministic*** algorithms in connection with the theory of NP-completeness (see Chapter 9). Roughly speaking, in the context of NP-completeness, we view a nondeterministic algorithm as a method of verifying the correctness of a given possible solution to a problem instance.

1.2 Definitions and Terminology

In this section, we review basic definitions relating to problems, algorithms and related concepts.

1.2.1 Problems

The purpose of an algorithm is to solve a specified problem, so we begin by defining the notion of a problem.

problem

A problem, say Π, is defined by specifying all possible problem instances. Usually there are an infinite number of problem instances. For each problem instance I, we need to carry out a particular computational task in order to obtain the problem solution.

problem instance

A problem instance I is the *input* for the specified problem Π.

problem solution

The problem solution is the *output* (i.e., the correct answer) for the specified problem instance I.

size of a problem instance

$\mathsf{size}(I)$ is a positive integer which is a measure of the size of the instance I. More precisely, $\mathsf{size}(I)$ is the number of *bits* required to specify the instance I, i.e., the *space* that is needed to write down the given instance.

There are some tricky aspects about defining the size of a problem instance that we will encounter at various points in this book. Suppose we have an array of n positive integers which comprises a problem instance. If each integer occupies one word of memory, then the total number of bits of storage for the array would be $32n$ bits. The constant factor 32 would not be relevant for any complexity analyses, so we would just say that the size of the problem instance is n.

On the other hand, there could be situations where we are performing arithmetic operations on large integers (sometimes this is called ***multiprecision*** arithmetic). The time required to add two large integers clearly depends on how big they are. The "size" of a positive integer, say x, would be the number of bits it occupies when written in binary. With this definition of size, we can express the complexity of multiprecision addition as a function of the size of the integers we are adding together. If x and y are m-bit positive integers, then it should not be too surprising that it requires time $\Theta(m)$ to compute the sum $x+y$. It is helpful to recognize that the binary representation of an m-bit integer has a leading (i.e., high-order) bit equal to 1, and $m-1$ other bits, each of which is 0 or 1. Thus we have the following lemma.

LEMMA 1.1 *If x is a positive integer and* $\mathsf{size}(x)=m$, *then*

$$2^{m-1} \leq x \leq 2^m - 1.$$

Expressing $m = \mathsf{size}(x)$ as a function of x, we have the following:

LEMMA 1.2 *If x is a positive integer, then*

$$\mathsf{size}(x) = \lceil \log_2(x+1) \rceil.$$

COROLLARY 1.3 *If x is a positive integer, then*

$$\mathsf{size}(x) \in \Theta(\log_2 x).$$

Observe that the value of a positive integer x is exponentially larger than its size.

1.2.2 Running Time and Complexity

When we have written an actual computer program, we can test its efficiency by seeing how long it takes to run on one or more problem instances. Of course the program must be run on a particular computer to obtain actual timings, and various factors such as the computer, computing environment, etc., will influence these timings.

running time of a program

$T_{\mathrm{M}}(I)$ denotes the running time (in seconds, say) of a particular program M on a problem instance I.

worst-case running time as a function of input size

$T_{\mathrm{M}}(n)$ denotes the ***maximum*** running time of program M on instances of size n:

$$T_{\mathrm{M}}(n) = \max\{T_{\mathrm{M}}(I) : \mathsf{size}(I) = n\}.$$

average-case running time as a function of input size

$T_{\mathrm{M}}^{avg}(n)$ denotes the ***average*** running time of program M over all instances of size n:

$$T_{\mathrm{M}}^{avg}(n) = \frac{1}{|\{I : \mathsf{size}(I) = n\}|} \sum_{\{I:\mathsf{size}(I)=n\}} T_{\mathrm{M}}(I).$$

The above formula specifies that M would be run on all possible instances of size n, the total time would be computed, and then the average would be computed by dividing by the number of different instances considered.

There are only finitely many instances of a given size, so the maximum and average running times are both guaranteed to be finite.

By analyzing algorithms, we are attempting to predict the time that would be required by a program that implements the algorithm, without actually implementing the algorithm. If we are given a program, we can run the program and compute its running time on inputs of various sizes. Empirical timing data may suggest a pattern, but it does not prove anything. In particular, it does not allow us to conclude that the program will run in any specified amount of time on some instance that we have not tested already.

However, by analyzing the structure of the underlying algorithm, we can determine the ***growth rate*** of the running time (i.e., the complexity) of the algorithm. For example, we might use mathematical techniques to prove that a particular sorting algorithm has complexity $\Theta(n^2)$ (where n is the number of elements in the array that is being sorted). An actual computer program that implements the algorithm will have a running time whose dominant term is a quadratic, i.e., an^2, for some positive constant a. The actual running time might be a more complicated function that also contains lower-order terms such as linear and constant terms, e.g.,

$$T_{\mathrm{M}}(n) = an^2 + bn + c,$$

for constants a, b and c.

In general, we cannot expect to determine the value of the constant a through analysis of the algorithm, because the value of a is highly dependent on the particular program that implements the algorithm, as well as the computer on which the program is run. However, once we know that the algorithm has quadratic complexity, we could then use empirical timing data to estimate a value of a, if we wished to do so.

Based on this discussion, in the following definitions, we relate the complexity of an algorithm to the running time of a program that implements the algorithm.

worst-case complexity of an algorithm

Let $f : \mathbb{Z}^+ \rightarrow \mathbb{R}$. An algorithm ALG has ***worst-case complexity*** $f(n)$ if there exists a program M implementing the algorithm ALG such that $T_{\mathrm{M}}(n) \in \Theta(f(n))$.

average-case complexity of an algorithm

Let $f : \mathbb{Z}^+ \rightarrow \mathbb{R}$. An algorithm ALG has ***average-case complexity*** $f(n)$ if there exists a program M implementing the algorithm ALG such that $T_{\mathrm{M}}^{avg}(n) \in \Theta(f(n))$.

To summarize, the running time of a program is influenced by many factors, including the programming language, processor, operating system, etc. Running time can only be determined by running the program on a specific computer. On the other hand, the complexity of an algorithm can be determined by high-level mathematical analysis. It is *independent* of the above-mentioned factors affecting running time.

Note the terminology: we speak of the *running time of a program* and the *complexity of an algorithm.*

Complexity is inherently a less precise measure than running time since it is asymptotic in nature and it incorporates unspecified constant factors and unspecified lower-order terms. However, if algorithm ALG has lower complexity than algorithm B, then a program implementing algorithm ALG will be faster than a program implementing algorithm B for *sufficiently large inputs.*

It is important to recognize that the complexity of an algorithm does not say anything about its behavior on small instances. For example, consider

$$T_1(n) = 75n + 500 \quad \text{and} \quad T_2(n) = 5n^2.$$

Clearly

$$T_1 \in \Theta(n) \quad \text{and} \quad T_2 \in \Theta(n^2),$$

so T_1 is faster asymptotically. However $T_2(n) \leq T_1(n)$ if $n \leq 20$, so T_1 it is not faster for small values of n. The value $n = 20$ is called the ***crossover point*** of the two algorithms. It is the value beyond which the asymptotically faster algorithm is truly faster than the asymptotically slower algorithm. See Figure 1.1.

1.3 Order Notation

The basic idea of order notation is to ignore lower-order terms and constant factors in complicated mathematical functions. This allows us to simplify expressions for the complexity of an algorithm. In general, we are only interested in functions that have some relation to the running time of a program. The running time of a program cannot be a negative number. Hence, we will make the somewhat less restrictive assumption when we use order notation that all functions under consideration take on positive values for sufficiently large integers n. So we are only interested in functions $f : \mathbb{Z}^* \rightarrow \mathbb{R}$ with the property that $f(n) > 0$ for all $n > n_0$, for some constant n_0.

The domain of f has to include the positive integers, at least, but it could be larger. For example, f might be defined on the set of all integers or the set of all real numbers.

For example, we would not use any kind of order notation to describe a function such as

$$f(n) = -n^2 + 100n,$$

since $f(n) < 0$ for all $n > 100$. This function f clearly cannot represent the running time of a program or the complexity of an algorithm.

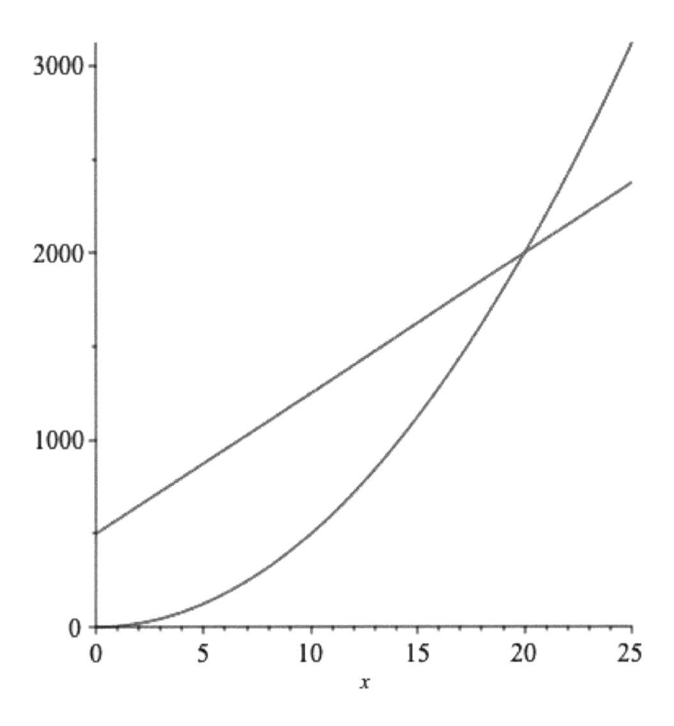

FIGURE 1.1: A graph of $T_1(x) = 75x + 500$ and $T_2(x) = 5x^2$.

We begin by defining five types of ***order notation***.

O-notation

$f(n) \in O(g(n))$ if *there exist* constants $c > 0$ and $n_0 > 0$ such that

$$0 \leq f(n) \leq cg(n)$$

for all $n \geq n_0$. This is expressed verbally by saying that $f(n)$ is *big-oh* of $g(n)$. Here the complexity of f is *not higher* than the complexity of g.

Ω-notation

$f(n) \in \Omega(g(n))$ if *there exist* constants $c > 0$ and $n_0 > 0$ such that

$$0 \leq cg(n) \leq f(n)$$

for all $n \geq n_0$. Here we say that $f(n)$ is *big-omega* of $g(n)$. This expresses the fact that the complexity of f is *not lower* than the complexity of g.

Θ-notation

$f(n) \in \Theta(g(n))$ if *there exist* constants $c_1, c_2 > 0$ and $n_0 > 0$ such that

$$0 \leq c_1 g(n) \leq f(n) \leq c_2 g(n)$$

for all $n \geq n_0$. Here we say that $f(n)$ is *big-theta* of $g(n)$; the functions f and g have the *same complexity*.

o-notation

$f(n) \in o(g(n))$ if, *for every* constant $c > 0$, there exists a constant $n_0 > 0$ such that

$$0 \leq f(n) \leq cg(n)$$

for all $n \geq n_0$. Here $f(n)$ is *little-oh* of $g(n)$ and f has *lower complexity* than g.

ω-notation

$f(n) \in \omega(g(n))$ if, *for every* constant $c > 0$, there exists a constant $n_0 > 0$ such that

$$0 \leq cg(n) \leq f(n)$$

for all $n \geq n_0$. Here $f(n)$ is *little-omega* of $g(n)$ and f has *higher complexity* than g.

Here are a few comments about the above definitions.

- We say $f(n) \in O(g(n))$ or $f(n)$ *is* $O(g(n))$. Formally, $O(g(n))$ is a *set* of functions and $f(n)$ is a *specific* function. So we do *not* say $f(n) = O(g(n))$, because this statement is mathematical nonsense. (But many books use this abuse of notation anyway.)

- There is an alternative definition of Ω-notation which is more commonly used in number theory. It is sometimes called the *Hardy-Littlewood definition*. The definition we are using is the *Knuth definition*. Here is the alternative definition: $f(n) \in \Omega(g(n))$ if *there exists* a constant $c > 0$ such that for every $n_0 > 0$, *there exists* $n \geq n_0$ such that $0 \leq cg(n) \leq f(n)$.

- There are unexpected subtleties in trying to extend the definition of O-notation to multiple variables. For our purposes, the following definition is probably sufficient: $f(n,m) \in O(g(n,m))$ if *there exist* constants $c > 0$ and $n_0 > 0$ such that

$$0 \leq f(n,m) \leq cg(n,m)$$

if $n \geq n_0$ *or* $m \geq n_0$.

Intuitively, we have the following correspondences between order notation and growth rates:

- $f(n) \in O(g(n))$ means the growth rate of f is *no higher* than the growth rate of g,

- $f(n) \in o(g(n))$ means the growth rate of f is *significantly less than* the growth rate of g ,

- $f(n) \in \Omega(g(n))$ means the growth rate of f is *at least as high* as the growth rate of g ,

- $f(n) \in \omega(g(n))$ means the growth rate of f is *significantly greater than* the growth rate of g, and
- $f(n) \in \Theta(g(n))$ means the growth rate of f is *the same* as the growth rate of g.

In general, when we use order notation, $f(n)$ may be a complicated function and we try to make $g(n)$ a simple function. We illustrate by working through some simple examples.

Example 1.1 Let $f(n) = n^2 - 7n - 30$. Prove from first principles that $f(n) \in O(n^2)$. □

PROOF We can factor $f(n)$ as

$$f(n) = (n - 10)(n + 3).$$

It is clear that $f(n) \geq 0$ if $n \geq 10$, because 10 is the larger of the two roots of the quadratic equation $f(n) = 0$. Also, it is obvious that $f(n) \leq n^2$ for all $n \geq 0$. So we can take $c = 1$ and $n_0 = 10$ and the definition is satisfied. ∎

REMARK In Example 1.1, we started with a "complicated" function, namely $n^2 - 7n - 30$, and showed that is big-oh of a "simple" function, n^2. It is of course also true that $n^2 \in O(n^2 - 7n - 30)$, but there seems to be little point in starting with the "simple" function n^2 and then showing that it is big-oh of a more complicated function. ∎

Example 1.2 Let $f(n) = n^2 - 7n - 30$. Prove from first principles that $f(n) \in \Omega(n^2)$. □

PROOF Here it is easiest to pick a suitable value of c first, and then figure out a value of n_0 that works. Note that any value of $c < 1$ will work. (Why won't $c \geq 1$ work?) Suppose we take $c = 1/2$. Now we want

$$\frac{n^2}{2} \leq n^2 - 7n - 30,$$

or

$$n^2 - 14n - 60 \geq 0.$$

There are many ways to find a suitable n_0. One way is to compute the two roots of the quadratic equation $n^2 - 14n - 60 = 0$ and take n_0 to be the larger root, rounded up to an integer. Another way is to observe that

$$n^2 - 14n = n(n - 14) \geq 18 \times 4 = 72 > 60$$

if $n \geq 18$, so $n_0 = 18$ works. ∎

Example 1.3 Suppose $f(n) = n^2 + n$. Prove from first principles that $f(n) \notin O(n)$.
□

PROOF Note that, in order to prove $f(n)$ is *not* $O(g(n))$, we must replace all occurrences of "there exists" by "for every," replace all occurrences of "for every" by "there exists," and reverse the direction of the inequality.

Therefore, we need to show, for every $c > 0$ and for every $n_0 \geq 0$, that there exists $n \geq n_0$ such that

$$n^2 + n > cn.$$

Suppose $n > 0$. Then

$$\begin{aligned} n^2 + n > cn &\Leftrightarrow n + 1 > c \\ &\Leftrightarrow n > c - 1. \end{aligned}$$

Let $n \geq \max\{c, n_0\}$. Then $n \geq n_0$ and $n^2 + n > cn$, as desired. ▮

1.3.1 Techniques for Order Notation

The following theorem provides a useful method for comparing the growth rates of functions.

THEOREM 1.4 *Suppose that $f(n) > 0$ and $g(n) > 0$ for all $n \geq n_0$. Suppose that*

$$L = \lim_{n \to \infty} \frac{f(n)}{g(n)}.$$

Then

$$f(n) \in \begin{cases} o(g(n)) & \text{if } L = 0 \\ \Theta(g(n)) & \text{if } 0 < L < \infty \\ \omega(g(n)) & \text{if } L = \infty. \end{cases}$$

In applying Theorem 1.4, it is sometimes useful to apply l'Hôpital's rule. This familiar result from calculus states that

$$\lim_{n \to \infty} \frac{f(n)}{g(n)} = \lim_{n \to \infty} \frac{f'(n)}{g'(n)}$$

whenever $f(n)$ and $g(n)$ are differentiable functions and

$$\lim_{n \to \infty} f(n) = \lim_{n \to \infty} g(n) = \infty.$$

We now look at some specific examples that illustrate the application of Theorem 1.4.

Example 1.4 Suppose we want to compare the growth rate of the functions $(\ln n)^2$ and $n^{1/2}$. We have

$$\begin{aligned}
\lim_{n\to\infty} \frac{(\ln n)^2}{n^{1/2}} &= \lim_{n\to\infty} \frac{2\ln n \times (1/n)}{(1/2)n^{-1/2}} && \text{from l'Hôpital's rule} \\
&= \lim_{n\to\infty} \frac{4\ln n}{n^{1/2}} \\
&= \lim_{n\to\infty} \frac{4 \times (1/n)}{(1/2)n^{-1/2}} && \text{from l'Hôpital's rule} \\
&= \lim_{n\to\infty} \frac{8}{n^{1/2}} \\
&= 0,
\end{aligned}$$

so $(\ln n)^2 \in o(n^{1/2})$. Note the use of l'Hôpital's rule twice in this example. □

More generally, it holds that $(\log n)^c \in o(n^d)$ for all $c, d > 0$. For example, $(\log n)^{1000} \in o(n^{.001})$. Also, Theorem 1.4 can be used to prove that $n^c \in o(d^n)$ for all $c > 0$, $d > 1$. For example, $n^{1000} \in o((1.001)^n)$.

Example 1.5 As another example, we can use Theorem 1.4 to compare the growth rate of the functions n^2 and $n^2 - 7n - 30$. We have

$$\begin{aligned}
\lim_{n\to\infty} \frac{n^2 - 7n - 30}{n^2} &= \lim_{n\to\infty} \left(1 - \frac{7}{n} - \frac{30}{n^2}\right) \\
&= 1,
\end{aligned}$$

so $n^2 - 7n - 30 \in \Theta(n^2)$. □

Example 1.5 can be generalized to prove the following theorem. The method of proof is similar and we leave the details for the reader to verify.

THEOREM 1.5 *Suppose*

$$f(n) = \sum_{i=0}^{t} a_i x^i$$

with $a_t > 0$. *Then* $f(n) \in \Theta(x^t)$.

Example 1.6 We compare the growth rate of the two functions $f(n) = (3 + (-1)^n)n$ and $g(n) = n$. Here the limit of the quotient $f(n)/g(n)$ of the two functions does not exist, so we cannot apply Theorem 1.4. However, it is nevertheless true that $(3 + (-1)^n)n \in \Theta(n)$. Note that

$$(3 + (-1)^n) = 4$$

if n is even and

$$(3 + (-1)^n) = 2$$

if n is odd. Therefore

$$2n \leq (3 + (-1)^n)n \leq 4n$$

for all $n \geq 0$. □

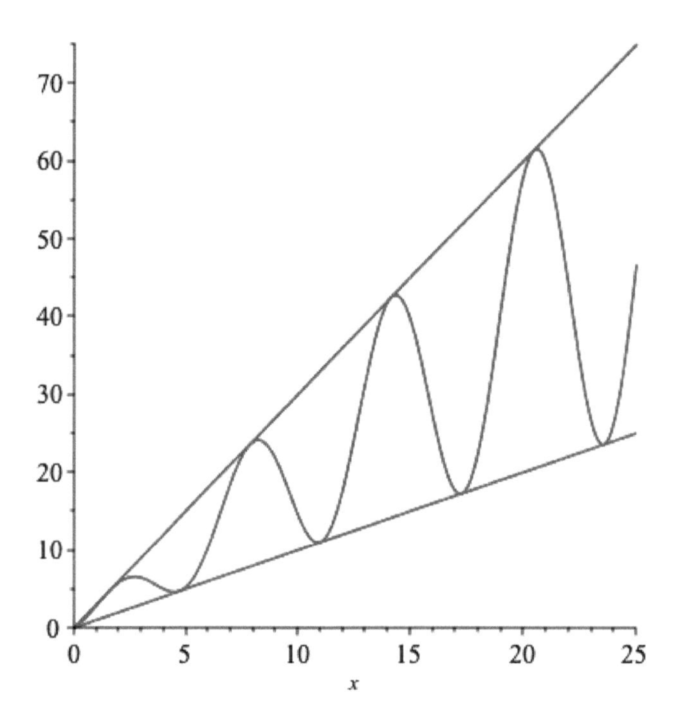

FIGURE 1.2: A graph of $x(2+\sin x)$, x and $3x$.

Example 1.7 It is possible to prove that

$$n\left(2+\sin\frac{n\pi}{2}\right) \in \Theta(n),$$

even though

$$\lim_{n\to\infty}\left(2+\sin\frac{n\pi}{2}\right)$$

does not exist. See Figure 1.2, which suggests that

$$n \leq n\left(2+\sin\frac{n\pi}{2}\right) \leq 3n.$$

This can be proven easily, using the fact that

$$-1 \leq \sin\frac{n\pi}{2} \leq 1.$$

□

Example 1.8 In order to compare the growth rates of the functions

$$f(n) = n\left|\sin\frac{n\pi}{2}\right| + 1$$

and

$$g(n) = \sqrt{n},$$

we observe that

$$f(n) = \begin{cases} 1 & \text{if } n \text{ is even} \\ n+1 & \text{if } n \text{ is odd.} \end{cases}$$

If follows that $f(n) \notin O(g(n))$ and $g(n) \notin O(f(n))$. Thus, there is no relation between the growth rates of $f(n)$ and $g(n)$. □

1.3.2 Relationships between Order Notations

We list several relationships between different flavors of order notation.

THEOREM 1.6

1. $f(n) \in \Theta(g(n)) \Leftrightarrow g(n) \in \Theta(f(n))$
2. *If* $f(n) \in \Theta(g(n))$ *and* $g(n) \in \Theta(h(n))$, *then* $f(n) \in \Theta(h(n))$
3. $f(n) \in O(g(n)) \Leftrightarrow g(n) \in \Omega(f(n))$
4. $f(n) \in o(g(n)) \Leftrightarrow g(n) \in \omega(f(n))$
5. $f(n) \in \Theta(g(n)) \Leftrightarrow f(n) \in O(g(n))$ *and* $f(n) \in \Omega(g(n))$
6. $f(n) \in o(g(n)) \Rightarrow f(n) \in O(g(n))$
7. $f(n) \in \omega(g(n)) \Rightarrow f(n) \in \Omega(g(n))$.

The above are all straightforward to prove. To illustrate, we provide a proof that $f(n) \in \Theta(g(n))$ implies $g(n) \in \Theta(f(n))$.

PROOF Suppose $f(n) \in \Theta(g(n))$. Then there exist positive constants c_1, c_2, n_0 such that

$$0 \leq c_1 g(n) \leq f(n) \leq c_2 g(n)$$

if $n \geq n_0$. Thus

$$0 \leq \left(\frac{1}{c_2}\right) f(n) \leq g(n) \leq \left(\frac{1}{c_1}\right) f(n)$$

if $n \geq n_0$. Define

$$c_1' = \frac{1}{c_2}, \quad c_2' = \frac{1}{c_1} \quad \text{and} \quad n_0' = n_0.$$

Then

$$0 \leq c_1' f(n) \leq g(n) \leq c_2' f(n)$$

if $n \geq n_0'$, as desired. ∎

1.3.3 Algebra of Order Notations

We record some rules that allow us to simplify complicated functions when using order notation. The proofs of these formulae are straightforward applications of the definitions.

First, we have three ***"maximum" rules***.

THEOREM 1.7 *Suppose that $f(n) > 0$ and $g(n) > 0$ for all $n \geq n_0$. Then:*

1. $O(f(n) + g(n)) = O(\max\{f(n), g(n)\})$
2. $\Theta(f(n) + g(n)) = \Theta(\max\{f(n), g(n)\})$
3. $\Omega(f(n) + g(n)) = \Omega(\max\{f(n), g(n)\})$.

The above "maximum" rules are also valid for a finite number of functions (at least two). Also, note that it is acceptable to use "=" signs here because we are referring to *sets*.

We also have ***"summation" rules.***

THEOREM 1.8 *Suppose $\mathcal{I}$ is a* finite *set. Then*

1. $O\left(\sum_{i \in \mathcal{I}} f(i)\right) = \sum_{i \in I} O(f(i))$
2. $\Theta\left(\sum_{i \in \mathcal{I}} f(i)\right) = \sum_{i \in I} \Theta(f(i))$
3. $\Omega\left(\sum_{i \in \mathcal{I}} f(i)\right) = \sum_{i \in I} \Omega(f(i))$.

Here is a small example.

Example 1.9

$$
\begin{aligned}
\sum_{i=1}^{n} O(i) &= O\left(\sum_{i=1}^{n} i\right) \\
&= O\left(\frac{n(n+1)}{2}\right) \qquad \text{(sum of an arithmetic sequence)} \\
&= O(n^2).
\end{aligned}
$$

□

We will see a bit later that summation rules are commonly used in loop analysis.

It is also interesting to consider products and quotients of functions. For products, we have the following.

THEOREM 1.9

1. *If* $f_1(n) \in O(g_1(n))$ *and* $f_2(n) \in O(g_2(n))$, *then*
$$f_1(n)f_2(n) \in O(g_1(n)g_2(n)).$$
2. *If* $f_1(n) \in \Omega(g_1(n))$ *and* $f_2(n) \in \Omega(g_2(n))$, *then*
$$f_1(n)f_2(n) \in \Omega(g_1(n)g_2(n)).$$
3. *If* $f_1(n) \in \Theta(g_1(n))$ *and* $f_2(n) \in \Theta(g_2(n))$, *then*
$$f_1(n)f_2(n) \in \Theta(g_1(n)g_2(n)).$$

The situation is a bit more complicated for quotients. Observe that, if we want an upper bound on the fraction x/y, we require an upper bound on x and a lower bound on y. Similarly, if we want a lower bound on the fraction x/y, we require a lower bound on x and an upper bound on y. We have the following three properties.

THEOREM 1.10

1. *If* $f_1(n) \in O(g_1(n))$ *and* $f_2(n) \in \Omega(g_2(n))$, *then*
$$\frac{f_1(n)}{f_2(n)} \in O\left(\frac{g_1(n)}{g_2(n)}\right).$$
2. *If* $f_1(n) \in \Omega(g_1(n))$ *and* $f_2(n) \in O(g_2(n))$, *then*
$$\frac{f_1(n)}{f_2(n)} \in \Omega\left(\frac{g_1(n)}{g_2(n)}\right).$$
3. *If* $f_1(n) \in \Theta(g_1(n))$ *and* $f_2(n) \in \Theta(g_2(n))$, *then*
$$\frac{f_1(n)}{f_2(n)} \in \Theta\left(\frac{g_1(n)}{g_2(n)}\right).$$

1.3.4 Common Growth Rates

We present in Table 1.1 a list of some common growth rates (in increasing order). Notice that we make a distinction between ***polynomial*** and ***exponential*** growth. A polynomial growth rate is anything having the form $\Theta(n^c)$ for a positive constant c, whereas an exponential growth rate is an expression of the form $\Theta(c^n)$ for a positive constant $c > 1$. Interestingly, there are growth rates that arise in practice that are between polynomial and exponential. Two examples of such "intermediate" growth rates are presented in Table 1.1.

First, the 1983 algorithm for ***graph isomorphism*** due to Babai and Luks has

TABLE 1.1: Some common growth rates (in increasing order).

polynomial	$\Theta(1)$	constant growth rate
	$\Theta(\log n)$	logarithmic growth rate
	$\Theta(\sqrt{n})$	
	$\Theta(n)$	linear growth rate
	$\Theta(n^2)$	quadratic growth rate
	$\Theta(n^{1000})$	
	$\Theta\left(e^{cn^{1/3}(\log n)^{2/3}}\right)$	number field sieve
	$\Theta\left(n^{c\sqrt{n}\log_2 n}\right)$	graph isomorphism
exponential	$\Theta(1.001^n)$	
	$\Theta(2^n)$	
	$\Theta(e^n)$	
	$\Theta(n!)$	
	$\Theta(n^n)$	

complexity $\Theta\left(n^{c\sqrt{n}\log_2 n}\right)$. (Here we are given two graphs on n vertices and we wish to determine if they are isomorphic.)

Second, the complexity of the ***number field sieve***, which is a factoring algorithm, is $\Theta\left(e^{cn^{1/3}(\log n)^{2/3}}\right)$. (Here we are given an integer having n bits in its binary representation and we are trying to factor it into primes.)

An algorithm having polynomial-time complexity is known as a ***polynomial-time algorithm*** and an algorithm having exponential-time complexity is known as an ***exponential-time algorithm***.

It is interesting to see how the running time is affected when the size of the problem instance doubles (i.e., if we replace n by $2n$).

- For logarithmic complexity, we have $T(n) = c\log_2 n$ and

$$\begin{aligned} T(2n) &= c\log_2(2n) \\ &= c + c\log n \\ &= T(n) + c. \end{aligned}$$

- For linear complexity, we have $T(n) = cn$ and

$$\begin{aligned} T(2n) &= c(2n) \\ &= 2T(n). \end{aligned}$$

- For a growth rate of $\Theta(n\log n)$, we have $T(n) = cn\log_2 n$ and

$$T(2n) = c(2n)\log_2(2n)$$

$$= 2cn + 2cn\log_2 n$$
$$= 2T(n) + \frac{2T(n)}{\log_2 n}$$
$$= 2T(n)\left(1 + \frac{1}{\log_2 n}\right).$$

- For quadratic complexity, we have $T(n) = cn^2$ and

$$\begin{aligned} T(2n) &= c(2n)^2 \\ &= 4T(n). \end{aligned}$$

- For cubic complexity, we have $T(n) = cn^3$ and

$$\begin{aligned} T(2n) &= c(2n)^3 \\ &= 8T(n). \end{aligned}$$

- For exponential complexity, we have $T(n) = c2^n$ and

$$\begin{aligned} T(2n) &= c2^{2n} \\ &= c(2^n)^2 \\ &= \frac{(T(n))^2}{c}. \end{aligned}$$

1.4 Mathematical Formulae

In this section, we give several useful mathematical formulae. Some of the formulae will be presented without proof.

1.4.1 Sequences

An ***arithmetic sequence*** with n terms has the form

$$a, a+d, a+2d, \ldots, a+(n-1)d.$$

The value d is the ***difference*** of the sequence. The sum of the sequence is

$$\sum_{i=0}^{n-1}(a+di) = na + \frac{dn(n-1)}{2}. \tag{1.1}$$

When $d > 0$, the sum is $\Theta(n^2)$. That is, the sum of n terms of an arithmetic sequence with $d > 0$ has a quadratic growth rate.

A ***geometric sequence*** with n terms has the form

$$a, ar, ar^2, \ldots, ar^{n-1}.$$

The value r is the ***ratio*** of the sequence. We will assume $r > 0$. The sum of the sequence is

$$\sum_{i=0}^{n-1} ar^i = \begin{cases} a\frac{r^n - 1}{r - 1} \in \Theta(r^n) & \text{if } r > 1 \\ na \in \Theta(n) & \text{if } r = 1 \\ a\frac{1 - r^n}{1 - r} \in \Theta(1) & \text{if } 0 < r < 1. \end{cases} \tag{1.2}$$

Note that $f(n) = r^n$ is an *exponential function* of n if $r > 1$ and $\lim_{n\to\infty} f(n) = 0$ if $0 < r < 1$. Hence, the sum of n terms of a geometric sequence has an exponential, linear, or constant growth rate, depending on the value of the ratio r.

Further, in the case where $0 < r < 1$, we see that

$$\lim_{n\to\infty} r^n = 0,$$

so the infinite sum is finite and

$$\sum_{i=0}^{\infty} ar^i = \frac{a}{1 - r} \in \Theta(1). \tag{1.3}$$

An ***arithmetic-geometric sequence*** with n terms has the form

$$a, a + dr, (a + 2d)r^2, \ldots, (a + (n-1)d)r^{n-1}.$$

We will assume that $d > 0$ and $r > 0$. The sum of the sequence is

$$\sum_{i=0}^{n-1} (a + di)r^i = \frac{a}{1 - r} - \frac{(a + (n-1)d)r^n}{1 - r} + \frac{dr(1 - r^{n-1})}{(1 - r)^2}, \tag{1.4}$$

provided that $r \neq 1$.

Observe that the growth rate of the sum is $\Theta(nr^n)$ if $r > 1$. When $0 < r < 1$, we can take the limit in (1.4) as $n \to \infty$ and obtain the following formula for the infinite sum:

$$\sum_{i=0}^{\infty} (a + di)r^i = \frac{a}{1 - r} + \frac{dr}{(1 - r)^2} \qquad \text{if } 0 < r < 1. \tag{1.5}$$

Example 1.10 We give an example of an arithmetic-geometric sequence and provide an illustration of how its sum can be computed without referring to the formula (1.4). Consider the following sum:

$$\sum_{i=1}^{n-1} (2i - 1)2^i.$$

This sum can be evaluated as follows:

1×2	3×2^2	5×2^3	$\cdots$	$(2n-1)2^n$	sum
2	2^2	2^3	$\cdots$	2^n	$2^{n+1} - 2$
	2×2^2	2×2^3	$\cdots$	2×2^n	$2(2^{n+1} - 4)$
		2×2^3	$\cdots$	2×2^n	$2(2^{n+1} - 8)$
				$\vdots$	$\vdots$
				2×2^n	$2(2^{n+1} - 2^n)$

Summing the last column, we get

$$(2n-1)2^{n+1} - 2(2^{n+1} - 4) - 2 = (2n-3)2^{n+1} + 6.$$

Note that this agrees with the formula (1.4). ☐

The ***harmonic sequence*** with n terms is the sequence

$$1, \frac{1}{2}, \frac{1}{3}, \ldots, \frac{1}{n}.$$

The sum of these n terms, which we write as

$$H_n = \sum_{i=1}^{n} \frac{1}{i},$$

is called the ***nth harmonic number***. It can be shown that

$$H_n \in \Theta(\log n). \tag{1.6}$$

Thus, the harmonic numbers have a logarithmic growth rate.

It is not difficult to prove (1.6) by upper-bounding and lower-bounding the sum $\sum_{i=1}^{n}(1/i)$ using terms of the form $1/2^j$. We illustrate the idea with $n = 8$. First we consider the Ω-bound. We have

$$\begin{aligned} H_8 &= 1 + \frac{1}{2} + \frac{1}{3} + \frac{1}{4} + \frac{1}{5} + \frac{1}{6} + \frac{1}{7} + \frac{1}{8} \\ &\geq 1 + \frac{1}{2} + \frac{1}{4} + \frac{1}{4} + \frac{1}{8} + \frac{1}{8} + \frac{1}{8} + \frac{1}{8} \\ &= 1 + \frac{1}{2} + \frac{1}{2} + \frac{1}{2} \\ &= \frac{5}{2}. \end{aligned}$$

More generally, using the same technique, we can prove that

$$H_{2^k} \geq 1 + \frac{k}{2} = \frac{k+2}{2}.$$

Hence, if $n = 2^k$, then we have $k = \log_2 n$ and $H_n \in \Omega(\log n)$.

The upper bound uses a similar idea:

$$\begin{aligned} H_8 &= 1 + \frac{1}{2} + \frac{1}{3} + \frac{1}{4} + \frac{1}{5} + \frac{1}{6} + \frac{1}{7} + \frac{1}{8} \\ &\leq 1 + \frac{1}{2} + \frac{1}{2} + \frac{1}{4} + \frac{1}{4} + \frac{1}{4} + \frac{1}{4} + \frac{1}{8} \\ &= 1 + 1 + 1 + \frac{1}{8} \\ &= \frac{25}{8}. \end{aligned}$$

In general, using the same technique, we can prove that

$$H_{2^k} \leq k + \frac{1}{2^k}.$$

Hence, if $n = 2^k$, then $H_n \in O(\log n)$.

For an arbitrary positive integer n, we can write $2^k \leq n < 2^{k+1}$ for a unique integer k. Then, from the above formulas, we have

$$\begin{aligned} \frac{k+2}{2} &\leq H_{2^k} \\ &\leq H(n) \\ &< H_{2^{k+1}} \\ &\leq k + 1 + \frac{1}{2^{k+1}}. \end{aligned}$$

Using the fact that $k \in \Theta(\log n)$, we see that $H_n \in \Theta(\log n)$ for all positive integers n.

The above argument is sufficient to establish the growth rate of the harmonic sequence. However, it is possible to prove the more precise result that

$$\lim_{n \to \infty} (H_n - \ln n) = \gamma, \tag{1.7}$$

where $\gamma \approx 0.57721$ is ***Euler's constant***.

Actually, the convergence in (1.7) is quite rapid. For all integers $n \geq 5$, it holds that

$$0 \leq H_n - \ln n - \gamma \leq 0.1.$$

The formula (1.7) also leads to an approximation of the difference between two harmonic numbers. For positive integers n and m with $n > m$,

$$H_n - H_m \approx \ln(n/m). \tag{1.8}$$

It is quite easy to see how (1.8) follows from (1.7). We have

$$\begin{aligned} H_n &\approx \ln n + \gamma \quad \text{and} \\ H_m &\approx \ln m + \gamma, \quad \text{so} \\ H_n - H_m &\approx \ln n - \ln m \\ &= \ln(n/m). \end{aligned}$$

1.4.2 Logarithm Formulae

Recall that $\log_b x = i$ if and only if $b^i = x$. Here, i is the ***logarithm*** of x to the base b. We present several useful formulae involving logarithms. In all these formulae, we assume that bases of logarithms are positive real numbers not equal to 1, and we only take the logarithm of a positive real number (because $\log_b x$ is not defined if $x < 0$).

THEOREM 1.11 *Suppose $b > 0$, $b \neq 1$, and $x, y > 0$. Then the following hold.*

$$\log_b xy = \log_b x + \log_b y \tag{1.9}$$

$$\log_b(x/y) = \log_b x - \log_b y \tag{1.10}$$

$$\log_b(1/x) = -\log_b x \tag{1.11}$$

$$\log_b x^y = y \log_b x \tag{1.12}$$

$$\log_b a = \frac{1}{\log_a b} \tag{1.13}$$

$$\log_b a = \frac{\log_c a}{\log_c b} \tag{1.14}$$

$$a^{\log_b c} = c^{\log_b a} \tag{1.15}$$

The first four formulae in Theorem 1.11 would be considered "basic" properties of logarithms. The last three formulae are perhaps less obvious, but they can be proven using familiar properties of exponents. To illustrate, we provide proofs of two of these formulae.

PROOF **[of (1.13)]** We have

$$\begin{aligned}\log_b a = x &\Rightarrow b^x = a \\ &\Rightarrow (b^x)^{1/x} = a^{1/x} \\ &\Rightarrow b = a^{1/x} \\ &\Rightarrow \log_a b = \frac{1}{x} \\ &\Rightarrow \log_a b = \frac{1}{\log_b a}.\end{aligned}$$

∎

PROOF **[of (1.14)]** Let $\log_c a = x$ and $\log_c b = y$. Then $\log_a c = 1/x$ and $\log_b c = 1/y$ from (1.13). Hence, $c = a^{1/x}$ and $c = b^{1/y}$. So we have

$$\begin{aligned}a^{1/x} = b^{1/y} &\Rightarrow a = b^{x/y} \\ &\Rightarrow \log_b a = \frac{x}{y} \\ &\Rightarrow \log_b a = \frac{\log_c a}{\log_c b}.\end{aligned}$$

■

From (1.14), we see that *logarithmic growth* is independent of the base of the logarithms. Fix constants $a, b > 1$; then

$$\log_b x = \log_b a \times \log_a x = c \log_a x,$$

where $c = \log_b a$ is a positive constant. Hence $\log_b x \in \Theta(\log_a x)$. So we often just write $\Theta(\log n)$ to denote logarithmic growth (as a function of n), without specifying a base for the logarithms.

1.4.3 Miscellaneous Formulae

We present several more useful formulae in this section.

- For a positive integer n, define

$$n! = n \times (n-1) \times \cdots \times 1.$$

$n!$ is referred to as "n factorial." Then it can be proven that

$$n! \in \Theta\left(n^{n+1/2} e^{-n}\right). \tag{1.16}$$

Upper and lower bounds on $n!$ are stated more precisely in ***Stirling's Formula***:

$$\sqrt{2\pi}\, n^{n+\frac{1}{2}} e^{-n} \leq n! \leq e\, n^{n+\frac{1}{2}} e^{-n}. \tag{1.17}$$

- We have the following formula for the growth rate of a logarithm of a factorial:

$$\log_2 n! \in \Theta(n \log n). \tag{1.18}$$

A direct proof of this formula is not difficult. First,

$$\begin{aligned}
\log_2 n! &= \sum_{i=1}^{n} \log_2 i \qquad \text{from (1.9)} \\
&\leq \sum_{i=1}^{n} \log_2 n \\
&= n \log_2 n,
\end{aligned}$$

so $\log_2 n! \in O(n \log n)$.

For the lower bound, we have

$$\begin{aligned}
\log_2 n! &= \sum_{i=1}^{n} \log_2 i \\
&\geq \sum_{i=n/2}^{n} \log_2 \left(\frac{n}{2}\right) \\
&= \left(\frac{n}{2}\right) \log_2 \left(\frac{n}{2}\right) \\
&= \left(\frac{n}{2}\right) (\log_2 n - 1),
\end{aligned}$$

so $\log_2 n! \in \Omega(n \log n)$.

- Here is an alternative way to prove (1.18). From (1.16), it follows that

$$\ln n! \in \Theta\left(\ln\left(n^{n+1/2}e^{-n}\right)\right).$$

However,

$$\begin{aligned}\ln\left(n^{n+1/2}e^{-n}\right) &= \left(n+\frac{1}{2}\right)\ln n - n\\ &\in \Theta(n \log n),\end{aligned}$$

so we conclude that $\ln n! \in \Theta(n \log n)$.

- Here is another useful observation. Using the fact that

$$\sum_{i=1}^{n} \log_b n = \log_b n!,$$

we have

$$\sum_{i=1}^{n} \log_b n \in \Theta(n \log n). \tag{1.19}$$

- Another useful formula is

$$\sum_{i=1}^{\infty} \frac{1}{i^2} = \frac{\pi^2}{6},$$

which implies that

$$\sum_{i=1}^{n} \frac{1}{i^2} \in \Theta(1).$$

- Here is a formula for the growth rate of the sum of the cth powers of the first n integers when $c \geq 1$:

$$\sum_{i=1}^{n} i^c \in \Theta(n^{c+1}). \tag{1.20}$$

When $c = 1$, we just have an arithmetic sequence, and the exact sum is given by (1.1). For $c = 2$ and $c = 3$, it is possible to prove the following exact formulae:

$$\sum_{i=1}^{n} i^2 = \frac{n(n+1)(2n+1)}{6} \tag{1.21}$$

and

$$\sum_{i=1}^{n} i^3 = \frac{n^2(n+1)^2}{4}. \tag{1.22}$$

- The ***Binomial Theorem*** provides a formula to expand $(x+y)^n$ as a sum of powers of x and y:
$$(x+y)^n = \sum_{i=0}^{n} \binom{n}{i} x^i y^{n-i}, \tag{1.23}$$
where
$$\binom{n}{i} = \frac{n!}{i!(n-i)!} \tag{1.24}$$
is a ***binomial coefficient***.
Setting $x = y = 1$, we obtain the following formula:
$$\sum_{i=0}^{n} \binom{n}{i} = 2^n. \tag{1.25}$$
Another useful formula involving binomial coefficients is the following:
$$\sum_{i=0}^{n} i \binom{n}{i} = n2^{n-1}. \tag{1.26}$$

1.4.4 Equivalence Relations

The notion of an equivalence relation is useful in studying algorithms. Roughly speaking, whenever we have a set X that is partitioned into smaller subsets, we have an equivalence relation. These subsets are called equivalence classes, and two elements of X that are in the same equivalence class are said to be equivalent.

We now give a more formal mathematical treatment of these ideas. Let X be any set (finite or infinite). A ***binary relation*** (or more simply, a ***relation***) on X is a set of ordered pairs $R \subseteq X \times X$. If $(x, y) \in R$, then we say that x is ***related*** to y. We sometimes express this using alternate notation. For example, it is common to record the fact that $(x, y) \in R$ by writing $x \sim_R y$, or just $x \sim y$ if the relation R is fixed.

An arbitrary relation is not particular useful, since any collection of ordered pairs is a relation. We are mostly interested in relations that satisfy certain properties. In particular, a relation R on a set X is an ***equivalence relation*** if it satisfies the following three properties:

reflexivity

$x \sim_R x$ for all $x \in X$.

symmetry

$x \sim_R y$ if and only if $y \sim_R x$, for all $x, y \in X$.

transitivity

for all $x, y, z \in X$, if $x \sim_R y$ and $y \sim_R z$, then $x \sim_R z$.

For any $x \in X$, we define

$$[x] = \{y \in X : x \sim_R y\}.$$

The set $[x]$ is the ***equivalence class*** containing x. It is not hard to see that, for any two equivalence classes $[x]$ and $[y]$,

$$[x] = [y] \text{ or } [x] \cap [y] = \varnothing.$$

Thus the equivalence classes form a partition of the set X into smaller sets.

We give a few examples of simple equivalence relations.

1. Let $n \geq 2$ be an integer. For integers x and y, define $x \sim_n y$ if and only if $n \mid (x - y)$. Of course, this is more commonly written as $x \equiv y \bmod n$. This relation is an equivalence relation defined on $\mathbb{Z}$. There are in fact n equivalence classes. The equivalence classes are

$$\{x : x \equiv j \bmod n\},$$

   for $j = 0, 1, \ldots, n - 1$.

2. Let G be a graph with vertices V and edges E. For two vertices $u, v \in V$, define $u \sim v$ if there is a path from u to v consisting of edges in E (if $u = v$, we allow the empty path that contains no edges). This relation is an equivalence relation defined on V and the equivalence classes are the *connected components* of the graph G. Connected components are discussed further in Section 7.5.2.

3. Suppose $f(n)$ and $g(n)$ are real-valued functions, defined on the positive integers, that take on positive real values for sufficiently large n. Define $f(n) \sim g(n)$ if $f(n) \in \Theta(g(n))$. This defines an equivalence relation: reflexivity is obvious, and symmetry and transitivity follow from the properties stated in Section 1.3.2. Thus, the mathematical statement $f(n) \in \Theta(g(n))$ can be interpreted as saying that $f(n)$ and $g(n)$ are in the same equivalence class of functions.

1.5 Probability Theory and Random Variables

We will require some basic facts concerning probability. The main definitions are reviewed and then we prove some important properties. First, we define the notion of a (discrete) random variable.[1]

A ***discrete random variable***, say $\mathbf{X}$, consists of a finite (or countably infinite) set

[1]Because we are only considering random variables that are discrete in this book, we will often refer to them simply as ***random variables***.

X and a ***probability distribution*** defined on X. The probability that the random variable $\mathbf{X}$ takes on the value $x \in X$ is denoted $\mathbf{Pr}[\mathbf{X} = x]$; sometimes we will abbreviate this to $\mathbf{Pr}[x]$ if the random variable $\mathbf{X}$ is fixed. Since we have a probability distribution defined on X, it must be the case that $0 \leq \mathbf{Pr}[x]$ for all $x \in X$, and

$$\sum_{x \in X} \mathbf{Pr}[x] = 1.$$

As an example, we could consider a coin toss to be a random variable defined on the set $\{heads, tails\}$. The associated probability distribution would be

$$\mathbf{Pr}[heads] = \mathbf{Pr}[tails] = \frac{1}{2}.$$

Suppose we have random variable $\mathbf{X}$ defined on X, and suppose $E \subseteq X$. The probability that $\mathbf{X}$ takes on a value in the subset E is computed to be

$$\mathbf{Pr}[x \in E] = \sum_{x \in E} \mathbf{Pr}[x]. \tag{1.27}$$

The subset E is often called an ***event***.

Example 1.11 Suppose we consider a random throw of a pair of dice. This can be modeled by a random variable $\mathbf{Z}$ defined on the set

$$Z = \{1,2,3,4,5,6\} \times \{1,2,3,4,5,6\},$$

where $\mathbf{Pr}[(i,j)] = 1/36$ for all $(i,j) \in Z$. Let's consider the sum of the two dice. Each possible sum defines an event, and the probabilities of these events can be computed using equation (1.27). For example, suppose that we want to compute the probability that the sum is 4. This corresponds to the event

$$S_4 = \{(1,3),(2,2),(3,1)\},$$

and therefore $\mathbf{Pr}[S_4] = 3/36 = 1/12$.

The probabilities of all the possible sums can be computed in a similar fashion. If we denote by S_j the event that the sum is j, then we obtain the following:

$$\begin{aligned}
\mathbf{Pr}[S_2] &= \frac{1}{36} & \mathbf{Pr}[S_8] &= \frac{5}{36}\\
\mathbf{Pr}[S_3] &= \frac{1}{18} & \mathbf{Pr}[S_9] &= \frac{1}{9}\\
\mathbf{Pr}[S_4] &= \frac{1}{12} & \mathbf{Pr}[S_{10}] &= \frac{1}{12}\\
\mathbf{Pr}[S_5] &= \frac{1}{9} & \mathbf{Pr}[S_{11}] &= \frac{1}{18}\\
\mathbf{Pr}[S_6] &= \frac{5}{36} & \mathbf{Pr}[S_{12}] &= \frac{1}{36}\\
\mathbf{Pr}[S_7] &= \frac{1}{6}.
\end{aligned}$$

Since the events $S_2, \ldots, S_{12}$ partition the set Z, it follows that we can consider the value of the sum of a pair of dice to be a random variable in its own right, which has the probability distribution computed above. □

If a discrete random variable $\mathbf{X}$ takes on real values, then we can define the ***expectation*** of $\mathbf{X}$, which is denoted by $E[\mathbf{X}]$, by the following formula:

$$E[\mathbf{X}] = \sum_{x \in X} (\mathbf{Pr}[x] \times x). \tag{1.28}$$

The expectation of a random variable is thus the weighted average value of that random variable.

THEOREM 1.12 *Suppose we perform a sequence of trials of an experiment, where each trial succeeds with probability p. We assume that the trials are all independent. Then the expected number of trials until the first successful outcome is achieved is $1/p$.*

PROOF We define a random variable $\mathbf{X}$ that takes on positive integral values as follows:

$$\mathbf{X} = i \text{ if the first successful outcome is in the } i\text{th trial.}$$

Clearly $\mathbf{X} = i$ if and only if the first $i - 1$ trials fail and the ith trial succeeds. Since the trials are independent, we have

$$\mathbf{Pr}[\mathbf{X} = i] = (1-p)^{i-1}p.$$

Now we can compute

$$\begin{aligned} E[\mathbf{X}] &= \sum_{i=1}^{\infty} i(1-p)^{i-1}p \\ &= p \sum_{i=1}^{\infty} i(1-p)^{i-1} \end{aligned} \tag{1.29}$$

Suppose we take $a = 0$, $d = 1$ and $r = 1 - p$ in (1.5). Then we obtain

$$\sum_{i=0}^{\infty} i(1-p)^i = \frac{1-p}{p^2}.$$

Dividing by $1 - p$, we obtain

$$\sum_{i=0}^{\infty} i(1-p)^{i-1} = \frac{1}{p^2}.$$

Substituting into (1.29), we see that

$$E[\mathbf{X}] = p \times \frac{1}{p^2} = \frac{1}{p}.$$

Therefore the expected number of trials until the first success is obtained is $1/p$.
∎

It turns out that there is an easier way to prove Theorem 1.12. It is based on the following formula to compute an expectation of a random variable that takes on values that are non-negative integers.

THEOREM 1.13 *Suppose a random variable* $\mathbf{X}$ *takes on values* $0, 1, \ldots$. *Then*

$$E[\mathbf{X}] = \sum_{i=1}^{\infty} \mathbf{Pr}[\mathbf{X} \geq i]. \tag{1.30}$$

PROOF We have

$$\begin{aligned}
E[\mathbf{X}] &= \sum_{i=0}^{\infty} i\,\mathbf{Pr}[\mathbf{X} = i] \\
&= \sum_{i=0}^{\infty} i(\mathbf{Pr}[\mathbf{X} \geq i] - \mathbf{Pr}[\mathbf{X} \geq i+1]) \\
&= \sum_{i=0}^{\infty} i\,\mathbf{Pr}[\mathbf{X} \geq i] - \sum_{i=0}^{\infty} i\,\mathbf{Pr}[\mathbf{X} \geq i+1] \\
&= \sum_{i=0}^{\infty} i\,\mathbf{Pr}[\mathbf{X} \geq i] - \sum_{i=1}^{\infty} (i-1)\mathbf{Pr}[\mathbf{X} \geq i]) \\
&\qquad\qquad \text{replacing } i+1 \text{ by } i \text{ in the second sum} \\
&= \sum_{i=1}^{\infty} \mathbf{Pr}[\mathbf{X} \geq i].
\end{aligned}$$

∎

We can now give an alternative proof of Theorem 1.12.

PROOF It is clear that

$$\mathbf{Pr}[\mathbf{X} \geq i] = (1-p)^{i-1}.$$

Then, from (1.30), we have

$$\begin{aligned}
E[\mathbf{X}] &= \sum_{i=1}^{\infty} \mathbf{Pr}[\mathbf{X} \geq i] \\
&= \sum_{i=1}^{\infty} (1-p)^{i-1} \\
&= \frac{1}{1-(1-p)} \\
&= \frac{1}{p},
\end{aligned}$$

from (1.3), since the desired sum is a geometric series. ∎

We next consider the concept of joint and conditional probabilities. Suppose $\mathbf{X}$ and $\mathbf{Y}$ are random variables defined on sets X and Y, respectively. Let $x \in X$ and $y \in Y$. The ***joint probability,***

$$\mathbf{Pr}[\mathbf{X} = x, \mathbf{Y} = y],$$

is the probability that $\mathbf{X}$ takes on the value x *and* $\mathbf{Y}$ takes on the value y.

The ***conditional probability,***

$$\mathbf{Pr}[\mathbf{X} = x | \mathbf{Y} = y],$$

denotes the probability that $\mathbf{X}$ takes on the value x, given that $\mathbf{Y}$ takes on the value y.

Joint probability can be related to conditional probability by the formula

$$\mathbf{Pr}[\mathbf{X} = x, \mathbf{Y} = y] = \mathbf{Pr}[\mathbf{X} = x | \mathbf{Y} = y]\mathbf{Pr}[\mathbf{Y} = y].$$

Interchanging x and y, we also have that

$$\mathbf{Pr}[\mathbf{X} = x, \mathbf{Y} = y] = \mathbf{Pr}[\mathbf{Y} = y | \mathbf{X} = x]\mathbf{Pr}[\mathbf{X} = x].$$

From these two expressions, we immediately obtain the following result, which is known as Bayes' Theorem.

THEOREM 1.14 (Bayes' Theorem) *If* $\mathbf{Pr}[y] > 0$, *then*

$$\mathbf{Pr}[\mathbf{X} = x | \mathbf{Y} = y] = \frac{\mathbf{Pr}[\mathbf{X} = x]\mathbf{Pr}[\mathbf{Y} = y | \mathbf{X} = x]}{\mathbf{Pr}[\mathbf{Y} = y]}.$$

The random variables $\mathbf{X}$ and $\mathbf{Y}$ are said to be ***independent random variables*** if

$$\mathbf{Pr}[\mathbf{X} = x, \mathbf{Y} = y] = \mathbf{Pr}[\mathbf{X} = x] \times \mathbf{Pr}[\mathbf{Y} = y]$$

for all $x \in X$ and $y \in Y$.

The following is an immediate corollary of Theorem 1.14.

COROLLARY 1.15 $\mathbf{X}$ *and* $\mathbf{Y}$ *are independent random variables if and only if*

$$\mathbf{Pr}[\mathbf{X} = x | \mathbf{Y} = y] = \mathbf{Pr}[\mathbf{X} = x]$$

for all $x \in X$ *and* $y \in Y$.

It is also useful to observe the relation between the probability distributions on X and Y, and the joint probability distribution:

$$\begin{aligned} \mathbf{Pr}[\mathbf{X} = x] &= \sum_{y \in Y} \mathbf{Pr}[\mathbf{X} = x, \mathbf{Y} = y] \\ \mathbf{Pr}[\mathbf{Y} = y] &= \sum_{x \in X} \mathbf{Pr}[\mathbf{X} = x, \mathbf{Y} = y]. \end{aligned} \tag{1.31}$$

Suppose that $\mathbf{X}$ and $\mathbf{Y}$ are real-valued random variables. Then the sum of $\mathbf{X}$ and $\mathbf{Y}$ is again a random variable. This random variable $\mathbf{Z} = \mathbf{X} + \mathbf{Y}$ is defined in terms of the joint probability distribution of $\mathbf{X}$ and $\mathbf{Y}$ as follows:

$$\mathbf{Pr}[\mathbf{Z} = z] = \sum_{\{(x,y):x+y=z\}} \mathbf{Pr}[\mathbf{X} = x, \mathbf{Y} = y]. \tag{1.32}$$

Observe that the formula (1.32) considers all combinations of x and y that sum to z in order to define the random variable $\mathbf{Z} = \mathbf{X} + \mathbf{Y}$.

One important and very useful property of expectation is called ***linearity of expectation***. Informally, this says that the expectation of the sum of random variables is equal to the sum of their expectations. This property would be considered to be "intuitively obvious" in the case that the random variables are independent. However, perhaps surprisingly, linearity of expectation holds for *any* random variables, independent or not.

THEOREM 1.16 (linearity of expectation) *Suppose that* $\mathbf{X}$ *and* $\mathbf{Y}$ *are real-valued random variables. Then* $E[\mathbf{X} + \mathbf{Y}] = E[\mathbf{X}] + E[\mathbf{Y}]$.

PROOF Let $\mathbf{Z} = \mathbf{X} + \mathbf{Y}$. We compute as follows:

$$\begin{aligned}
E[\mathbf{Z}] &= \sum_z (z \times \mathbf{Pr}[\mathbf{Z} = z]) \\
&= \sum_z \left(z \times \sum_{\{(x,y):x+y=z\}} \mathbf{Pr}[\mathbf{X} = x, \mathbf{Y} = y] \right) && \text{from (1.32)} \\
&= \sum_x \sum_y ((x+y) \times \mathbf{Pr}[\mathbf{X} = x, \mathbf{Y} = y]) \\
&= \sum_x x \sum_y \mathbf{Pr}[\mathbf{X} = x, \mathbf{Y} = y] + \sum_y y \sum_x \mathbf{Pr}[\mathbf{X} = x, \mathbf{Y} = y] \\
&= \sum_x (x \times \mathbf{Pr}[\mathbf{X} = x]) + \sum_y (y \times \mathbf{Pr}[\mathbf{Y} = y]) && \text{from (1.31)} \\
&= E[\mathbf{X}] + E[\mathbf{Y}].
\end{aligned}$$

∎

Linearity of expectation also holds for the sum of any $r \geq 2$ real-valued random variables.

THEOREM 1.17 (generalized linearity of expectation) *Suppose that* $\mathbf{X_i}$ *is a real-valued random variable, for* $1 \leq i \leq r$. *Then*

$$E\left[\sum_{i=1}^{r} \mathbf{X_i}\right] = \sum_{i=1}^{r} E[\mathbf{X_i}].$$

Let's return to Example 1.11 and compute the expected value of the sum of two dice.

Example 1.12 A random throw of a die corresponds to a random variable $\mathbf{X}$ where

$$\mathbf{Pr}[\mathbf{X} = i] = \frac{1}{6}$$

for $i = 1, 2, 3, 4, 5, 6$. It is therefore easy to compute

$$\begin{aligned} E[\mathbf{X}] &= \sum_{i=1}^{6} \left(\frac{1}{6} \times i \right) \\ &= \frac{1}{6} \times (1 + 2 + 3 + 4 + 5 + 6) \\ &= \frac{21}{6} \\ &= 3.5. \end{aligned}$$

Now, suppose that we have a pair of dice, which correspond to two random variables $\mathbf{X_1}$ and $\mathbf{X_2}$. Let $\mathbf{X} = \mathbf{X_1} + \mathbf{X_2}$. We have

$$E[\mathbf{X_1}] = E[\mathbf{X_2}] = 3.5,$$

so $E[\mathbf{X}] = 7$ by Theorem 1.16. In other words, the expected value of the sum of two dice is equal to 7.

We would reach the same result using the probability distribution that we computed in Example 1.11, but the calculation would be more complicated. It would amount to computing

$$\sum_{j=2}^{12} (j \times \mathbf{Pr}[S_j])$$

using the values $\mathbf{Pr}[S_j]$ that were determined in Example 1.11. The reader can verify that the same result (namely, 7) is obtained. □

The two random variables defined in Example 1.12 are independent. We now present an example involving linearity of expectation of random variables that are not independent.

Example 1.13 Suppose that n children, say $C_1, \dots, C_n$, have n different books, say $B_1, \dots, B_n$. The books are collected and then randomly redistributed among the n children (one book per child). What is the expected number of children who are given the correct book?

Suppose we define random variables $\mathbf{X_i}$ (for $i = 1, \dots, n$), where

$$\begin{aligned} \mathbf{X_i} &= 1 \quad \text{if } C_i \text{ receives book } B_i \\ \mathbf{X_i} &= 0 \quad \text{if } C_i \text{ doesn't receive book } B_i. \end{aligned}$$

These are sometimes called ***indicator*** random variables. Since the books are distributed randomly, it is clear that

$$\begin{aligned} \mathbf{Pr}[\mathbf{X_i} = 1] &= \frac{1}{n} \quad \text{and} \\ \mathbf{Pr}[\mathbf{X_i} = 0] &= 1 - \frac{1}{n}, \end{aligned}$$

for $i = 1, \ldots, n$. Then we compute

$$\begin{aligned} E[\mathbf{X_i}] &= 1 \times \mathbf{Pr}[\mathbf{X_i} = 1] + 0 \times \mathbf{Pr}[\mathbf{X_i} = 0] \\ &= \mathbf{Pr}[\mathbf{X_i} = 1] \\ &= \frac{1}{n}. \end{aligned}$$

Now, suppose we define

$$\mathbf{X} = \sum_{i=1}^{n} \mathbf{X_i}.$$

It is clear that $E[\mathbf{X}]$ is the expected number of books that are returned to their correct owners. From Theorem 1.17, we have

$$\begin{aligned} E[\mathbf{X}] &= \sum_{i=1}^{n} E[\mathbf{X_i}] \\ &= n \times \frac{1}{n} \\ &= 1. \end{aligned}$$

Hence, on average, one child receives the correct book. □

1.6 Notes and References

Knuth [42] is a good reference for many of the mathematical concepts discussed in this chapter. For a discussion on the use of asymptotic notation with multiple variables, see Howell [35].

There are now a wide variety of books on design and analysis of algorithms. Recommended books include the following:

- *Introduction to Algorithms. Third Edition,* by Cormen, Leiserson, Rivest and Stein [16].
- *Fundamental of Algorithmics,* by Brassard and Bratley [11].
- *Algorithms,* by Dasgupta, Papadimitriou and Vazirani [18].
- *Algorithm Design,* by Kleinberg and Tardos [39].
- *Algorithms. Fourth Edition,* by Sedgewick and Wayne [54].

Exercises

1.1 Give a proof from first principles (not using limits) that $n^3 - 100n + 1000 \in \Theta(n^3)$.

1.2 Compare the growth rates of the functions

$$n^{\log_2 \log_2 n} \quad \text{and} \quad (\log_2 n)^{\log_2 n}.$$

1.3 Compare the growth rates of the following three functions:

$$n!, \quad n^{n+1} \quad \text{and} \quad n^{n+1/2}.$$

1.4 For each pair of functions $f(n)$ and $g(n)$, fill in the correct asymptotic notation among Θ, o, and ω in the statement

$$f(n) \in \_\_(g(n)).$$

Formal proofs are not necessary, but provide brief justifications for all of your answers.

(a) $f(n) = \sum_{i=1}^{n-1}(i+1)/i^2$ vs. $g(n) = \log_2(n^{100})$
(b) $f(n) = n^{3/2}$ vs. $g(n) = (n+1)^9/(n^3-1)^2$.
(c) $f(n) = 3^{\log_9 n}$ vs. $g(n) = n^{1/4} + \sqrt{n} + \log n$
(d) $f(n) = (32768)^{n/5}$ vs. $g(n) = (6561)^{n/4}$

1.5 Suppose that $f(n)$ is $O(g(n))$. Give a formal proof that $f(n) + 100$ is $O(g(n) + 100)$.

1.6 Determine the growth rate of

$$\sum_{i=1}^{n} H_n,$$

where H_n is the nth harmonic number.

HINT First, prove that

$$\sum_{i=1}^{n} H_n = (n+1)H_n - n.$$

1.7 Use induction to prove the formulae (1.1) and (1.2).

1.8 Use the formula (1.4) to compute the sum

$$3 + 5 \times 2 + 7 \times 2^2 + \cdots + 21 \times 2^9.$$

1.9 For $c \geq 1$, prove that

$$\sum_{i=1}^{n} i^c \in \Theta(n^{c+1})$$

by computing upper and lower bounds on the given sum.

1.10 Use (1.23) and a bit of calculus to prove (1.26), as follows. First, let $y = 1$ in (1.23), then take the derivative with respect to x, and finally set $x = 1$.

1.11 Use Stirling's Formula (1.17) to show that

$$\binom{2n}{n} \in \Theta\left(\frac{4^n}{\sqrt{n}}\right).$$

1.12 Consider the following relation $\sim$ that is defined on the set $\{n \in \mathbb{Z} : n \geq 2\}$:

$$n \sim m \text{ if and only if } \gcd(n, m) > 1.$$

Which of the properties reflexivity, symmetry, and transitivity does $\sim$ satisfy?

1.13 Consider the following relation $\sim$ that is defined on the set $\{n \in \mathbb{Z} : n \geq 2\}$:

$$n \sim m \text{ if and only if } n | m.$$

Which of the properties reflexivity, symmetry, and transitivity does $\sim$ satisfy?

1.14 Prove that the following generalization of Theorem 1.16 holds: Suppose that $\mathbf{X}$ and $\mathbf{Y}$ are real-valued random variables and α and β are real numbers. Then $E[\alpha\mathbf{X} + \beta\mathbf{Y}] = \alpha E[\mathbf{X}] + \beta E[\mathbf{Y}]$.

1.15 Compute the expected value of the sum of a pair of dice directly from the definition, i.e., using the formula (1.28).

1.16 Suppose there are four pairs of socks in a drawer, having the colors red, black, blue and brown. Suppose that five of the eight socks are chosen randomly.

(a) What is the probability that the red pair is among the five chosen socks?

(b) Let $\mathbf{X}$ be a random variable that takes on the value 1 if the red pair is among five randomly chosen socks; and 0, otherwise. Compute $E[\mathbf{X}]$.

(c) Use linearity of expectation to compute the expected number of pairs of socks among five randomly chosen socks.

Chapter 2

Algorithm Analysis and Reductions

This chapter begins with basic techniques of loop analysis. Then we present a case study of designing simple algorithms for the famous **3SUM** problem. We introduce the idea of reductions, which will play an important role in the study of intractability. Finally, we provide some examples of exact analysis of certain algorithms.

2.1 Loop Analysis Techniques

Many simple algorithms involve nested loops, so it is helpful to develop some techniques to analyze the complexity of loop structures. The following are the main techniques used in loop analysis.

- Identify ***elementary operations*** that require constant time (which we denote as $\Theta(1)$ time). Elementary operations are discussed in Section 2.1.1.
- The complexity of a loop is expressed as the *sum* of the complexities of each iteration of the loop.
- Analyze independent loops *separately*, and then *add* the results: use "maximum" rules (Theorem 1.7) and simplify whenever possible.
- If loops are nested, start with the *innermost loop* and proceed outwards. In general, this kind of analysis requires evaluation of *nested summations* using "summation" rules (Theorem 1.8).

Sometimes a O-bound is sufficient. However, we often want a more precise Θ-bound. Two general strategies used in loop analysis to obtain Θ-bounds are as follows:

1. Use Θ-bounds *throughout the analysis* and thereby obtain a Θ-bound for the complexity of the algorithm.
2. Prove a O-bound and a *matching* Ω-bound *separately* to get a Θ-bound. Sometimes this technique is easier because arguments for O-bounds may use simpler upper bounds (and arguments for Ω-bounds may use simpler lower bounds) than arguments for Θ-bounds do.

We will illustrate both of these strategies with various examples.

DOI: 10.1201/9780429277412-2

Algorithm 2.1: LOOPANALYSIS1(n)

```
comment: n is a positive integer
(1) sum ← 0
(2) for i ← 1 to n
      do { for j ← 1 to i
             do { sum ← sum + (i − j)^2
                  sum ← ⌊sum/i⌋
(3) return (sum)
```

2.1.1 Elementary Operations in the Unit Cost Model

For now, we will discuss elementary operations in the ***unit cost model***. In this model, we assume that arithmetic operations such as $+$, $-$, $\times$ and integer division take time $\Theta(1)$. This is a reasonable assumption for integers of *bounded size* (e.g., integers that fit into one word of memory).

If we want to consider the complexity of arithmetic operation on integers of arbitrary size, we need to consider ***bit complexity***, where we express the complexity as a function of the length of the integers (as measured in bits). We will see some examples later, such as **Multiprecision Multiplication** (see Section 4.9).

In general, we use common sense in deciding which operations take time $\Theta(1)$. For example, sorting an array or copying an array are obviously *not* unit cost operations, as the time required for these kinds of operations depends on the length of the array. On the other hand, concatenating two linked lists *could* be a unit cost operation, if pointers are maintained to the beginning and ends of the lists. So, in this example, we would need to know some details about how the list data structure is implemented.

2.1.2 Examples of Loop Analysis

We begin by performing a simple loop analysis of Algorithm 2.1. First, we do a Θ-bound analysis. We consider the three main steps in Algorithm 2.1, which are labeled (1), (2) and (3).

(1) This is just an assignment statement, which has complexity $\Theta(1)$.

(2) The complexity of the inner **for** loop is $\Theta(i)$. The complexity of the outer **for** loop can be computed as

$$\begin{aligned}\sum_{i=1}^{n} \Theta(i) &= \Theta\left(\sum_{i=1}^{n} i\right) && \text{from Theorem 1.8}\\ &= \Theta\left(\frac{n(n+1)}{2}\right) && \text{from (1.1)}\\ &= \Theta(n^2).\end{aligned}$$

In the above, we are using the formula for the sum of an arithmetic sequence.

(3) This statement just returns a value to the calling program, so it has complexity $\Theta(1)$.

Finally, using Theorem 1.8, the total of (1) + (2) + (3) is

$$\Theta(1) + \Theta(n^2) + \Theta(1) = \Theta(n^2).$$

The alternative approach is to prove separate O- and Ω-bounds. We focus here on the two nested **for** loops (i.e., (2)), since (1) and (3) both have complexity $\Theta(1)$.

upper bound

$$\begin{aligned}\sum_{i=1}^{n} O(i) &\leq \sum_{i=1}^{n} O(n) \\ &= O\left(\sum_{i=1}^{n} n\right) \\ &= O(n^2).\end{aligned}$$

Here, we make use of the observation that the inner loop takes time $O(n)$ for all values of $i \leq n$.

lower bound

$$\begin{aligned}\sum_{i=1}^{n} \Omega(i) &\geq \sum_{i=n/2}^{n} \Omega(i) \\ &\geq \sum_{i=n/2}^{n} \Omega\left(\frac{n}{2}\right) \\ &= \Omega\left(\frac{n^2}{4}\right) \\ &= \Omega(n^2).\end{aligned}$$

This bound makes use of the fact that the inner loop takes time $\Omega(n/2) = \Omega(n)$ whenever $i \geq n/2$.

combined bound

Finally, since the upper and lower bounds match, the complexity is $\Theta(n^2)$.

We next analyze a slightly more complicated example given in Algorithm 2.2.

```
Algorithm 2.2: LOOPANALYSIS2(A)

comment: A is an array of length n
max ← 0
for i ← 1 to n
  do { for j ← i to n
         do { sum ← 0
              for k ← i to j
                do { sum ← sum + A[k]
                     if sum > max
                       then max ← sum
return (max)
```

First, we do a Θ-bound analysis. The innermost loop (the **for** loop on k) has complexity $\Theta(j-i+1)$, since there are $j-i+1$ iterations, each taking $\Theta(1)$ time. The next loop (**for** j) has complexity

$$\begin{aligned}\sum_{j=i}^{n}\Theta(j-i+1) &= \Theta\left(\sum_{j=i}^{n}(j-i+1)\right)\\ &= \Theta\left(1+2+\cdots+(n-i+1)\right)\\ &= \Theta\left(\frac{(n-i+1)(n-i+2)}{2}\right) \qquad \text{from (1.1)}\\ &= \Theta\left((n-i+1)(n-i+2)\right),\end{aligned}$$

since this is an arithmetic sequence. The outer loop (**for** i) has complexity

$$\begin{aligned}&\sum_{i=1}^{n}\Theta((n-i+1)(n-i+2))\\ &\quad= \Theta\left(\sum_{i=1}^{n}(n-i+1)(n-i+2)\right) \qquad \text{from Theorem 1.8}\\ &\quad= \Theta\left(1\times 2+2\times 3+\cdots+n(n+1)\right)\\ &\quad= \Theta\left(\frac{n^3+3n^2+2n}{3}\right)\\ &\quad= \Theta(n^3).\end{aligned}$$

In the above computation, we use the non-obvious fact that

$$1\times 2+2\times 3+\cdots+n(n+1) = \frac{n^3+3n^2+2n}{3}.$$

This could be proven by induction, for example.

$\mathcal{L}_1$	$\mathcal{L}_2$
$i = 1, \dots, \frac{n}{3}$	$i = 1, \dots, n$
$j = 1 + \frac{2n}{3}, \dots, n$	$j = i + 1, \dots, n$
$k = 1 + \frac{n}{3}, \dots, \frac{2n}{3}$	$k = i, \dots, j$

FIGURE 2.1: Two loop structures.

The above proof was a bit tedious, so it might be of interest to observe that there is a quicker analysis. The number of times the **if** statement is executed is exactly

$$|\{(i,k,j) : 1 \leq i \leq k \leq j \leq n\}| = \binom{n+2}{3}.$$

But again, this depends on a formula (in this case, a combinatorial formula) that might not be familiar to many people.

The alternative approach is to prove separate O- and Ω-bounds, as follows.

upper bound

It is rather obvious that the algorithm has complexity $O(n^3)$; the reader can fill in the details.

lower bound

Consider the two loop structures shown in Figure 2.1. $\mathcal{L}_2$ is the actual loop structure of the algorithm. $\mathcal{L}_1$ is an alternative loop structure with the property that $\mathcal{L}_1 \subset \mathcal{L}_2$. This property holds because the algorithm examines all triples (i,k,j) with $1 \leq i \leq k \leq j \leq n$, and $\mathcal{L}_1$ just considers a subset of these triples.

In $\mathcal{L}_1$, each of i, j, and k take on exactly $n/3$ values, and $\mathcal{L}_1$ considers all the possible combinations. So the total number of iterations is just the product

$$\frac{n}{3} \times \frac{n}{3} \times \frac{n}{3},$$

and therefore there are

$$\left(\frac{n}{3}\right)^3 = \frac{n^3}{27}$$

iterations in $\mathcal{L}_1$. Hence, the number of iterations in L_2 is $\Omega(n^3)$.

In the lower bound analysis, we note that we are not required to choose three equal-length intervals to define the loop structure $\mathcal{L}_1$. For example, we could have

Algorithm 2.3: LOOPANALYSIS3(n)

```
comment: n is a positive integer
sum ← 0
for i ← 1 to n
  do { j ← i
       while j ≥ 1
         do { sum ← sum + i/j
              j ← ⌊j/2⌋
return (sum)
```

instead let i range from 1 to $n/2$, k range from $n/2+1$ to $3n/4$, and j range from $3n/4+1$ to n. Then the total number of iterations would be

$$\frac{n}{2} \times \frac{n}{4} \times \frac{n}{4} = \frac{n^3}{32} \in \Omega(n^3).$$

Finally, we perform a loop analysis where the result includes a "log" term. Consider Algorithm 2.3. Our first task is to compute the number of iterations of the **while** loop. Observe that j takes on the values

$$i, \frac{i}{2}, \frac{i}{4}, \dots,$$

until

$$\frac{i}{2^t} < 1$$

for some integer t. This condition is satisfied if and only if $2^t > i$, or $t > \log_2 i$. So the number of iterations of the **while** loop is $\Theta(\log i)$.

Then, it is easy to see that the overall complexity of Algorithm 2.3 is

$$\begin{aligned} \sum_{i=1}^{n} \Theta(\log i) &= \Theta\left(\sum_{i=1}^{n} \log i\right) && \text{from Theorem 1.8} \\ &= \Theta(\log n!) \\ &= \Theta(n \log n) && \text{from (1.18).} \end{aligned}$$

2.2 Algorithms for the 3SUM Problem

Sometimes it is useful to "jump right in" and so we consider some examples to illustrate the concepts and techniques we have been discussing in previous sections. We introduce a simple problem known as **3SUM** and we provide various algorithms to solve it. This allows us to exemplify some of the ideas of design, complexity analysis and correctness of algorithms that we have been discussing.

Problem 2.1: 3SUM

Instance: An array A of n distinct integers, $A = [A[1], \ldots, A[n]]$.
Question: Do there exist three distinct elements in A whose sum equals 0?

We have not been very specific about how the output is to be presented. We are asking a question with a yes/no answer, so it might be sufficient just to output "yes" or "no," without any additional information. However, in the case where the desired three distinct elements exist, it might be more useful to output distinct indices i, j and k such that $A[i] + A[j] + A[k] = 0$.

2.2.1 An Obvious Algorithm

The **3SUM** problem has an obvious algorithm to solve it. We basically just compute all possible sums $A[i] + A[j] + A[k]$, where $1 \leq i < j < k \leq n$, and see if any of them are equal to 0. The following algorithm determines *all* such triples (i, j, k), in the case where there is more than one. An alternative would be to terminate the algorithm as soon as a single such triple is found.

Algorithm 2.4: TRIVIAL3SUM(A)

```
comment: A is an array of length n
for i ← 1 to n − 2
  do { for j ← i + 1 to n − 1
         do { for k ← j + 1 to n
                do { if A[i] + A[j] + A[k] = 0
                       then output (i, j, k)
```

Here are a few comments:

- Algorithm 2.4 outputs all solutions (i, j, k) to the equation $A[i] + A[j] + A[k] = 0$. If there are no solutions, then Algorithm 2.4 does not produce any output.
- Throughout this book, array indices will always start at 1.
- We only want to generate each solution once. This is achieved in Algorithm 2.4 because the loop structure ensures that $i < j < k$.
- The complexity of TRIVIAL3SUM is obviously $O(n^3)$, because there are three nested loops, all of which involve indices between 1 and n, and each iteration takes only $O(1)$ time.

- The complexity is actually $\Theta(n^3)$. An easy way to see this is to observe that we consider all possible subsets of three distinct indices i, j and k chosen from the set $\{1, \ldots, n\}$. The number of sums that are tested is therefore exactly $\binom{n}{3}$, because we have $i < j < k$ (as mentioned above). Then we can compute

$$\binom{n}{3} = \frac{n(n-1)(n-2)}{6} \in \Theta(n^3),$$

and this tells us that the algorithm has complexity $\Theta(n^3)$.

2.2.2 An Improvement

Instead of having three nested loops, suppose we have *two* nested loops (with indices i and j, say). For each pair of indices (i, j), we *search* for an array element $A[k]$ such that $A[i] + A[j] + A[k] = 0$. We can still ensure that $i < j < k$, so we do not find multiple occurrences of each solution.

If we sequentially try all the possible k-values (for given values of i and j), then we basically end up with the previous algorithm. However, if we *pre-sort* the array A, then we can replace the linear search by a binary search (see Section 4.5 for details of the binary search algorithm).

Algorithm 2.5: IMPROVED3SUM(A)

```
comment: A is an array of length n
external SORT
(B, π) ← SORT(A)
comment: π is the permutation such that A[π(i)] = B[i] for all i
for i ← 1 to n − 2
  do { for j ← i + 1 to n − 1
         do { perform a binary search for B[k] = −B[i] − B[j]
                  in the subarray [B[j + 1], ..., B[n]]
              if the search is successful
                then output (π(i), π(j), π(k))
```

Therefore, Algorithm 2.5 incorporates a ***preprocessing step*** in which we sort the array A. This can be done using any desired sorting algorithm. However, it would make sense to use an efficient sorting algorithm (such as HEAPSORT; see Section 3.4.5) that has complexity $O(n \log n)$. The pre-processing step is only done *once*. We assume that, after sorting A, the result is stored in the array B. Also, we assume that a permutation π is returned by the sorting algorithm, where $A[\pi(i)] = B[i]$ for all i.

With these modifications, the innermost loop of the algorithm will have complexity $O(\log n)$ instead of $O(n)$. Overall, the complexity is reduced from $O(n^3)$ to $O(n^2 \log n)$. We obtain Algorithm 2.5.

The complexity of Algorithm 2.5 must take into account the cost of the preprocessing step. Therefore, the overall complexity is

$$O(n \log n + n^2 \log n) = O(n^2 \log n),$$

from Theorem 1.7. The "$n \log n$" term accounts for the SORT and the $n^2 \log n$ term is the complexity of the n^2 binary searches. It is also possible to prove the lower bound $\Omega(n^2 \log n)$ for the complexity of this algorithm.

Note that we want to ensure that $i < j < k$, so we only perform a binary search of the subarray $[B[j+1], \ldots, B[n]]$.

In Algorithm 2.5, we have not written out the pseudocode for SORT or the binary searches, as these may be considered standard algorithms that we can call as subroutines, if desired. We also note that $\pi(i), \pi(j)$ and $\pi(k)$ are indices in the *original array A*, before it was sorted. An alternative to returning indices of the solution is to return their actual *values*; in this case, we would not need the permutation π.

Example 2.1 We give a small example to show how the required permutation π can be obtained in a simple way. Suppose

$$A = [2, -8, 6 - 3].$$

We construct the list of ordered pairs

$$(2, 1), (-8, 2), (6, 3), (-3, 4)$$

and then sort them by their first coordinates, obtaining

$$(-8, 2), (-3, 4), (2, 1), (6, 3).$$

Thus

$$B = [-8, -3, 2, 6] \quad \text{and} \quad \pi = [2, 4, 1, 3].$$

Applying Algorithm 2.5, we find that

$$B[1] + B[3] + B[4] = 0,$$

so the triple

$$(\pi(1), \pi(3), \pi(4)) = (2, 1, 3)$$

is returned by the algorithm. This indicates that

$$A[2] + A[1] + A[3] = 0.$$

□

2.2.3 A Further Improvement

The SORT used in Algorithm 2.5 modifies the input to permit a more efficient algorithm to be used (a binary search as opposed to linear search). However, there is a better way to make use of the sorted array B. Namely, for a given $B[i]$, we *simultaneously scan from both ends* of A, looking for indices j and k such that $j < k$ and

$$B[j] + B[k] = -B[i].$$

Initially, we have $j = i + 1$ and $k = n$. At any stage of the scan, we either *increment* j or *decrement* k (or both, if it happens that $B[i] + B[j] + B[k] = 0$). See Algorithm 2.6.

```
Algorithm 2.6: QUADRATIC3SUM(A)

comment: A is an array of length n
external SORT
(B, π) ← SORT(A)
comment: π is the permutation such that A[π(i)] = B[i] for all i
for i ← 1 to n − 2
  do { j ← i + 1
       k ← n
       while j < k
         do { S ← B[i] + B[j] + B[k]
              if S < 0
                then j ← j + 1
              else if S > 0
                then k ← k − 1
              else { output (π(i), π(j), π(k))
                     j ← j + 1
                     k ← k − 1
```

The resulting Algorithm 2.6 has complexity

$$O(n \log n + n^2) = O(n^2),$$

from Theorem 1.7. As before, the SORT corresponds to the "$n \log n$" term.

We observe that the two nested loops have complexity $O(n^2)$, as follows. There are $O(n)$ iterations of the **for** loop with index i. In each iteration of the **while** loop, we increment j or decrement k (or both). Thus the value of $k - j$ is decreased by at least 1 in each iteration of the **while** loop. Initially we have $j = i + 1 \geq 2$ and $k = n$, so $k - j \leq n - 2$. Since the **while** loop terminates when $k - j \leq 0$, there are $O(n)$ iterations of the **while** loop. Therefore, each iteration of the **for** loop takes $O(n)$ time and the overall complexity of the algorithm is $O(n^2)$.

Recently, subquadratic algorithms for **3SUM** have been found. An algorithm with complexity $O(n^2(\log \log n)^{O(1)}/(\log n)^2)$ was developed in 2018.

Example 2.2 For simplicity, suppose we have the following array, which is already sorted:

$$-11 \quad -10 \quad -7 \quad -3 \quad 2 \quad 4 \quad 8 \quad 10.$$

Algorithm 2.6 will not find any solutions when $i = 1$:

$\boxed{-11}$	$\boxed{-10}$	-7	-3	2	4	8	$\boxed{10}$	$S = -11$
$\boxed{-11}$	-10	$\boxed{-7}$	-3	2	4	8	$\boxed{10}$	$S = -8$
$\boxed{-11}$	-10	-7	$\boxed{-3}$	2	4	8	$\boxed{10}$	$S = -4$
$\boxed{-11}$	-10	-7	-3	$\boxed{2}$	4	8	$\boxed{10}$	$S = 1$
$\boxed{-11}$	-10	-7	-3	$\boxed{2}$	4	$\boxed{8}$	10	$S = -1$
$\boxed{-11}$	-10	-7	-3	2	$\boxed{4}$	$\boxed{8}$	10	$S = 1$

However, when $i = 2$, a solution is found:

-11	$\boxed{-10}$	$\boxed{-7}$	-3	2	4	8	$\boxed{10}$	$S = -7$
-11	$\boxed{-10}$	-7	$\boxed{-3}$	2	4	8	$\boxed{10}$	$S = -3$
-11	$\boxed{-10}$	-7	-3	$\boxed{2}$	4	8	$\boxed{10}$	$S = 2$
-11	$\boxed{-10}$	-7	-3	$\boxed{2}$	4	$\boxed{8}$	10	$S = 0$

□

The correctness of Algorithm 2.6 is perhaps not so obvious, so it is probably useful to give a detailed proof of correctness. We are assuming that the elements in the array are distinct, and for simplicity, we will also assume that the array is already sorted.

It is clear that any solution found by Algorithm 2.6 is valid, so we just need to show that the algorithm will find all the solutions. Suppose that

$$A[i_0] + A[j_0] + A[k_0] = 0,$$

where $i_0 < j_0 < k_0$. The algorithm should find this solution when $i = i_0$ in the **for** loop. We assume that the algorithm does not find this solution, and then derive a contradiction.

When the iteration $i = i_0$ of the algorithm terminates, we have $j = k = \ell$, say. We consider three cases.

case 1

Suppose $\ell \leq j_0$. At some point in time before the termination of this iteration, k must have been decremented from k_0 to $k_0 - 1$ since $\ell \leq j_0 < k_0$. At that point in time, $i = i_0$, $j \leq \ell \leq j_0$ and $k = k_0$. Then

$$\begin{aligned} A[i] + A[j] + A[k] &\leq A[i_0] + A[j_0] + A[k_0] \\ &= 0. \end{aligned}$$

Since k is decremented only when $A[i] + A[j] + A[k] \geq 0$, it must be the case that $A[i] + A[j] + A[k] = 0$. This happens if and only if $(i, j, k) = (i_0, j_0, k_0)$. In this case, the solution (i_0, j_0, k_0) would have been reported by the algorithm, which is a contradiction.

case 2

Suppose $\ell \geq k_0$. This case is handled by an argument similar to case 1.

case 3

Suppose $j_0 < \ell < k_0$. At some point in time before the termination of this iteration, k must have been decremented from k_0 to $k_0 - 1$ since $\ell < k_0$. In order for k to be decremented, $A[i] + A[j] + A[k] \geq 0$ at this point in time. We have $i = i_0$ and $k = k_0$, so $A[i] + A[j] + A[k] \geq 0$ only if $j \geq j_0$. If $j = j_0$, then $A[i] + A[j] + A[k] = 0$ and the solution (i_0, j_0, k_0) would have been reported by the algorithm. If $j > j_0$, then there was a time earlier in this iteration when j was incremented from j_0 to $j_0 + 1$. At this time, $k \geq k_0$, so

$$\begin{aligned} A[i] + A[j] + A[k] &\geq A[i_0] + A[j_0] + A[k_0] \\ &= 0. \end{aligned}$$

However, j is not incremented unless the sum is ≤ 0. Hence we conclude that $(i, j, k) = (i_0, j_0, k_0)$ and the solution (i_0, j_0, k_0) would have been reported by the algorithm.

2.3 Reductions

Reductions are one of the most important concepts in the study of algorithms and complexity. We begin by introducing the basic idea of a reduction. Suppose Π_1 and Π_2 are two problems and we use a *hypothetical algorithm* solving Π_2 as a subroutine to solve Π_1. Then we say that we have a ***Turing reduction*** (or more simply, a ***reduction***) from Π_1 to Π_2; this is denoted $\Pi_1 \leq \Pi_2$. The hypothetical algorithm solving Π_2 is called an ***oracle***.

The notation "$\leq$" is used to suggest (roughly speaking) that solving Π_1 is not substantially more difficult than solving Π_2. However, this is certainly *not* a mathematically precise statement.

Here are a few basic facts about reductions and how the oracle is used in a reduction.

- The reduction must treat the oracle as a ***black box***. This means that the input-output behavior of the oracle is completely specified, but we have no information and make no assumptions about the internal computations that are performed by the oracle.
- There might be more than one call made to the oracle in the reduction.

- For the purposes of a reduction, it does not matter if the oracle is a "real" algorithm.
- However, if $\Pi_1 \leq \Pi_2$ *and* we also have an algorithm A_2 that solves Π_2, then we can plug A_2 into the reduction and obtain an actual algorithm that solves Π_1.

We present a simple example of a reduction to start. Consider the algebraic identity

$$xy = \frac{(x+y)^2 - (x-y)^2}{4}, \tag{2.1}$$

where x and y are positive integers.

We note that the identity (2.1) is elementary to prove:

$$\begin{aligned} \frac{(x+y)^2 - (x-y)^2}{4} &= \frac{x^2 + y^2 + 2xy - (x^2 + y^2 - 2xy)}{4} \\ &= \frac{4xy}{4} \\ &= xy. \end{aligned}$$

We show that **Multiplication** $\leq$ **Squaring**, using the formula (2.1), as follows.

```
Algorithm 2.7: MULTIPLICATIONTOSQUARING(x, y)

external COMPUTESQUARE
u ← x + y
v ← x − y
s ← COMPUTESQUARE(u)
t ← COMPUTESQUARE(v)
z ← (s − t)/4
return (z)
```

In the reduction, we are assuming the existence of an oracle COMPUTESQUARE that computes the square of its input. There are two calls to the oracle; the first call computes $(x+y)^2$ and the second call computes $(x-y)^2$. The final "division by 4" just consists of truncating the two low-order bits, which are guaranteed to be 00, so it is not a "real" division.

The existence of this reduction shows that a multiplication costs no more than two squarings + one addition + two subtractions.

2.3.1 The Target 3SUM Problem

In this section, we look at some reductions involving variations of the **3SUM** problem. First, we present a variation called **Target3SUM**. The original problem **3SUM** is the special case of **Target3SUM** in which $T = 0$.

Problem 2.2: Target3SUM

Instance: An array A of n distinct integers, $A = [A[1], \ldots, A[n]]$, and a target T.
Question: Do there exist three distinct elements in A whose sum equals T?

Suppose we are asked to design an algorithm for **Target3SUM**. It is straightforward to *modify* any algorithm solving the **3SUM** problem so it solves the **Target3SUM** problem. Another approach is to find a reduction **Target3SUM** $\leq$ **3SUM**. This would allow us to *re-use* code as opposed to *modifying* code. (We are not claiming that re-using code, by finding a reduction, is preferable from a programming point of view to modifying code. We are just using this example to illustrate how a reduction between the two problems can be found.)

The reduction is based on the observation that

$$A[i] + A[j] + A[k] = T$$

if and only if

$$\left(A[i] - \frac{T}{3}\right) + \left(A[j] - \frac{T}{3}\right) + \left(A[k] - \frac{T}{3}\right) = 0.$$

This suggests the following approach, which works if T is divisible by three.

Algorithm 2.8: TARGET3SUMTO3SUM(A, T)

```
comment: A is an array of length n
precondition: assume T is divisible by 3
external 3SUM-SOLVER
for i ← 1 to n
  do B[i] ← A[i] − T/3
return (3SUM-SOLVER(B))
```

The above approach might not be satisfactory if T is not divisible by three, because the modified array B would not consist of integers in this case. However, the transformation $B[i] \leftarrow 3A[i] - T$ works for *any* integer T.

Algorithm 2.9: GENERALTARGET3SUMTO3SUM(A, T)

```
comment: A is an array of length n
external 3SUM-SOLVER
for i ← 1 to n
  do B[i] ← 3A[i] − T
return (3SUM-SOLVER(B))
```

In the above algorithms, we have used 3SUM-SOLVER as an oracle to solve

Target3SUM. This gives the desired reduction **Target3SUM** $\leq$ **3SUM**. We also note that there is a "trivial" reduction **3SUM** $\leq$ **Target3SUM**; see Algorithm 2.10.

Algorithm 2.10: 3SUMTOTARGET3SUM(A)

comment: A is an array of length n
external TARGET3SUM-SOLVER
$T \leftarrow 0$
return (3SUM-SOLVER(A, T))

This reduction just sets $T = 0$ and calls the 3SUM-SOLVER oracle.

2.3.2 Complexity of a Reduction

When we analyze the complexity of a reduction, we just consider the work done "outside" the oracle. More precisely, we consider each oracle call to have complexity $O(1)$. For example, the reduction TARGET3SUMTO3SUM has complexity $O(n)$, because we do $O(n)$ work to construct the array B, followed by one call to the oracle 3SUM-SOLVER.

However, we can also consider the complexity of the algorithm that is obtained from the reduction by replacing the oracle 3SUM-SOLVER by a "real" algorithm. In this complexity analysis, we take into account the complexity of the reduction, the complexity of the algorithm that we substitute for the oracle, and the number of calls that are made to the oracle (there is only one call to the oracle in this example). Using Theorem 1.7, we obtain the following.

- If we plug in TRIVIAL3SUM, the complexity is
$$O(n + n^3) = O(n^3).$$
- If we plug in IMPROVED3SUM, the complexity is
$$O(n + n^2 \log n) = O(n^2 \log n).$$
- If we plug in QUADRATIC3SUM, the complexity is
$$O(n + n^2) = O(n^2).$$

In each case, it turns out that the "n" term (which is the complexity of the reduction) is subsumed by the second term (which is the complexity of the algorithm that replaces the oracle).

2.3.3 Many-one Reductions

The reduction TARGET3SUMTO3SUM had a very special structure:

1. We *transformed* an instance I of the first problem, **Target3SUM**, to an instance I' of the second problem, **3SUM**.

2. Then we called the oracle *once*, on the transformed instance I'.
3. It was not required in this case, but the solution to the transformed instance I' might need to be modified to obtain the solution to the original instance I.

Reductions of this form, in the context of decision problems (i.e., problems where we ask a question that has a yes-no answer), are called ***many-one reductions*** (if the reduction is polynomial-time, it is also known as ***polynomial transformation*** or ***Karp reduction***).

Polynomial transformations are an essential tool in proving that certain problems are "probably intractable" (more precisely, NP-complete). A polynomial transformation from a known intractable (or probably intractable) problem to a problem of unknown difficulty proves that the *second problem* is also intractable (or probably intractable, resp.). We will see many examples of reductions of this type in Chapter 9.

2.3.4 The Three Array 3SUM Problem

We now consider another reduction involving another variant of the **3SUM** problem.

Problem 2.3: 3array3SUM

Instance: Three arrays of n distinct integers, A, B and C.
Question: Do there exist array elements, one from each of A, B and C, whose sum equals 0?

It turns out that there is a reduction **3array3SUM** $\leq$ **3SUM**, but it is a bit trickier than the previous reductions we considered. As a first attempt, given an instance $I = (A, B, C)$ of **3array3SUM**, we might concatenate the three arrays A, B and C and then call an oracle 3SUM-SOLVER. However, this might not work correctly, as 3SUM-SOLVER might find a solution consisting of two elements from A and one element from B, say. This would not be a valid solution for the original instance of **3array3SUM**.

So we have to transform the instance $I = (A, B, C)$ of **3array3SUM** to an instance I' of **3Sum** in such a way that I' has a solution *if and only if* I has a solution. Thus, a proof of correctness of the reduction will be an "if and only if" proof.[1]

In this case, finding a transformation that yields a correct reduction requires a bit of cleverness. See Algorithm 2.11 for one possible solution.

To prove that Algorithm 2.11 is a correct reduction, we show that (i, j, k) is a solution to the instance $I' = A'$ of **3SUM** if and only if $(i, j - n, k - 2n)$ is a solution to the instance $I = (A, B, C)$ of **3array3SUM**. Assume first that (i', j', k') is a solution to **3array3SUM**. Then

$$A[i'] + B[j'] + C[k'] = 0.$$

[1]We should stress that any proof of correctness of any reduction is inherently an "if and only if" proof.

Algorithm 2.11: 3ARRAY3SUMTO3SUM(A, B, C)

comment: A, B and C are arrays of length n
external 3SUM-SOLVER
for $i \leftarrow 1$ **to** n
do $\begin{cases} D[i] \leftarrow 10A[i] + 1 \\ E[i] \leftarrow 10B[i] + 2 \\ F[i] \leftarrow 10C[i] - 3 \end{cases}$
let A' denote the concatenation of D, E and F
comment: A' has length $3n$
if 3SUM-SOLVER$(A') = (i, j, k)$
  then return $(i, j - n, k - 2n)$

Hence,

$$\begin{aligned} D[i'] + E[j'] + F[k'] &= 10A[i'] + 1 + 10B[j'] + 2 + 10C[k'] - 3 \\ &= 10(A[i'] + B[j'] + C[k']) \\ &= 0, \end{aligned}$$

and thus

$$A'[i'] + A'[j' + n] + A'[k' + 2n] = 0.$$

This is the "easy" direction of the proof.

Conversely, suppose that $A'[i] + A'[j] + A'[k] = 0$. We claim that this sum must consist of one element from each of D, E and F. This can be proven by considering the sum modulo 10 and observing that the only way to get a sum of three elements that is divisible by 10 is

$$1 + 2 - 3 \bmod 10 = 0$$

(observe that the desired sum, 0, is divisible by 10). The result follows by translating back to the original arrays A, B and C.

To prove the claim, we look at all the combinations of sums of three elements of A', reduced modulo 10. All elements of D are congruent to 1 modulo 10, all elements of E are congruent to 2 modulo 10, and all elements of F are congruent to -3 modulo 10. So we have the following combinations (modulo 10) to consider:

$$\begin{array}{rlrl} 1+1+1 & = 3 & 2+2+2 & = 6 \\ -3-3-3 & = 1 & 1+2+2 & = 5 \\ 1-3-3 & = 5 & 2+1+1 & = 4 \\ 2-3-3 & = 6 & -3+1+1 & = 9 \\ -3+2+2 & = 1 & 1+2-3 & = 0. \end{array}$$

From the above computations, we see that the only way to get a sum equal to 0 (modulo 10) is to have one element from each of D, E and F. Then, we have

$$D[i'] + E[j'] + F[k'] = 0$$

for some indices i', j' and k', which implies

$$A[i'] + B[j'] + C[k'] = 0.$$

Example 2.3 Suppose

$$A = [3, 6, -2], \quad B = [8, 5, -4] \quad \text{and} \quad C = [7, -1, 1].$$

Then

$$A' = [31, 61, -19, 82, 52, -38, 67, -13, 7].$$

Suppose 3SUM-SOLVER(A') finds the solution $(1, 6, 9)$. These are indices of A', so

$$\begin{aligned} A'[1] + A'[6] + A'[9] &= 31 - 38 + 7 \\ &= 0. \end{aligned}$$

Then

$$\text{3ARRAY3SUMTO3SUM}(A, B, C) = (1, 6 - 3, 9 - 6) = (1, 3, 3),$$

which indicates that

$$\begin{aligned} A[1] + B[3] + C[3] &= 3 - 4 + 1 \\ &= 0. \end{aligned}$$

□

2.4 Exact Analysis

Sometimes it is possible to give an exact analysis of an algorithm, say if we want to count the exact number of times a certain operation is performed during a specific algorithm. Exact analysis of an algorithm can consider the best, worst or average cases. We illustrate this approach with a few simple problems in this section.

2.4.1 Maximum

We begin with the **Maximum** problem, which is defined as Problem 2.4. This problem asks us to find the maximum element in an array.

Problem 2.4: Maximum

Instance: An array A of n integers, $A = [A[1], \ldots, A[n]]$.
Find: The maximum element in A.

Algorithm 2.12: FINDMAX(A)

```
comment: A is an array of length n
max ← A[1]
for i ← 2 to n
  do { if A[i] > max
       then max ← A[i]
return (max)
```

The **Maximum** problem has an obvious simple solution, where we just examine all the elements in A one at a time, keeping track of the largest element we have encountered as we proceed. See Algorithm 2.12.

First, it might be of interest to formally prove that FINDMAX is correct. This is done easily by defining a suitable ***loop invariant***:

> At the end of iteration i (for $i = 2, \ldots, n$), the current value of *max* is the maximum element in $[A[1], \ldots, A[i]]$.

The correctness of the loop invariant can be proven by induction. The base case, when $i = 2$, is obvious. Now we make an induction assumption that the claim is true for $i = j$, where $2 \leq j \leq n - 1$, and then we would prove that the claim is true for $i = j + 1$ (the reader should fill in the details!). When $j = n$, we are done and the correctness of Algorithm 2.12 is proven.

It is obvious that the complexity of Algorithm 2.12 is $\Theta(n)$. However, more precisely, we can observe that the number of comparisons of array elements done by Algorithm 2.12 is exactly $n - 1$. This is because we perform one such comparison (namely, a comparison of $A[i]$ to *max*) in each of the $n - 1$ iterations of the **for** loop.

It turns out that Algorithm 2.12 is *optimal* with respect to the number of comparisons of array elements. That is, any algorithm that correctly solves the **Maximum** problem for an array of n elements requires *at least* $n - 1$ comparisons of array elements.

We describe a quick way to see that the maximum element in the array A can never be computed using fewer than $n - 1$ comparisons. This analysis is based on modeling the execution of the algorithm using a certain graph (we note that graphs and graph algorithms are discussed in detail in Chapter 7). Let ALG be any algorithm that solves the **Maximum** problem and suppose we run ALG on an array A consisting of n distinct elements. Suppose we construct a graph G on vertex set $\{1, \ldots, n\}$, in which $\{i, j\}$ is an edge of G if and only if ALG compares $A[i]$ and $A[j]$ at some time during the execution of the algorithm ALG.

Suppose ALG performs at most $n - 2$ comparisons. Then G contains at most $n - 2$ edges and hence G is not a connected graph (in order for a graph to be connected, it must contain at least $n - 1$ edges). Choose any two connected components C_1 and C_2 of G. Clearly, there is no way for ALG to determine if the maximum element in C_1 is greater than or less than the maximum element in C_2. Therefore ALG cannot guarantee that it finds the maximum element in A.

2.4.2 Maximum and Minimum

Next, we investigate a variation of the **Maximum** problem, where we want to find the maximum and minimum elements in an array. This problem is defined formally as Problem 2.5.

Problem 2.5: Max-Min

Instance: An array A of n integers, $A = [A[1], \ldots, A[n]]$.
Find: The maximum and the minimum element in A.

The **Max-Min** problem also has an obvious simple solution, which we present as Algorithm 2.13. This algorithm is based on examining the elements of A, one at a time, updating the maximum and minimum as we proceed.

The complexity of Algorithm 2.13 is $\Theta(n)$. However, we can observe, more precisely, that Algorithm 2.13 requires $2n - 2$ comparisons of array elements given an array of size n.

The complexity of Algorithm 2.13 is obviously optimal, but it turns out that there are algorithms to solve the **Max-Min** problem that require fewer comparisons of array elements than Algorithm 2.13.

Here is one simple improvement, based on the observation that it is unnecessary to compare $A[i]$ to min whenever $A[i] > max$. The only change is to insert "else" before the second "if" statement. See Algorithm 2.14.

The number of comparisons that are done in Algorithm 2.14 is no longer constant. It is an interesting exercise to compute the number of comparisons in the best, worst, and average cases. First, we note that we perform either one or two

Algorithm 2.13: FINDMAXMIN(A)

```
comment: A is an array of length n
max ← A[1]
min ← A[1]
for i ← 2 to n
     ⎧ if A[i] > max
     ⎪   then max ← A[i]
  do ⎨ if A[i] < min
     ⎩   then min ← A[i]
return (max, min)
```

Algorithm 2.14: IMPROVEDFINDMAXMIN(A)

```
comment: A is an array of length n
max ← A[1]
min ← A[1]
for i ← 2 to n
        if A[i] > max
          then max ← A[i]
  do    
          else  if A[i] < min
                  then min ← A[i]
return (max, min)
```

comparisons in each iteration of the **for** loop, so the total number of comparisons is at least $n-1$ and at most $2n-2$.

The best case would be $n-1$ comparisons. This happens if and only if the array A is sorted in increasing order.

The worst case, $2n-2$ comparisons, occurs whenever $A[1]$ is the maximum element in the array.

To analyze the average case, we consider all $n!$ permutations of the n elements in A (for simplicity, we assume these n elements are distinct). In iteration i of the **for** loop (for $i = 2, \dots, n$), one comparison is performed if and only if

$$A[i] > \max\{A[1], \dots, A[i-1]\}.$$

Any of the first i elements of A is equally likely to be the largest of these i elements, where this probability is computed over all $n!$ permutations of the elements in A. Thus, we have

$$\mathbf{Pr}[A[i] > \max\{A[1], \dots, A[i-1]\}] = \frac{1}{i}.$$

Suppose $\mathbf{X_i}$ is the random variable that indicates the number of comparisons required in iteration i of the **for** loop. Then

$$\mathbf{Pr}[\mathbf{X_i} = 1] = \frac{1}{i}$$
$$\mathbf{Pr}[\mathbf{X_i} = 2] = 1 - \frac{1}{i}.$$

Hence, the expected number of comparisons required in iteration i is

$$\begin{aligned} E[\mathbf{X_i}] &= \frac{1}{i} \times 1 + \left(1 - \frac{1}{i}\right) \times 2 \\ &= 2 - \frac{1}{i}. \end{aligned}$$

Now define

$$\mathbf{X} = \mathbf{X_2} + \mathbf{X_3} + \cdots + \mathbf{X_n}.$$

X is a random variable that records the total number of comparisons that are required. The expected number of comparisons is $E[\mathbf{X}]$. By linearity of expectation (Theorem 1.17), we have

$$\begin{aligned} E[\mathbf{X}] &= \sum_{i=2}^{n} E[\mathbf{X_i}] \\ &= \sum_{i=2}^{n} \left(2 - \frac{1}{i}\right) \\ &= 2(n-1) - (H_n - 1) \\ &= 2n - 1 - H_n, \end{aligned}$$

where H_n is the nth harmonic number. Since $H_n \in \Theta(\log n)$ from (1.6), the expected number of comparisons, $E[\mathbf{X}]$, is still (roughly) $2n$, since $\log n \in o(n)$.

With some ingenuity, we can actually reduce the number of comparisons of array elements by (roughly) 25%. Suppose n is even and we consider the elements *two at a time*. Initially, we compare the first two elements and initialize maximum and minimum values based on the outcome of this comparison. One comparison is required to do this.

Then, whenever we compare two new elements, we subsequently compare the larger of the two elements to the current maximum and the smaller of the two to the current minimum. Thus, three additional comparisons are done to process two additional array elements. This yields Algorithm 2.15, which only requires a total of

$$1 + \frac{n-2}{2} \times 3 = \frac{3n}{2} - 2$$

comparisons.

It is possible to prove that any algorithm that solves the **Max-Min** problem requires at least $3n/2 - 2$ comparisons of array elements in the worst case. Therefore the algorithm BESTFINDMAXMIN is in fact optimal with respect to the number of comparisons of array elements that the algorithm performs.

2.4.3 Insertion Sort

INSERTIONSORT is a simple sorting algorithm. Suppose we have an array

$$A = [A[1], \ldots, A[n]]$$

consisting of n distinct integers. In the ith iteration (for $i = 2, \ldots, n$), we insert $A[i]$ into its correct position in the sorted subarray

$$[A[1], \ldots, A[i-1]].$$

Algorithm 2.15: BESTFINDMAXMIN(A)

```
comment: A is an array of length n
precondition: n is even
if A[1] > A[2]
  then { max ← A[1]
       { min ← A[2]
  else { max ← A[2]
       { min ← A[1]
for i ← 2 to n/2
  do { if A[2i − 1] > A[2i]
     {   then { if A[2i − 1] > max
     {        {   then max ← A[2i − 1]
     {        { if A[2i] < min
     {        {   then min ← A[2i]
     {   else { if A[2i] > max
     {        {   then max ← A[2i]
     {        { if A[2i − 1] < min
     {        {   then min ← A[2i − 1]
return (max, min)
```

This is done by shifting all elements of the subarray $[A[1], \ldots, A[i-1]]$ that are larger than $A[i]$ one position to the right, and then dropping $A[i]$ into its correct position. See Algorithm 2.16.

Algorithm 2.16: INSERTIONSORT(A)

```
comment: A is an array of length n
for i ← 2 to n
  do { x ← A[i]
     { j ← i − 1
     { while (j > 0) and (A[j] > x)
     {   do { A[j + 1] ← A[j]
     {      { j ← j − 1
     { A[j + 1] ← x
return (A)
```

It should be noted that, in the **while** statement, we assume that the second condition (namely, $A[j] > x$) is *not* evaluated if the first condition ($j > 0$) fails.

An exact analysis of Algorithm 2.16 can be done by counting the number of times two array elements are compared. That is, we determine the number of times that the condition $A[j] > x$ (in the **while** statement) is evaluated, where $x = A[i]$. Again, we can consider the best, worst and average cases.

We look first at a specific iteration, say iteration i of the **for** loop, where $2 \leq i \leq n$. The number of comparisons performed in this iteration is at least 1 and at most $i-1$. Exactly one comparison is done when $A[i] > A[i-1]$, and $i-1$ comparisons are performed when $A[i] < A[2]$. When $i = 2$, one comparison is always performed.

Over the $n-1$ iterations of the algorithm, the total number of comparisons is at least $n-1$ and at most

$$\sum_{i=2}^{n}(i-1) = \frac{n(n-1)}{2},$$

from (1.1).

The best case is $n-1$ comparisons. This occurs if and only if the array A is already sorted in increasing order (e.g., $[1,2,3,4,5]$), or if the A is sorted in increasing order and then the first two elements are interchanged (e.g., $[2,1,3,4,5]$).

It is a bit more complicated to determine when the worst case of $n(n-1)/2$ comparisons can arise. One situation is when the array A is initially sorted in decreasing order, but there are other possibilities when the worst case occurs. The worst case corresponds to $i-1$ comparisons being done in iteration i, for all i. However, the last (i.e., the $(i-1)$st) comparison in iteration i can succeed or fail without affecting the number of comparisons that are done in this iteration. That is, for the worst case to occur, it is necessary and sufficient that $A[i]$ is the smallest or second smallest among the i values $A[1], \dots, A[i]$, for all i.

Now, let's analyze the average case. First, we focus on a specific iteration of the **for** loop. Let $\mathbf{X_i}$ be the random variable indicating how may comparisons are done during iteration i. We see that

$$\begin{aligned}
\mathbf{X_i} &= 1 && \text{if } A[i] > A[i-1] \\
\mathbf{X_i} &= 2 && \text{if } A[i-1] > A[i] > A[i-2] \\
\mathbf{X_i} &= 3 && \text{if } A[i-2] > A[i] > A[i-3] \\
&\vdots && \vdots \\
\mathbf{X_i} &= i-2 && \text{if } A[3] > A[i] > A[2] \\
\mathbf{X_i} &= i-1 && \text{if } A[2] > A[i] > A[1] \text{ or } A[i] < A[1].
\end{aligned}$$

There are $i-1$ possible values of $\mathbf{X_i}$. In order to compute the expected value, $E[\mathbf{X_i}]$, we need to compute the probability of each possible outcome. This probability is computed over all $n!$ permutations of the elements in A. It is not hard to see that

$$\begin{aligned}
\mathbf{Pr}[\mathbf{X_i} = j] &= \frac{1}{i} && \text{for } j = 1, \dots, i-2 \\
\mathbf{Pr}[\mathbf{X_i} = i-1] &= \frac{2}{i}.
\end{aligned}$$

Hence, we can compute

$$\begin{aligned}
E[\mathbf{X_i}] &= \sum_{j=1}^{i-1} (j \times \mathbf{Pr}[X_i = j]) \\
&= \sum_{j=1}^{i-2} \left(j \times \frac{1}{i}\right) + (i-1) \times \frac{2}{i} \\
&= \frac{(i-1)(i-2)}{2i} + \frac{2(i-1)}{i} \qquad \text{from (1.1)} \\
&= \frac{(i-1)(i+2)}{2i} \\
&= \frac{i+1}{2} - \frac{1}{i}.
\end{aligned}$$

Now define

$$\mathbf{X} = \mathbf{X_2} + \mathbf{X_3} + \cdots + \mathbf{X_n}.$$

$\mathbf{X}$ is a random variable that records the total number of comparisons that are required. The expected number of comparisons is $E[\mathbf{X}]$. By linearity of expectation (Theorem 1.17), we have

$$\begin{aligned}
E[\mathbf{X}] &= \sum_{i=2}^{n} E[\mathbf{X_i}] \\
&= \sum_{i=2}^{n} \left(\frac{i+1}{2} - \frac{1}{i}\right) \\
&= \frac{(n+1)(n+2)}{4} - \left(1 + \frac{1}{2}\right) - (H_n - 1) \qquad \text{from (1.1)} \\
&= \frac{n^2 + 3n}{4} - H_n,
\end{aligned}$$

where H_n is the nth harmonic number. Since $H_n \in \Theta(\log n)$ from (1.6), the expected number of comparisons is $\Theta(n^2)$.

2.5 Notes and References

This chapter discussed basic analysis techniques such as loop analysis. It also gives some examples of exact analysis, average case analysis and proofs of correctness. Reductions are also introduced in this chapter. Reductions of course play a leading role in Chapter 9. However, several of my colleagues have suggested that it is beneficial to introduce the basic idea of a reduction well before they are required in the context of proofs of NP-completeness. Thus, we have given some examples of "positive" uses of reductions in this chapter, i.e., to demonstrate how an algorithm for one problem might be used to solve a different problem.

The **3SUM** problem is seemingly very elementary, but it is not known how efficiently it can be solved. It had been conjectured that any algorithm solving **3SUM** would necessarily have complexity $\Omega(n^2)$. This conjecture was disproved in 2014 with the discovery of an algorithm solving **3SUM** that has complexity $O(n^2/(\log n/\log\log n)^{3/2})$; see Grønlund and Petti [29] for a journal paper presenting the proof. There have been some improvements since then, and the most efficient algorithm known at the present time is due to Chan [12]; it has complexity $O(n^2(\log\log n)^{O(1)}/(\log n)^2)$. It has been conjectured that **3SUM** cannot be solved in time $O(n^{2-\epsilon})$ if $\epsilon > 0$.

Exercises

2.1 Analyze the complexity of the following algorithm by proving a Θ-bound.

Algorithm 2.17: LOOPANALYSIS4(n)

comment: assume n is a positive integer
$t \leftarrow 0$
for $i \leftarrow 1$ **to** n
do $\begin{cases} j \leftarrow i \\ \textbf{while } j \leq n^2 \\ \quad \textbf{do } \begin{cases} j \leftarrow j + i \\ t \leftarrow t + j \end{cases} \end{cases}$
return (t)

2.2 Analyze the complexity of the following algorithm by proving a Θ-bound. You can assume here that computing $\lceil\sqrt{i}\rceil$ is an "elementary operation" that takes $O(1)$ time.

Algorithm 2.18: LOOPANALYSIS5(n)

comment: assume n is a positive integer
$s \leftarrow 0$
for $i \leftarrow 1$ **to** n^2
do $\begin{cases} \textbf{for } j \leftarrow 1 \textbf{ to } \lceil\sqrt{i}\rceil \\ \quad \textbf{do } s \leftarrow s + (j-i)^2 \end{cases}$
return (s)

2.3 Analyze the complexity of the following algorithm by proving a Θ-bound.

Algorithm 2.19: LOOPANALYSIS6(n)

```
comment: assume n is a positive integer
S ← 0
for i ← 1 to n
      ⎧ j ← n
      ⎪ while j > 0
  do  ⎨
      ⎪      ⎧ S ← S + j²
      ⎩  do  ⎨
             ⎩ j ← j − i²
return (S)
```

2.4 Consider the problem **M3SUM**, which is defined as follows.

Problem 2.6: M3SUM

Instance: An array of n distinct positive integers,

$$A = [A[1], \ldots, A[n]].$$

Question: Do there exist three array elements $A[i]$, $A[j]$ and $A[k]$ such that

$$A[i] + A[j] = A[k],$$

where $1 \leq i, j, k \leq n$ and i, j, k are all distinct?

Define

$$B[\ell] = 4A[\ell] - 1$$

for $1 \leq \ell \leq n$ and define

$$B[\ell + n] = -4A[\ell] + 2$$

for $1 \leq \ell \leq n$. Show that solving **3SUM** on the array B (of length $2n$) will solve **M3SUM** on the array A (so this is a many-one reduction from **M3SUM** to **3SUM**).

2.5 The problem **ThreeCollinearPoints** is defined as follows.

Problem 2.7: ThreeCollinearPoints

Instance: An array of n distinct points in the Euclidean plane,

$$B = [(x_1, y_1), \ldots, (x_n, y_n)].$$

We assume that all the x_i's and y_i's are integers.

Question: Do there exist three collinear points in the array B?

Given an instance of **3SUM**, say $A = [A[1], \ldots, A[n]]$, construct the list B of n points in the Euclidean plane, where

$$B[i] = (A[i], (A[i])^3)$$

for $1 \leq i \leq n$. Show that this mapping $A \mapsto B$ can be used to construct a many-one reduction **3SUM** $\leq$ **ThreeCollinearPoints**.

HINT Three points $(x_1, y_1), (x_2, y_2), (x_3, y_3)$ with distinct x-coordinates are collinear if and only if

$$\frac{y_3 - y_1}{x_3 - x_1} = \frac{y_3 - y_2}{x_3 - x_2}.$$

2.6 **algorithm design**

Suppose we are given an m by n array of integers, where each row is sorted in increasing order from left to right and each column is sorted in increasing order from top to bottom. The following would be an example of such an array with $m = 4$ and $n = 7$:

10	12	13	21	32	34	43
16	21	23	26	40	54	65
21	23	31	33	54	58	74
32	46	59	65	74	88	99

(a) Design an $O(m \log n)$-time algorithm that determines whether or not a given integer x is contained in the array.

HINT Make use of a standard binary search.

(b) Design an $O(m + n)$-time algorithm to solve the same problem.

HINT Follow a path, beginning in the upper left corner of the array.

2.7 **algorithm design**

Suppose Alice spends a_i dollars on the ith day and Bob spends b_i dollars on the ith day, for $1 \leq i \leq n$. We want to determine whether there exists some set of t consecutive days during which the total amount spent by Alice is exactly the same as the total amount spent by Bob in some (possibly different) set of t consecutive days. That is, we want to determine if there exist i, j, t (with $0 \leq i, j \leq n - t$ and $1 \leq t \leq n$) such that

$$a_{i+1} + a_{i+2} + \cdots + a_{i+t} = b_{j+1} + b_{j+2} + \cdots + b_{j+t}.$$

For example, for the inputs

$$10, 21, 11, 12, 19, 15 \quad \text{and} \quad 12, 9, 2, 31, 21, 8,$$

the answer is "*yes*" because

$$11 + 12 + 19 = 9 + 2 + 31.$$

(a) Design and analyze an algorithm that solves the problem in $\Theta(n^3)$ time by brute force.

(b) Design and analyze a better algorithm that solves the problem in $\Theta(n^2 \log n)$ time.

HINT Use sorting.

2.8 **algorithm design**

Suppose we are given a set of n rectangles, denoted $\mathcal{R}_1, \dots, \mathcal{R}_n$. Each rectangle $\mathcal{R}_i$ is specified by four integers b_i, h_i, ℓ_i, w_i:

- b_i is the y-coordinate of the base of $\mathcal{R}_i$,
- $h_i > 0$ is the height of $\mathcal{R}_i$,
- ℓ_i is the x-coordinate of the left side of $\mathcal{R}_i$, and
- $w_i > 0$ is the width of $\mathcal{R}_i$.

We want to find a new rectangle $\mathcal{R}$ with the smallest area such that it encloses all n rectangles $\mathcal{R}_i, \dots, \mathcal{R}_n$. The sides of the rectangle $\mathcal{R}$ will be parallel to the x- and y-axes. Design and analyze an algorithm that solves this problem using at most $3n - 4$ comparisons.

2.9 **programming question**

Write a program implementing the algorithm described below, which finds the largest and second largest elements of an array A of n elements using $n + \Theta(\log n)$ comparisons. Here is the basic idea.

Assume for convenience that n is a power of 2, i.e., $n = 2^m$. Place the n elements of A in the leaves of a binary tree T (any convenient implementation of a binary tree can be used). Starting at the leaf nodes, compare every node of T with its sibling, and put the larger value in the parent node. Also keep track of which child of the parent node contains the larger value, e.g., by including an extra bit to denote *left* or *right*. Continue to do this, proceeding up the tree, until the root of T contains the maximum element, denoted max_1. At this point, $n - 1$ comparisons of array elements have been done.

Now the second largest element in A, denoted by max_2, must be a sibling of some node in T that contains max_1. The value of max_2 can be determined by following the path of nodes containing max_1, starting at the root. The number of additional comparisons required is exactly $m - 1 = \log_2 n - 1$ and the total number of comparisons is $n - 2 + m$.

REMARK The above-described algorithm turns out to be optimal, in terms of the number of comparisons of array elements required to compute max_1 and max_2. ∎

Chapter 3

Data Structures

We assume that most students studying design and analysis of algorithms have taken a prior course in basic data structures. However, we thought it would be helpful to provide a treatment of some basic data structures, for the purpose of review and reference. Therefore, in this chapter, we discuss various aspects of linked lists, stacks, queues, priority queues, binary search trees and hash tables.

3.1 Abstract Data Types and Data Structures

In our algorithms, we make use of various kinds of abstract data types, such as stacks, queues, priority queues, etc. We usually think of an ***abstract data type*** (or ***ADT***) as consisting of a mathematical model (which typically describes a certain collection of data) together with one or more operations defined on that model. We can define an ADT and then use its operations in an algorithm as needed. On the other hand, when we write a computer program that implements a particular algorithm, we would also have to implement the ADT using a suitable data structure.

Typical examples of data structures include arrays, linked lists, binary trees, etc. We study these in the following sections of this chapter. A data structure is generally more closely tied to a specific programming language than an ADT is. However, as usual, we present algorithms associated with data structures using structured pseudocode, which could subsequently be implemented using various high-level computing languages.

If we want to analyze the complexity of an algorithm, then we need to be aware of the complexity of the operations in any ADTs used by the algorithm. Of course the complexity of ADT operations may depend on the specific data structure used to implement the ADT. However, many commonly used ADTs also have standard implementations in which the complexities of the operations in the ADT are predetermined.

For example, there are multiple ways to implement a stack ADT, but we would generally assume that PUSH and POP operations both can be done in $O(1)$ time (see Section 3.3). For some common ADTs, it is therefore sometimes required as part of the definition of the ADT that the operations have prespecified complexities.

DOI: 10.1201/9780429277412-3

3.2 Arrays, Linked Lists and Sets

Two of the most common data structures are ***arrays*** and ***linked lists***. These topics are covered in every introductory programming course, so we just review a few essential features of these data structures here. We also discuss the use of bit strings to store sets whose elements are all "small" integer values.

An array consists of a fixed number (say n) of entries of a given type, e.g., integers. The entries are denoted $A[1], A[2], \ldots, A[n]$. The size of an array may be specified at compile time or run time, but once an array is defined, its size is fixed. (Of course, given an array of length n, it is not required that all of its entries are defined at any given point in time.) An array occupies a contiguous block of memory, which means that we can access any entry $A[i]$ of an array A in $O(1)$ time, simply by computing a suitable offset from the beginning of the array.

Here is a depiction of an array containing four integer values:

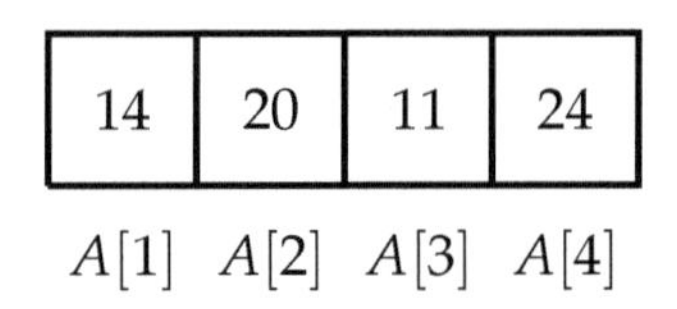

A linked list consists of a sequence of ***nodes***. Each node contains a value and one or more ***links***. In its simplest form, each node is linked to the next node in the list. A link is actually a ***pointer*** that specifies the address in memory of a node. We access a linked list through its ***head***, which is a pointer to the first node in the list. The last node in the list, the ***tail***, is not linked to any other node.

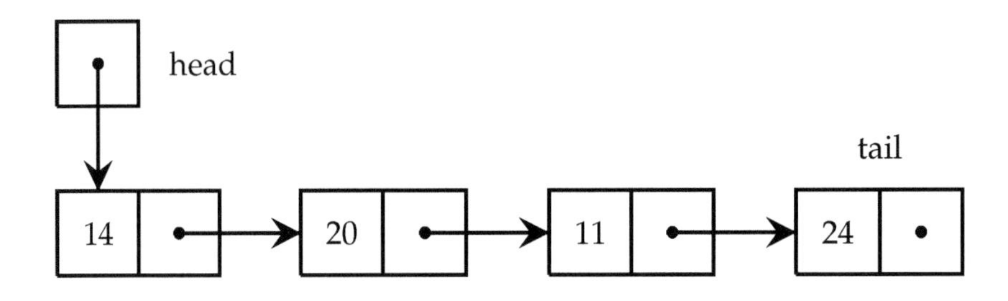

The only way to access the ith node in a linked list is to start at the head and follow i links one by one until the desired node is reached. This is obviously much less efficient than the corresponding operation in an array.

A linked list does not have a fixed length and hence it can grow or shrink dynamically. Whenever we want to add a new node to a linked list, we can allocate a new node from a special piece of reserved memory. Linked lists occupy more memory than arrays because space has to be allocated in each node for one or more pointers.

There are many variations of linked lists. Sometimes it is useful to maintain pointers to both the head and the tail of the list. We might choose to have an additional pointer from each node to the previous node in the list (this is sometimes called a ***back-pointer***). Such a list is called a ***doubly-linked list***. Finally, a list might be ***circular***, where there is a pointer from the end (i.e., the tail) of the list back to the head. The kind of pointers included in a list will depend on the operations we might wish to perform on the list, for example, as specified in an ADT.

As we have mentioned, we access a node in a linked list via a pointer. A pointer N is a memory address. If we wish to access that memory address, we ***dereference*** the pointer. For example, in the C programming language, $*N$ denotes the contents of the memory to which N points. A node in a linked list typically contains data as well as a pointer to the next node. In this chapter, we often refer to this data as a ***key***. The key and the pointer in a node could be indicated as $(*N).key$ and $(*N).next$, respectively. However, this is awkward syntax, so C defines the operator ->, so that N->*key* means the same thing as $(*N).key$.

The "last" node in a linked list is not linked to any other node. The pointer in such a node will have a special value, which is often denoted by *null*.

In any pseudocode we present, we will implicitly assume that dereferencing is done automatically, so we will just write $N.key$ to denote the key in the node pointed to by N. Similarly, $N.next$ denotes the (next) node in the list, to which N points. If $N.next = null$, then N does not point to any other node.

It is interesting to compare various operations that we might wish to perform on arrays and linked lists. As usual, we will consider worst-case complexity. We consider several operations, which are typical operations specified in common ADTs. Three of the most fundamental operations are SEARCH, INSERT and DELETE. In a SEARCH, we wish to determine if a particular key value occurs in the data structure. An INSERT operation is used to insert a new key value into the data structure, and a DELETE operation is used to delete an existing key value from the data structure.

We will examine arrays and linked lists containing integer values. We also distinguish between ***sorted*** and ***unsorted*** arrays and linked lists. Unsorted structures may contain keys in an arbitrary order, while sorted structures are required to maintain all the keys in sorted order (it doesn't really matter if the keys are in increasing or decreasing order).

We assume that there are n items stored in the data structure. In the case of an array, we also assume that we keep track of the number of items currently in the array by using an auxiliary variable *size*. This will allow us to easily add a new item to the end of the array if we wish to do so.

First, let's suppose that we want to SEARCH for a particular value. In a sorted array, we can use a BINARY SEARCH (see Section 4.5) that has complexity $O(\log n)$. For an unsorted array or linked list, the complexity of a search is $O(n)$. Note that a sorted linked list does not permit a BINARY SEARCH to take place, because there is no way to access arbitrary items in the list in $O(1)$ time.

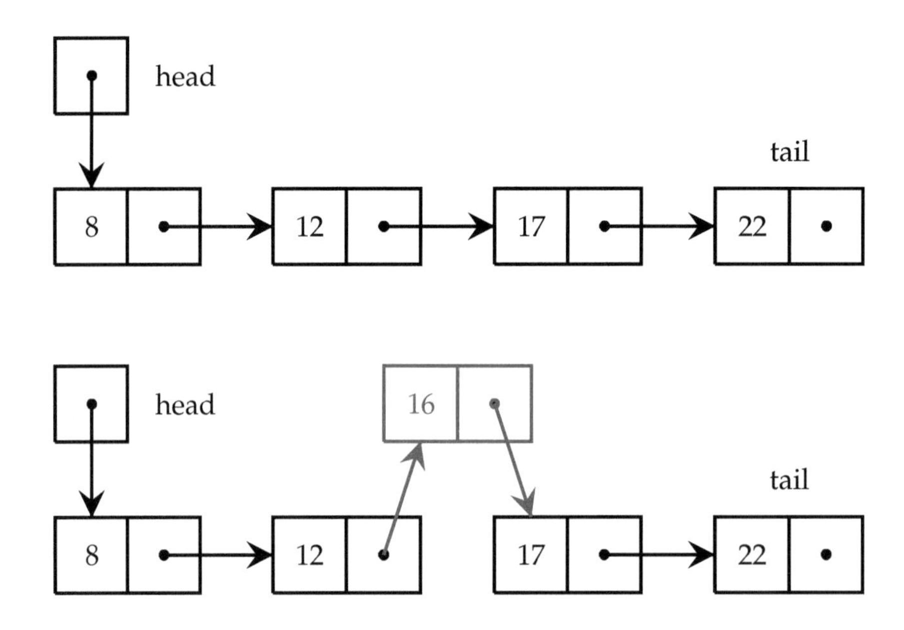

FIGURE 3.1: Inserting a node into a sorted linked list.

Searching through a linked list basically involves traversing the list from beginning to end, comparing each key value to a desired value x. This is usually termed a ***linear search***. Algorithm 3.1 is a basic linear search. It returns a pointer to the node that contains the value x (if it exists), and *null*, otherwise.

Algorithm 3.1: LINEARSEARCH(*Head*, x)

> **comment:** *Head* is the head of a linked list
> *CurNode* ← *Head*
> **while** *CurNode* ≠ *null*
> **do** {
>   **if** *CurNode.key* = x
>   **then return** (*CurNode*)
>   **else** *CurNode* ← *CurNode.next*
> **return** (*CurNode*)

If the linked list is sorted (say in increasing order), then we could potentially halt early if the desired value x is not in the list. Specifically, we could abort the search as soon as we encounter a node containing a value $y > x$.

As another simple operation, suppose we want to INSERT a new item into a list. We outline how this would be done for each of the four structures we are considering:

- For an unsorted linked list, we can easily insert the new item at the *beginning*

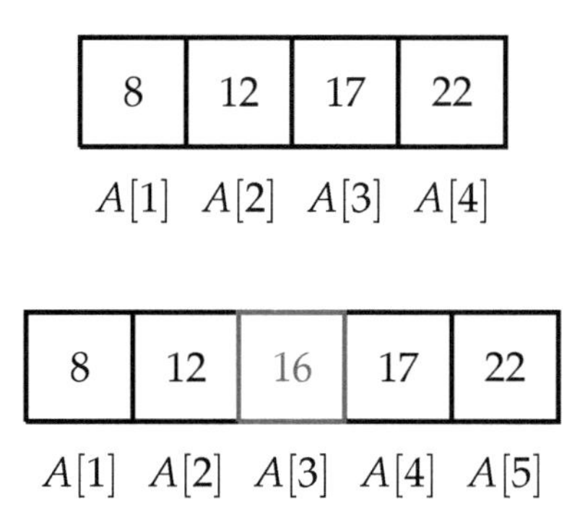

FIGURE 3.2: Inserting a node into a sorted array.

of the list in $O(1)$ time. Suppose *New* is the node that we are inserting. Let *Head* be the pointer to the first node in the list. Note that $Head = null$ when the list is empty. Then the pointer updates would be as follows:

1. *New.next* $\leftarrow$ *Head*
2. *Head* $\leftarrow$ *New*.

After these updates, *Head* points to the new node, which is linked to the previous first node in the list.

- For an unsorted array, we can easily insert the new item at the *end* of the array in $O(1)$ time. When we insert the item, we also increase *size* by one.

- For a sorted linked list, we first need to find the correct place to insert a new node. This SEARCH takes time $O(n)$. However, after having found the desired position of the new item, it only takes $O(1)$ time to insert the new node into the list.

  In a bit more detail, we require a pointer N to the node that will *precede* the new node after it is inserted. The linear search algorithm (Algorithm 3.1) can easily be modified to incorporate this extra feature.

  As an example, suppose we want to insert a new node having the value 16 into the first list in Figure 3.1. Then we need the pointer to the node containing the value 12. If *New* is the pointer to the new node, then pointers would be updated as follows:

  1. *New.next* $\leftarrow$ *N.next*
  2. *N.next* $\leftarrow$ *New*.

  If we want to insert a node at the beginning of the list, it is done in the same manner as insertion at the beginning of an unsorted linked list, which was described above.

- For a sorted array, we again SEARCH for the correct place to insert a new node, which takes time $O(\log n)$ using BINARY SEARCH. However, after having found the desired position of the new item, it takes $O(n)$ time to insert the new node into the list, because all the later items in the array have to be shifted one position to the right. We also increase *size* by one.

Another fundamental operation is to DELETE an item. We might first have to find the relevant item in the array or list, which requires a SEARCH. After finding this item, performing the actual deletion of that item is done as follows:

- For an unsorted or sorted linked list, deleting a specified item (once it has been found) can be done in $O(1)$ time. Note, however, that we need the pointer to the node *before* the node that is being deleted, in order that the pointers in the linked list can be updated correctly. For example, in Figure 3.1, if we wanted to delete the node containing the value 17, we would need a pointer to the node containing the value 16. Suppose that this pointer is denoted *Prev*. Then we update as follows:

  1. *Old* $\leftarrow$ *Prev.next*
  2. *Prev.next* $\leftarrow$ *Prev.next.next*
  3. discard *Old*.

  Deleting the first node in the list is similar:

  1. *Old* $\leftarrow$ *Head*
  2. *Head* $\leftarrow$ *Head.next*
  3. discard *Old*.

- For an unsorted array, we can move the last item in the array to occupy the position of the element that is being deleted. No other items are moved. The *size* of the array is then decreased by one. This takes time $O(1)$.

- For a sorted array, we need to shift all items following the item being deleted to the left by one position. The *size* of the array is then decreased by one. This takes time $O(n)$.

3.2.1 Sets and Bit Strings

If we are dealing with items whose values are small positive integers, for example, then a very different approach can be considered. That is, suppose that all the values under consideration are elements of the ***universe*** $\mathcal{U} = \{1, \ldots, m\}$, where m is a "small" integer. Any subset $X \subseteq \mathcal{U}$ can be represented as a ***bit string*** of length m, say $(x_1, \ldots, x_m)$, where $x_i = 1$ if $i \in X$ and $x_i = 0$ if $i \notin X$. (Bit strings are also known as ***bit arrays*** or ***bit vectors***.)

Any subset of $\mathcal{U}$ has the same length when considered as a bit string, namely m bits. SEARCH(i) just checks the value of the bit x_i. INSERT(i) sets $x_i \leftarrow 1$ and DELETE(i) sets $x_i \leftarrow 0$. So these operations are very efficient, requiring $O(1)$ time. Other set operations that are easy to perform include the following:

- the ***union*** of two sets X and Y is computed as the logical "or" of the two associated bit strings;
- the ***intersection*** of two sets X and Y is computed as the logical "and" of the two associated bit strings;
- the ***complement*** of the set X with respect to $\mathcal{U}$ is computed by complementing every bit in the bit string associated with X.

The complexity of these operations is $\Theta(m)$. However, m is a pre-specified constant, so these operations all require $O(1)$ time.

We observe that if we require an ADT that supports arbitrary INSERT and DELETE operations, then none of these array, linked list or bit string data structures is ideal. This is because some array and linked list operations take time $\Omega(n)$ to carry out in the worst case, and bit string data structures are limited by their space requirements (i.e., the size of the universe $\mathcal{U}$). This motivates the study of more complicated data structures, such as those based on ***binary search trees***, for which all the desired operations only require $O(\log n)$ time. Another approach is to use a ***hash table***. (We will discuss both of these topics a bit later, in Sections 3.5 and 3.6, resp.)

3.3 Stacks and Queues

We first define stacks and queues as abstract data types.

stack

A stack is an ADT consisting of a collection of items with operations PUSH and POP.

PUSH(S, x) inserts a new item x into the stack S and POP(S) deletes an item from the stack S.

Items are removed in ***LIFO*** (or ***last-in first-out***) order. That is when we POP a stack, the item that is removed is the item that was most recently PUSHed onto the stack.

queue

A queue is an ADT consisting of a collection of items with operations ENQUEUE and DEQUEUE.

ENQUEUE(Q, x) inserts a new item x into the queue Q and DEQUEUE(Q) deletes an item from the queue Q.

Items are removed in FIFO (***first-in first-out***) order. That is, when we DEQUEUE a queue, the item that is removed is the first item that was ENQUEUEd.

We usually think of a queue as having a ***front*** and ***rear***; items enter at the rear of the queue and are deleted from the front of the queue.

In a stack or queue, we only need to implement insertion and deletion operations of a special type. It turns out that we can use either linked lists or arrays as data structures to implement these ADTs so that all operations can be done in $O(1)$ time.[1]

3.3.1 Stacks

First, we consider the stack ADT. If we implement a stack using a linked list, a bit of thought shows that we can PUSH and POP efficiently from the head of the linked list (this is usually called the ***top*** of the stack). This enables both of these operations to be carried out in $O(1)$ time.

If we use an array, then the top of the stack would be the *end* of the array. Thus, we simply insert and delete elements from the end of the array, which takes $O(1)$ time. However, the array has a fixed size, say n, so, as we mentioned above, the stack cannot hold more than n items at any time.

3.3.2 Resizing and Amortized Analysis

If we want to PUSH an item onto a "full" stack, then a commonly used solution is to allocate a new array having size $2n$ (i.e., double the size of the current array) and copy the n current items to the new array. Then the new item can be inserted into the stack. We call this ***resizing*** the stack.

It is clear that resizing a stack of size n will require time $\Theta(n)$, so we no longer have worst-case complexity $O(1)$ for every PUSH operation. But the situation is not so bad, as this resizing is only required "occasionally." Suppose we start with an initial array of size 1 and we consider a sequence of n PUSH operations, with no POP operations. After the initial PUSH operation, the stack is full and we would need to resize the array when we do a second PUSH operation. The resized array can store two items, so resizing would be required again when the third PUSH operation is carried out. Now we have an array of size 4, so the next resizing would happen with the fifth PUSH operation.

Let P_i denote the ith PUSH operation. In general, we see that resizing is required when $i = 2^j + 1$ for an integer $j \geq 0$, because the "current" array of size 2^j would be completely filled. Now suppose we consider the total complexity required for a sequence of n PUSH operations. The complexity of P_i is $\Theta(i)$ if $i = 2^j + 1$ and $\Theta(1)$ if $i \neq 2^j + 1$.

It is now easy to compute the total complexity of the sequence of operations $P_1, \dots, P_n$. For convenience, assume $n = 2^t + 1$ for a positive integer t. Then the

[1] In the case of array implementations, we assume that we never need to add an item to an array that is completely filled. However, we will briefly discuss how to mitigate this situation in Section 3.3.2.

complexity is given by the sum

$$\sum_{i=1}^{n} \Theta(1) + \sum_{j=0}^{t} \Theta(2^j) = \Theta(n) + \Theta(2^{t+1})$$
$$= \Theta(n) + \Theta(n)$$
$$= \Theta(n).$$

In the above analysis, we are using the formula (1.2) for the sum of a geometric sequence.

We have shown that a sequence of n PUSH operations requires time $\Theta(n)$, even when resizing is taken into account. Therefore we say that the ***amortized complexity*** of a PUSH operation is $\Theta(1)$. Amortized complexity can be considered as a kind of average-case complexity. However, amortized complexity measures the average complexity of an operation that is contained in an arbitrary sequence of operations. In contrast, the average-case complexity defined in Section 1.2.2 refers to the average complexity over all possible inputs of a given size.

3.3.3 Queues

Now we look at queues. First, we consider a linked list implementation of a queue. We need to insert (i.e., ENQUEUE) at one end of a queue and delete (i.e., DEQUEUE) at the other end. Thus it would be useful to have pointers to both the head and tail of the linked list. Then we could ENQUEUE at the tail of the list and DEQUEUE at the head of the list in $O(1)$ time. So the front of the queue is the beginning of the list and the rear of the queue is the end of the list.

We can also implement a queue Q using an array A. It is most convenient to ENQUEUE at the end of the array and DEQUEUE from the beginning. Suppose we explicitly keep track of the position of the front of the queue, say $A[\mathit{front}]$, as well as the size of the queue, which we denote by *size*. The rear of the queue is $A[\mathit{front} + \mathit{size} - 1]$. Thus, the queue consists of the array elements

$$A[\mathit{front}], A[\mathit{front} + 1], \dots, A[\mathit{front} + \mathit{size} - 1].$$

Then a DEQUEUE(Q) operation does the following:

1. return the value $A[\mathit{front}]$
2. $\mathit{front} \leftarrow \mathit{front} + 1$
3. $\mathit{size} \leftarrow \mathit{size} - 1$.

If we ENQUEUE(Q, x), then the following steps are carried out:

1. $\mathit{size} \leftarrow \mathit{size} + 1$.
2. $A[\mathit{front} + \mathit{size} - 1] \leftarrow x$.

There is one subtle point we need to address. As we perform a sequence of ENQUEUE and DEQUEUE operations, the position of the queue within the array A gradually migrates to the right. Since the array has a fixed size n, we could reach the end of the array A even if $size \leq n$.[2]

The solution when the queue reaches the end of the array is to treat the queue as being ***circular***. That is, we allow the queue Q to wrap around to the beginning of the array. For example, suppose $n = 10$, $front = 7$ and $size = 4$. Then Q would consist of the array elements

$$A[7], A[8], A[9], A[10].$$

If we ENQUEUE(Q, x), the new element x can be placed in $A[1]$. Thus Q would now consist of the array elements

$$A[7], A[8], A[9], A[10], A[1].$$

This modification amounts to evaluating array subscripts modulo n (more precisely, reducing subscripts modulo n to the range $1, \dots, n$). We leave the details for the reader to fill in.

3.4 Priority Queues and Heaps

We define priority queues.

priority queue

A priority queue is an ADT consisting of a collection of items with operations INSERT and DELETEMIN.

Items are removed according to their ***priority***, which (typically) is a positive integer. That is, at any given time, only the item of *highest priority* can be removed from the queue. We will follow the convention that the item having the *minimum* value has the highest priority.

The operation CONSTRUCTQUEUE builds a priority queue from a specified collection of items.

Other possible operations include INITIALIZEEMPTYQUEUE, ISEMPTY, and SIZEOFQUEUE. Note that ISEMPTY(PQ) = **true** if and only if SIZEOFQUEUE$(PQ) = 0$.

The above definition is for a ***minimum-oriented*** priority queue. A ***maximum-oriented*** priority queue is defined in the natural way, replacing the operation DELETEMIN by DELETEMAX.

---

[2]If $size = n$, then the array is completely filled and we can no longer perform an ENQUEUE operation. However, in this situation, we could resize the array, in a similar fashion to what we did in the case of a stack.

Suppose we want to construct a priority queue, say PQ, consisting of n items. One way to implement the operation CONSTRUCTQUEUE(PQ) is to perform INITIALIZEEMPTYQUEUE(PQ) followed by n INSERT operations. However, depending on the implementation, it might be more efficient to construct the priority queue "all at once" instead of reducing it to n INSERT operations. We will discuss this in detail in Section 3.4.1.

We observe that any priority queue can be used to sort. First, we would build a priority queue of n items, and then we would perform n DELETEMIN operations, storing the results in an array. This array is sorted in increasing order.

Suppose we use an unsorted array to implement a priority queue. The required operations can be implemented as follows.

- CONSTRUCTQUEUE($PQ, x_1, \ldots, x_n$) requires us to place n items (denoted $x_1, \ldots, x_n$) into an array. This takes time $O(n)$.
- INSERT(PQ, x) inserts a new item x into an unsorted array, exactly as in Section 3.2. We would increment *size* by 1 and insert the item at the end of the array. This takes time $O(1)$.
- DELETEMIN(PQ) requires a linear search to find the minimum item. Then we delete it, as described in Section 3.2. That is, we exchange the minimum element with the last item in the array, return the last item, and decrement *size* by 1. This takes time $O(n)$ (to perform the linear search).
- SIZEOFQUEUE(PQ) returns the value *size*. This takes time $O(1)$.
- INITIALIZEEMPTYQUEUE(PQ) sets $size \leftarrow 0$. This takes time $O(1)$.

The sorting algorithm based on this implementation is a well-known sorting algorithm that is known as SELECTIONSORT. It has complexity $\Theta(n^2)$.

We could instead use an unsorted linked list to implement a priority queue and the complexities of the operations would be similar. The reader can fill in the details.

Next, suppose we use an array sorted in *decreasing* order to implement our priority queue. Then the priority queue operations would be carried out at follows:

- SIZEOFQUEUE(PQ) and INITIALIZEEMPTYQUEUE(PQ) are the same as they were for an unordered array implementation.
- INSERT(PQ, x) inserts an item x into the priority queue PQ. This is the same as inserting an element into a sorted array, which we discussed in Section 3.2; it takes time $O(n)$.
- DELETEMIN(PQ) returns the last item in the array, and decrements *size* by 1. This takes time $O(1)$.

- CONSTRUCTQUEUE($PQ, x_1, \ldots, x_n$) is the same as sorting the array containing the elements $x_1, \ldots, x_n$ in decreasing order. So this takes time $O(n \log n)$ if we use an efficient sorting algorithm.

The sorting algorithm based on this implementation of a priority queue is INSERTIONSORT. This sorting algorithm was studied in detail in Section 2.4.3; it has complexity $\Theta(n^2)$.

We could instead use an unsorted linked list to implement the priority queue operations and the complexities of the operations would be similar. The reader can fill in the details.

3.4.1 Heaps

All of the implementations described above have some operations that take $\Theta(n)$ time. We would prefer an implementation where all operations (except CONSTRUCTQUEUE) can be performed in $O(\log n)$ time. There are various ***heap*** implementations that achieve these goals. We will focus on the simplest version of a heap, which is called a ***binary heap***. The (binary) heap implementation of a priority queue yields INSERT and DELETEMIN operations that both have complexity $\Theta(\log n)$. The resulting sorting algorithm, called HEAPSORT, has complexity $\Theta(n \log n)$.

Conceptually, a heap is a certain type of binary tree (however, the term "heap" is generally used to refer to a very specific array-based data structure). We begin with some basic terminology related to ***binary trees***. A ***node*** in a binary tree $\mathcal{T}$ has at most two children (which are named the ***left child*** and the ***right child***). A node is joined to each of its children by an ***edge***. A node with no children is a ***leaf node***. Every node has a unique ***parent node***, except for a special node called the ***root node***.

For any node N in a binary tree $\mathcal{T}$, there is a unique path from the root node to N. The number of edges in this path is the ***depth*** of the node N. Thus the root node has depth 0, its children have depth 1, etc. The nodes of depth i comprise the ith ***level*** of the tree $\mathcal{T}$, for $i = 0, 1, \ldots$. The ***depth of the tree***, denoted as $\mathsf{depth}(\mathcal{T})$, is the maximum depth of any node in the tree $\mathcal{T}$.

Considering a heap as a binary tree, all the levels of a heap are completely filled, except (possibly) for the last level. Furthermore, the filled items in the last level are ***left-justified***. A binary tree of this type is sometimes called a ***complete binary tree***.

A heap also satisfies the following additional property: if N is any node in a heap (except for the root node) and P is the parent of N, then $P.key \leq N.key$ (where $N.key$ denotes the value stored in any node N).

Let H be a heap (i.e., a binary tree, as described above) having n nodes and let A be an array of size n. The items of level i of H ($i \geq 0$) are stored in ***left-to-right order*** in the array elements

$$A[2^i], A[2^i + 1], \ldots, A[2^{i+1} - 1].$$

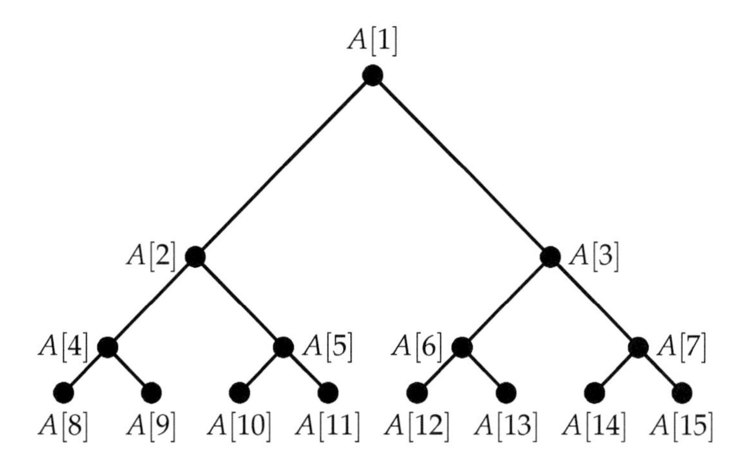

FIGURE 3.3: An array of 15 elements, depicted as a binary tree.

The items in the last level of H are stored in the array elements

$$A[2^i], A[2^i + 1], \ldots, A[n],$$

where $i = \lfloor \log_2 n \rfloor$. Figure 3.3 indicates the correspondence between array elements and nodes in the associated binary tree.

It is easy to find parents and children using this array representation, as follows:

- the left child of $A[i]$ (if it exists) is $A[2i]$,
- the right child of $A[i]$ (if it exists) is $A[2i + 1]$,
- the parent of $A[i]$ (where $i \neq 1$) is $A[\lfloor i/2 \rfloor]$ (note that $A[1]$ is the root node).

Thus we can follow paths, both "up" and "down" in the tree, without having to store and maintain pointers.

The following simple lemma relates the depth of a heap to the number of nodes in the heap. We leave the proof details to the reader.

LEMMA 3.1 *A heap (or any complete binary tree) of size n has depth*

$$\lceil \log_2(n + 1) \rceil - 1 \in \Theta(\log n).$$

In the next subsections, we look at how the priority queue operations can be performed on a heap. For convenience, our descriptions are given in terms of binary trees, even though the actual pseudocode will be presented using arrays.

3.4.2 Insertion into a Heap

HEAPINSERT(A, x) is done by a ***bubble-up*** process:

1. Place the new item x in the first available position of the heap.
2. As long as the new item is less than its parent, exchange it with its parent.
3. Increase $size(A)$ by 1.

Thus the new item "bubbles up" until it reaches its correct place in the heap. Since the new item follows a path from the bottom level of the heap towards the root, the number of comparisons and exchanges that take place is $O(\log n)$. The complexity of HEAPINSERT is therefore $O(\log n)$ since we only require $O(1)$ time for each comparison/exchange. A pseudocode description of HEAPINSERT is given in Algorithm 3.2.

Algorithm 3.2: HEAPINSERT(A, x)

```
size(A) ← size(A) + 1
A[size(A)] ← x
i ← size(A)
j ← ⌊i/2⌋
while i > 1 and A[j] < A[i]
       ⎧ temp ← A[i]
       ⎪ A[i] ← A[j]
   do  ⎨ A[j] ← temp
       ⎪ i ← j
       ⎩ j ← ⌊i/2⌋
```

Example 3.1 A heap of size 12 is depicted in Figure 3.4. Suppose we now perform the operation HEAPINSERT$(A, 3)$. We would create a new node containing the value 3 as the right child of the node that contains the value 6. Then we follow the path back to the root node. The value 3 is exchanged with 6, 5 and 4 (in this example, we bubble up all the way to the root node). The result of the bubbling-up process is the second heap shown in Figure 3.4. The red nodes are the ones that have been altered. ◻

3.4.3 Deleting the Minimum Element from a Heap

HEAPDELETEMIN(A) requires us to delete the minimum element from the heap. The minimum element of a heap A is just the root node, $A[1]$, which we return as HEAPDELETEMIN(A). We then replace this minimum element (i.e., the root node) with the last element in the heap and decrease $size(A)$ by 1. Then we

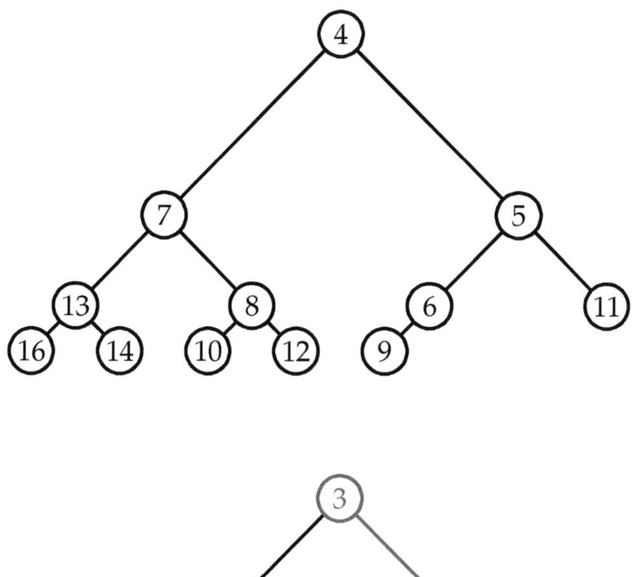

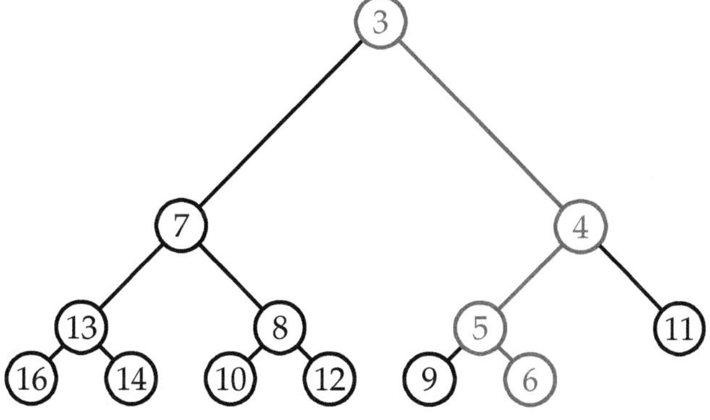

FIGURE 3.4: Inserting a new node into a heap.

restore the heap property by a ***bubble-down*** process, as follows. Starting at the root, we exchange this item with the smaller child, until it is smaller than all of its children or we reach the bottom of the tree. This operation will have complexity $O(\log n)$.

A pseudocode description of HEAPDELETEMIN is given in Algorithm 3.3.

Example 3.2 A heap is depicted in Figure 3.5. The operation HEAPDELETEMIN(A) would perform the following operations:

1. Return the minimum element, 2, which is the value in the root node.

2. Move the value 16 to the root node.

3. Exchange 16 with 4 (the smaller child).

4. Exchange 16 with 8 (the smaller child).

5. Exchange 16 with 10 (the smaller child).

Algorithm 3.3: HEAPDELETEMIN(A)

```
m ← A[1]
A[1] ← A[size(A)]
size(A) ← size(A) − 1
i ← 1
loop
  j ← 2i
  k ← 2i + 1
  if j ≤ size(A) and A[j] < A[i]
    then L ← j
    else L ← i
  if k ≤ size(A) and A[k] < A[L]
    then L ← k
  if L = i  then break
           ⎧ temp ← A[i]
    else   ⎨ A[i] ← A[L]
           ⎪ A[L] ← temp
           ⎩ i ← L
end loop
return (m)
```

The resulting heap is also shown in Figure 3.5. The red nodes are the ones that have been altered. □

It turns out to be useful to define a separate operation BUBBLEDOWN, which is described in Algorithm 3.4. In this algorithm, i is a position in the array A. The element in position i in the heap bubbles down to its correct position. That is, we successively exchange it with the smaller child, until it is smaller than all of its children.

HEAPDELETEMIN can be described very simply using the BUBBLEDOWN operation. The modified presentation is given as Algorithm 3.5.

3.4.4 Constructing a Heap

We now present an algorithm due to Floyd that uses BUBBLEDOWN to construct a heap in an efficient manner. Of course we can build a heap of n items by performing n successive HEAPINSERT operations. This approach will have complexity $\Theta(n \log n)$. However, a different approach, based on BUBBLEDOWN, has *linear* complexity, $\Theta(n)$. The basic idea is to build the heap from the bottom up. We consider the nodes in each level from right to left, and each value bubbles down to its correct position. See Algorithm 3.6.

We provide a small example to illustrate the application of Algorithm 3.6.

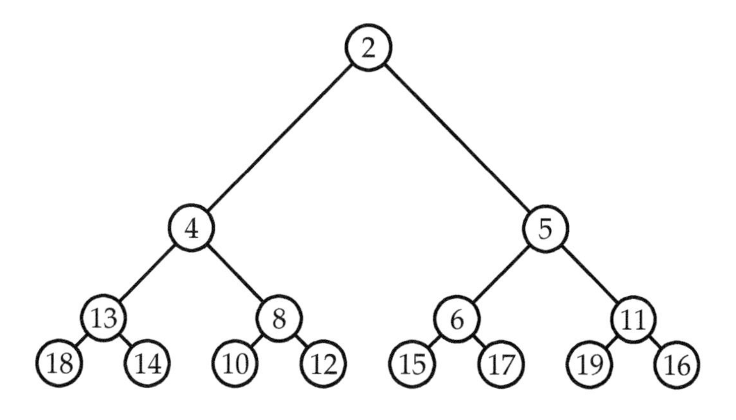

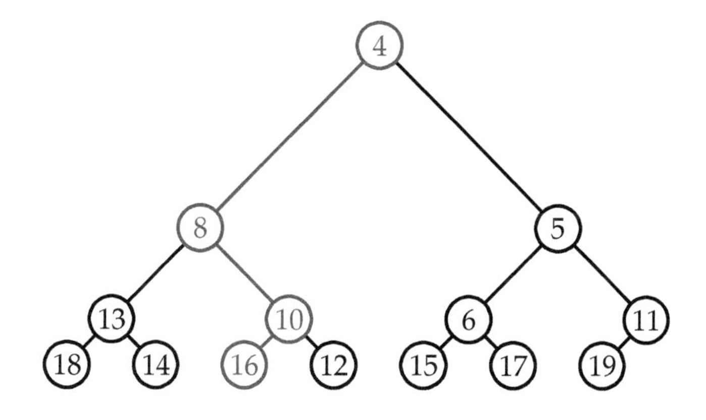

FIGURE 3.5: Deleting the minimum element from a heap.

Algorithm 3.4: BUBBLEDOWN(A, i)

loop
  $j \leftarrow 2i$
  $k \leftarrow 2i + 1$
  if $j \leq size(A)$ **and** $A[j] < A[i]$
    then $L \leftarrow j$
    else $L \leftarrow i$
  if $k \leq size(A)$ **and** $A[k] < A[L]$
    then $L \leftarrow k$
  if $L = i$
    then break
    else $\begin{cases} \text{EXCHANGE}(A[i], A[L]) \\ i \leftarrow L \end{cases}$
end loop

Algorithm 3.5: HEAPDELETEMIN2(A)

```
external BubbleDown
m ← A[1]
A[1] ← A[size(A)]
size(A) ← size(A) − 1
BubbleDown(1)
return (m)
```

Algorithm 3.6: BUILDHEAP(A)

```
comment: initially, A is an array not in "heap order"
n ← size(A)
for i ← ⌊n/2⌋ downto 1
  do BubbleDown(i)
comment: Now A is a heap
```

Example 3.3 Figure 3.6 shows how an array is transformed into a heap. The second tree is obtained after the four nodes on level 2 are bubbled down; the third tree results after the two nodes on level 1 are bubbled down; and the final tree (the heap) is produced when the root node is bubbled down.

Initially, the eight leaf nodes comprise eight heaps of size 1. The bottom two levels of the second tree constitute four heaps of size 3; the bottom three levels of the third tree comprise two heaps of size 7; and the four levels in the final tree form the desired heap of size 15. These are all indicated by red nodes in Figure 3.6. □

Let us analyze the complexity of BUILDHEAP. For simplicity, suppose that there are $2^m - 1$ nodes in the heap (so the last level is completely filled). There are 2^i nodes on level i, for $i = 0, 1, \ldots, m-1$. For each node on level i, the maximum number of exchanges that are required is $m - 1 - i$. The total number of exchanges required is at most

$$\begin{aligned}\sum_{i=0}^{m-1} (m-1-i)2^i &= \sum_{j=0}^{m-1} j\, 2^{m-1-j} \\ &= 2^{m-1} \sum_{j=0}^{m-1} j\, 2^{-j}.\end{aligned}$$

The last sum is an arithmetic-geometric sequence with ratio $r = 1/2$ (see Section 1.4). The finite sum is bounded above by the infinite sum, which converges, from

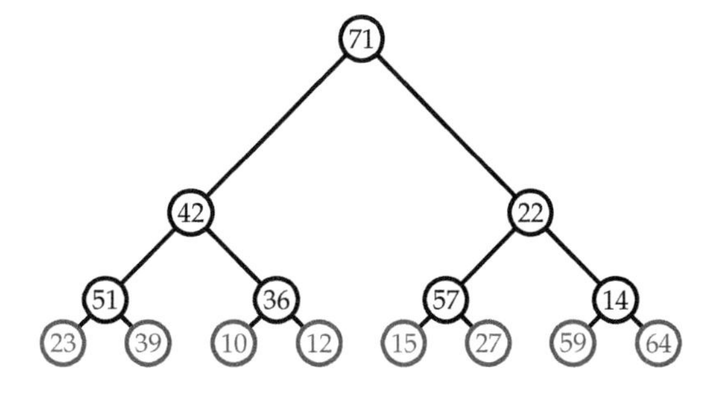

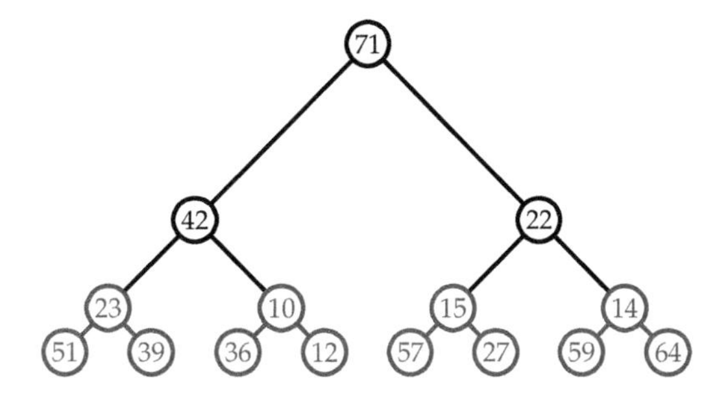

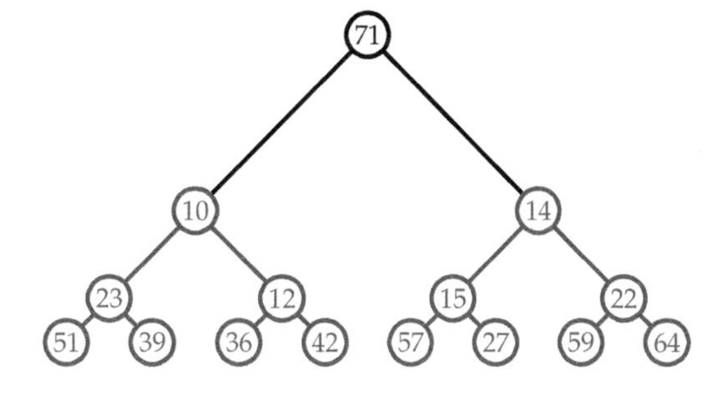

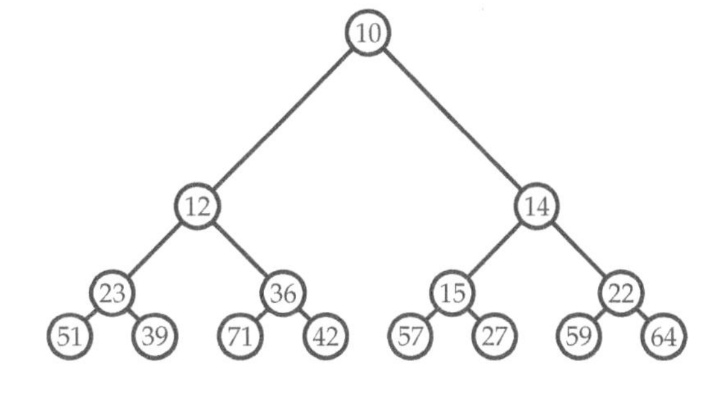

FIGURE 3.6: Building a heap.

(1.5). Thus, we have

$$\begin{aligned}\sum_{j=0}^{m-1} j2^{-j} &\leq \sum_{j=0}^{\infty} j2^{-j}\\ &\in O(1).\end{aligned}$$

Hence, the total number of exchanges is $O(2^{m-1}) \in O(n)$. Each exchange requires $O(1)$ time, so the complexity of Algorithm 3.6 is also $O(n)$.

3.4.5 HeapSort

HEAPSORT is a sorting algorithm that constructs a heap using BUILDHEAP and then performs n successive HEAPDELETEMIN operations. Each time a HEAPDELETEMIN operation is performed, the size of the heap decreases by 1. Therefore, the complexity is

$$\Theta\left(n + \sum_{i=1}^{n} \log i\right).$$

We showed in Section 1.4.3 that

$$\sum_{i=1}^{n} \log i \in \Theta(n \log n).$$

Therefore the complexity of HEAPSORT is $\Theta(n + n \log n) = \Theta(n \log n)$.

If we were just interested in the complexity of the sorting algorithm, we could build the heap using n HEAPINSERT operations instead of BUILDHEAP. The resulting sorting algorithm would still be a $\Theta(n \log n)$ sorting algorithm, though it would probably be slower by a constant factor.

3.5 Binary Search Trees

Suppose we want an ADT that supports arbitrary SEARCH, INSERT and DELETE operations that all have complexity $O(\log n)$, where n is the number of items that are stored. Such an ADT is often called a ***dictionary***. There are various ways to implement a dictionary so the three operations all have the desired complexity. Many of the popular implementations use a data structure based on a binary search tree, which we define shortly.

First we define the notion of a left and right subtree in a binary tree. Let N be a node in a binary tree $\mathcal{T}$. The ***left subtree*** of N consists of the left child, say N_L, of N (if it exists), along with any descendants of N_L. The ***right subtree*** of N consists of the right child, say N_R, of N (if it exists), along with any descendants of N_R.

A ***binary search tree*** is binary tree $\mathcal{T}$ in which every node $N \in \mathcal{T}$ contains a key such that the following properties hold:

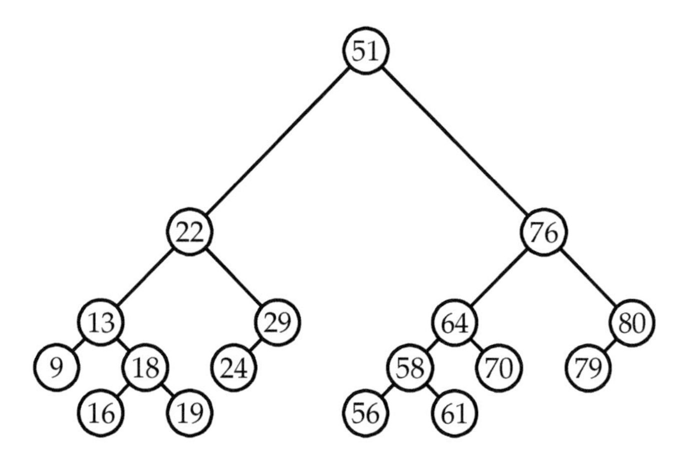

FIGURE 3.7: A binary search tree.

- for any node N' in the left subtree of N, $N'.key < N.key$, and
- for any node N' in the right subtree of N, $N'.key > N.key$.

Note that we are implicitly assuming that all the keys in the tree $\mathcal{T}$ are distinct.

Each node N in a binary search tree contains two pointers, namely, $N.left$ and $N.right$. These are pointers to the left child and right child, respectively. Of course, either or both of these pointers can have the value *null* if the child does not exist.

A binary search tree and a (binary) heap are both binary trees; however, the ordering of nodes (with respect to keys) in these two types of trees is different.

Algorithm 3.7: BSTBINARYSEARCH(*root*, *x*)

```
comment: we assume root is the root node of a binary search tree
CurNode ← root
while CurNode ≠ null
     ⎧ if CurNode.key = x
     ⎪   then return (CurNode)
  do ⎨ else if CurNode.key > x
     ⎪   then CurNode ← CurNode.left
     ⎩   else CurNode ← CurNode.right
return (CurNode)
```

Algorithm 3.7 determines if a specified key value x is in a given binary search tree. It returns a pointer to the node containing x (if it exists), and *null*, otherwise.

Example 3.4 A binary search tree is shown in Figure 3.7. If we searched for the

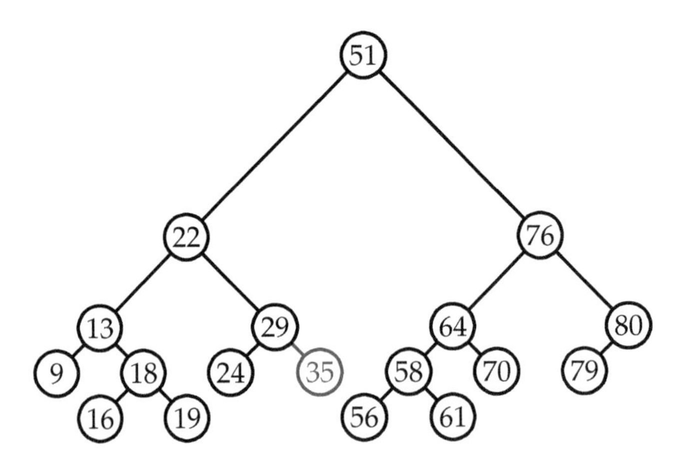

FIGURE 3.8: Inserting a node into a binary search tree.

value 16, we would follow the path 51, 22, 13, 18, 16. A search for the value 57 would follow the path 51, 76, 64, 58, 56, returning the value *null* since there is no node having the value 57 in the tree. ▯

Insertion in a binary search tree is not difficult. This basically consists of carrying out an unsuccessful search for an item not in the tree, and then creating a new leaf node where the search terminates, in order to hold the new item.

Example 3.5 Suppose we want to insert a node having the value 35 into the binary tree in Figure 3.7. A search for 35 would terminate at the right child of the node containing the value 29. After inserting the new node, we would obtain the tree shown in Figure 3.8. ▯

Deletion of an item that is currently in the tree can be more complicated. The first step is to find the node N containing the item to be deleted. There are various possible cases that can arise.

1. If N is a leaf node, then we simply delete it.
2. If N has one child, then replace N by its child.
3. Suppose N has two children. First, find the ***successor node*** of N, which is the smallest node that has a larger key than N. This successor node, say S, is the leftmost node in the right subtree of N. It is clear that S has no left child. Let R denote the parent of S.
   (a) $N.key \leftarrow S.key$
   (b) If the S is the right child of N, then

$$R.rightchild \leftarrow S.rightchild$$

otherwise

$$R.leftchild \leftarrow S.rightchild.$$

Alternatively, find the ***predecessor node*** of N, say P, which is the rightmost node in the left subtree of N. P has no right child. Attach the left subtree of P as the right subtree of the parent of P. Then replace N by P.

Example 3.6 Starting with the binary search tree in Figure 3.8, we indicate how various nodes would be deleted.

1. The nodes containing the key values 24 and 56 are leaf nodes, so they are simply deleted.

2. The node containing the key value 80 has one child (a left child, which has the value 79). We would replace this node (having the key value 80) by the node having the key value 79.

   The first tree in Figure 3.9 indicates the tree after these three deletions have been carried out.

3. Suppose we want to delete the node N containing the key value 51. We follow a path to the right child, and then to left children until a leaf node is reached. This node, S, which contains the key value 56, is the successor of 51. R is the node containing the key value 64. The right subtree of S becomes the *left* subtree of R (this is accomplished by updating one pointer). The result is the second tree shown in Figure 3.9.

4. Suppose we want to delete the node N containing the key value 22. The right child of N has no left child, so the right child of N is the successor S. S contains the key value 29. $N = R$ in this case, so the right subtree of S becomes the *right* subtree of R. The result is the third tree shown in Figure 3.9.

□

It should be clear that the (worst-case) complexity of searching, insertion or deletion is $O(\mathsf{depth}(\mathcal{T}))$. In the case of a complete binary tree, the complexity would be $\Theta(\log n)$ if the tree contains n nodes. However, if the binary tree $\mathcal{T}$ is "unbalanced," the complexity could be much worse. For example, in the worst case, all of the nodes in $\mathcal{T}$ would be in a single path. In this situation, the complexity would be $\Theta(n)$.

The main challenge in using binary search trees is to ensure that the depth of such a tree containing n nodes is $O(\log n)$. It is not guaranteed that this will be the case, especially after a number of insertions and deletions are performed. The insertion and deletion operations should also have complexity $O(\log n)$, while maintaining the desirable "balance" properties of the binary search tree. Over the

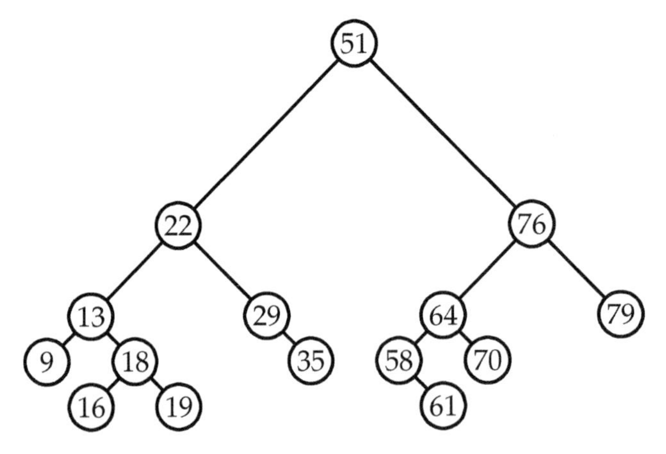

after deleting 24, 56 and 80

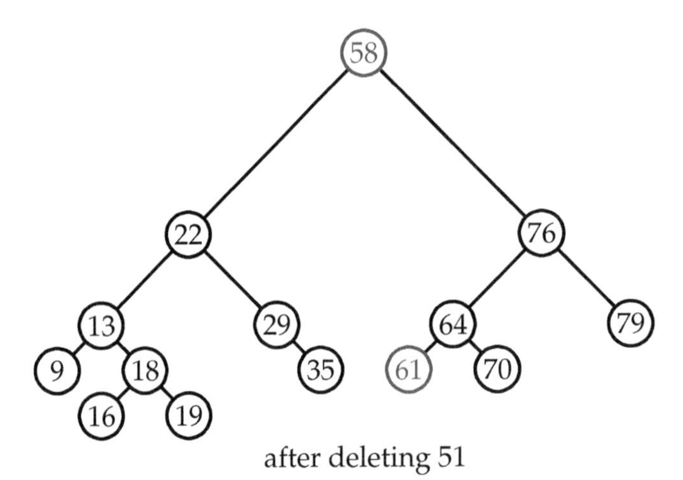

after deleting 51

after deleting 22

FIGURE 3.9: Deleting nodes from a binary search tree.

years, various tree-based solutions have been proposed to accomplish this goal, including ***AVL trees, red-black trees*** and ***B-trees***.

The first solution was the AVL tree, which was invented in 1962. An AVL tree is required to satisfy the following balance condition. For a node N in a binary search tree, define $\mathsf{balance}(N)$ to be the depth of the left subtree of N − the depth of the right subtree of N. Then, in an AVL tree, we require that

$$\mathsf{balance}(N) \in \{-1, 0, 1\}$$

for all nodes N. Every time a node is inserted or deleted in an AVL tree, it may be required to ***rebalance*** the tree, using operations that are termed ***left rotation, right rotation, left double rotation*** and ***right double rotation***. We do not describe these operations here, but they can be found in most data structures textbooks.

3.6 Hash Tables

Suppose $\mathcal{U}$ is a very large set, for example, the set of all integers. A ***hash table*** is table T of fixed size, say M, that is used to store items taking on values in $\mathcal{U}$. Each entry in the hash table is called a ***slot***. Because slots are often computed using modular arithmetic, it is convenient to assume that the slots are denoted $T[0], \dots, T[m-1]$. We are numbering the slots starting from $T[0]$ (whereas we usually denote the first element in an array A as $A[1]$.)

The value taken on by an item is often called its ***key***. The goal is that we should be able to accommodate a very large number of possible key values in a relatively small hash table.

A ***hash function*** h is used to determine where an item having a given key should be stored in the table T. So we have a function

$$h : \mathcal{U} \rightarrow \{0, \dots, M-1\},$$

where M is the size of the hash table T. Then an item with key k is stored in the slot $T[h(k)]$. The value $h(k)$ is called the ***hash value*** of the key k.

Two common methods of constructing hash functions are the ***division method*** and the ***multiplication method***. These are defined as follows.

division method

In the division method, the hash function is defined to be

$$h(k) = k \bmod M.$$

For this hash function, the modulus M is usually chosen to be a prime number that is not too close to a power of two. This serves to "spread out" the hash values in a fairly uniform way for a wide variety of data.

multiplication method

In the multiplication method, we choose a real-valued constant A where $0 < A < 1$. Then we define

$$h(k) = \lfloor M(kA - \lfloor kA \rfloor) \rfloor.$$

A popular choice for A is the ***golden ratio***, $A = (\sqrt{5} - 1)/2$.

In general, $|U| > M$, so there must exist ***collisions***, i.e., there are distinct keys $k, k' \in U$ such that $h(k) = h(k')$. If we want to insert two items, having keys k and k', into the hash table, then we have a problem because they are assigned to the same slot.

In the next sections, we will study two techniques to resolve collisions: ***chaining*** and ***open addressing***.

3.6.1 Chaining

Chaining (also known as ***separate chaining with linked lists***) is a method where each hash table slot $T[j]$ points to an unsorted linked list consisting of all the keys having hash value equal to j. This idea is credited to H.P. Luhn in 1953. The dictionary operations are straightforward:

search

To SEARCH for the item having key k, we compute the hash value $j = h(k)$, and then do a linear search of the linked list in slot $T[j]$.

insert

To INSERT a new item having key k into the table, compute the hash value $j = h(k)$, and then insert the new item at the beginning of the linked list in the slot $T[j]$.

delete

To DELETE an item having key k from the table, compute the hash value $j = h(k)$, and then delete the item from the linked list in the slot $T[j]$.

In the above, we use the basic operations defined on an unsorted linked list that we described in Section 3.2.

In order to analyze the complexity of various operations, we will treat the hash function $h : U \to \{0, \dots, M-1\}$ as a ***random function***. Formally, this means that h is chosen uniformly at random from the set of all $M^{|U|}$ hash functions with domain U and range $\{0, \dots, M-1\}$. When we speak of the expected complexity of an operation, e.g., SEARCH, we really mean the average over all possible hash functions with the specified domain and range.[3]

Intuitively, we are thinking of the evaluation of the "random" hash function h as a *random process*. This of course is an idealization of a "real" hash function. For

---

[3]This is sometimes called the ***uniform hashing assumption***.

any $k \in U$, we choose a value $j \in \{0, \ldots, M-1\}$ uniformly at random and define $h(k) = j$. Thus, the value of $h(k)$ is independent of any other values $h(k')$, where $k' \neq k$.

In the context of hash table operations, we often have a sequence of distinct keys, say $k_1, k_2, \ldots, k_r$ and we successively compute $h(k_1), h(k_2), \ldots, h(k_r)$. If we treat h as a random function, then the following property holds for any r:

$$\mathbf{Pr}[(h(k_1), h(k_2), \ldots, h(k_r)) = (j_1, j_2, \ldots, j_r)] = \frac{1}{M^r} \tag{3.1}$$

for all $(j_1, j_2, \ldots, j_r) \in \{0, \ldots, M-1\}^r$ and all r-tuples of *distinct* keys $(k_1, k_2, \ldots, k_r)$. This probability is computed over a random choice of the hash function h, as described above. The following independence property also holds for all $r \geq 2$:

$$\mathbf{Pr}[h(k_r) = j_r | (h(k_1), \ldots, h(k_{r-1})) = (j_1, \ldots, j_{r-1})] = \frac{1}{M} \tag{3.2}$$

for all $(j_1, j_2, \ldots, j_r) \in \{0, \ldots, M-1\}^r$ and all r-tuples of distinct keys $(k_1, k_2, \ldots, k_r)$.

Now, suppose we have a hash table containing n items. We define the ***load factor*** to be $\alpha = n/M$. There is no *a priori* restriction on the value of α: α can be < 1, $= 1$, or > 1. The complexity of the operations we analyze will, in general, depend on the load factor.

Let $k \in U$ and let $h(k) = j$. To analyze the complexity of SEARCH(k), we count the number of comparisons of k with the items in the slot $T[j]$. We consider two cases, depending on whether k is in the hash table (this is termed a ***successful search***) or k is not in the hash table (this is an ***unsuccessful search***).

For $0 \leq j \leq M-1$, let n_j denote the number of items in the slot $T[j]$. Observe that

$$\sum_{j=0}^{M-1} n_j = n.$$

First, we consider the complexity of an unsuccessful search. In an unsuccessful search for k, we have to traverse the entire list of elements in the slot $T[h(k)]$. Therefore, the number of comparisons required in an unsuccessful search for k is n_j, where $j = h(k)$. Then the expected number of comparisons in an unsuccessful search for a value k that is not in the hash table is

$$\begin{aligned} \sum_{j=0}^{M-1} \left(\mathbf{Pr}[h(k) = j] \times n_j\right) &= \frac{1}{M} \sum_{j=0}^{M-1} n_j \\ &= \frac{n}{M} \\ &= \alpha. \end{aligned}$$

In the above, we are assuming that $\mathbf{Pr}[h(k) = j] = 1/M$ for all j, which follows from (3.1) with $r = 1$.

Therefore, the expected complexity of an unsuccessful search is

$$\Theta(1) + \Theta(\alpha) = \Theta(1 + \alpha).$$

The $\Theta(1)$ term accounts for evaluating $h(k)$ and other overhead, especially in the case where α is very small.

Now, we look at a successful search. In a successful search for a key k, the number of comparisons that are done depends on the position of k in the list of values in the slot $T[j]$, where $j = h(k)$. The position of an item in a particular slot depends on the order that items are inserted into the hash table. Suppose that $k_1, \ldots, k_n$ are the n keys in the hash table, and they were inserted *in that particular order*.

Let $\mathbf{X}_i$ be a random variable whose value is the number of comparisons in a successful search for k_i. The value of the random variable $\mathbf{X}_i$ is the position where k_i occurs in the slot $T[j]$, where $j = h(k_i)$. Because items are inserted at the beginning of the appropriate linked list, we see that

$$\mathbf{X}_i = 1 + |\{\ell > i : h(k_\ell) = j\}|. \tag{3.3}$$

The value of $\mathbf{X}_i$ depends only on the hash values $h(k_{i+1}), \ldots, h(k_n)$. The $(n-i)$-tuple $(h(k_{i+1}), \ldots, h(k_n))$ is a random $(n-i)$-tuple from $\{0, \ldots, M-1\}^{n-i}$, by (3.1). Therefore, on average, $(n-i)/M$ of these $n-i$ hash values have the value j. That is,

$$E(\mathbf{X}_i) = 1 + \frac{n-i}{M}.$$

In a successful search, we randomly choose an index i between 1 and n. Therefore, the expected number of comparisons in a successful search is

$$\begin{aligned}
E\left[\frac{1}{n}\sum_{i=1}^{n}\mathbf{X}_i\right] &= \frac{1}{n}E\left[\sum_{i=1}^{n}\mathbf{X}_i\right] \\
&= \frac{1}{n}\sum_{i=1}^{n}E\left[\mathbf{X}_i\right] \quad \text{by linearity of expectation} \\
&= \frac{1}{n}\sum_{i=1}^{n}\left(1 + \frac{n-i}{M}\right) \qquad \text{from (3.3)} \\
&= \frac{1}{n}\left(n + \frac{n(n-1)}{2M}\right) \\
&= 1 + \frac{n-1}{2M} \\
&< 1 + \frac{\alpha}{2}.
\end{aligned}$$

Hence, the expected complexity of a successful search is also $\Theta(1 + \alpha)$.

We summarize the above analyses as follows.

THEOREM 3.2 *Suppose we use a random hash function in a hash table with chaining* *and suppose that the load factor of the hash table is α. The expected number of comparisons in a successful search is at most $1 + \alpha/2$, while the expected number of comparisons in an unsuccessful search is at most α.*

To INSERT a *new* item into the hash table, we need to first do a SEARCH (which will be unsuccessful). This search has expected complexity $\Theta(1 + \alpha)$. Then we insert the new item at the beginning of the relevant list, which only takes time $\Theta(1)$. Therefore the expected complexity of INSERT is $\Theta(1 + \alpha)$.

To DELETE an *existing* item from the hash table, we need to first do a SEARCH (which will be successful). As stated above, this search has expected complexity $\Theta(1 + \alpha)$. Then we delete the new item from the relevant list, which takes time $\Theta(1)$. Therefore the expected complexity of DELETE is $\Theta(1 + \alpha)$.

When $n \in O(M)$, (i.e., $n \leq cM$ for some constant c), we have that $\alpha \in O(1)$, so all operations take $O(1)$ expected time in this case.

3.6.2 Open Addressing

When using the technique of ***open addressing***, every item is stored in a different slot in the hash table (hence it is necessary that $n \leq M$, i.e., the load factor $\alpha \leq 1$). So the hash table is just an array in this case. An entry in this array (i.e., a slot) either is empty or holds one item.

To INSERT a new key k into the hash table, we follow a ***probe sequence*** $H(k, i)$, $i = 0, 1, \ldots,$ which specifies a sequence of possible slots, until we find an available slot. Here H is a hash function that maps an ordered pair (k, i) to a slot in the hash table.

Two common ways to construct probe sequences for a key k are the following:

linear probing

Given a hash function h that maps a key to a slot in the hash table, let

$$H(k, i) = h(k) + i \bmod M.$$

double hashing

Given two hash functions h_1 and h_2 that map keys to slots in the hash table, let

$$H(k, i) = h_1(k) + ih_2(k) \bmod M.$$

Clearly, linear probing is the simplest approach. If this method is used, we just probe successive slots starting from $h(k)$, until an available slot is found. If we reach the end of the table T, we wrap around to the beginning. So we might eventually examine all the slots, e.g., if they are all filled.

Linear probing can suffer from ***clustering***, where several contiguous slots become full. Unfortunately, once a cluster forms, it tends to grow. Suppose ℓ consecutive slots in a hash table are filled and we are inserting a new key k. If $h(k)$ takes on any one of these ℓ consecutive values, then the cluster is "extended" and its size is then (at least) $\ell + 1$. This happens with probability ℓ/M. That is, for any cluster of size ℓ, an INSERT operation will cause the cluster to be extended with probability at least ℓ/N. So the larger a cluster is, the more likely it is that the cluster will "grow."

Double hashing is a bit more complicated, but it better serves to "spread out" the slots occupied by colliding keys, as compared to linear probing. In the case of double hashing, we again have a random starting point, given by $h_1(k)$. However, we now also have a random "step size" that is given by the value $h_2(k)$. So we end up trying the slots

$$h_1(k), \quad h_1(k)+h_2(k), \quad h_1(k)+2h_2(k), \quad \ldots .$$

Again, we wrap around to the beginning of the table when we reach the end. In order to ensure that all slots are eventually considered in a given probe sequence, we should take M to be a prime number. If this is done, then $h_2(k)$ is guaranteed to be relatively prime to M, and there will be no repetitions in the probe sequence until we have looked at all possible slots.

We now describe how the basic dictionary operations would be carried out. To SEARCH for a key k in a hash table with open addressing, we examine the slots in the probe sequence $H(k,i)$, for $i = 0, 1, \ldots$, until we either find the item with key k (this constitutes a ***successful search***) or we encounter an empty slot (this is an ***unsuccessful search***).

To INSERT, we begin with a SEARCH, which will be an unsuccessful search. Then the new item can be inserted into the first available slot.

To DELETE, we first perform a SEARCH, which will be a successful search. Then we delete the item from the hash table. However, this creates a problem in that a subsequent SEARCH operation will not know that it should skip over the newly emptied slot. One possible fix is to mark a slot where an item has been deleted as *deleted* rather than *empty* and to terminate a SEARCH only when a slot marked *empty* is encountered. (Initially, every slot would be *empty*.) After an unsuccessful search, a new item can be inserted into a slot marked as *deleted*. So we should interpret the term "available slot" as meaning "deleted or empty slot."

After a period of time, the hash table might contain a large number of "deleted" slots. This will certainly have the effect of slowing down searches. One way to mitigate this problem is to ***re-hash*** the entire table once the number of "deleted" slots reaches some threshold. Of course this may require a considerable amount of work, so the threshold should be chosen carefully.

Example 3.7 As a toy example, suppose we have a hash table T with eleven slots, $T[0], \ldots, T[10]$. We use the trivial hash function $h(k) = k \bmod 11$ with linear probing. Suppose we insert keys $56, 62, 44, 25, 50, 90$ and 33 into T. The first six keys do not give rise to any collisions. They are inserted into the table as follows:

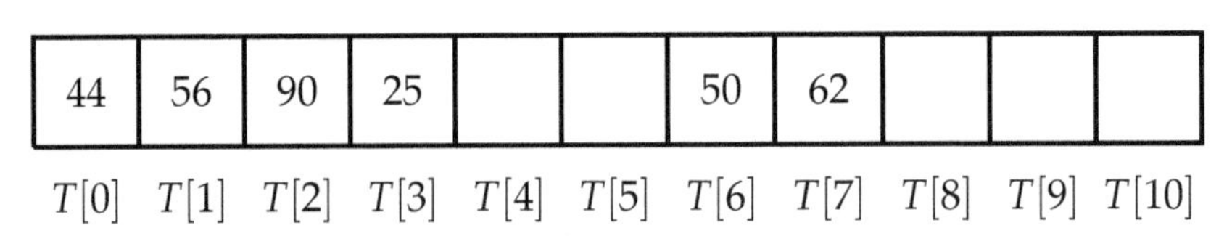

44	56	90	25			50	62			
$T[0]$	$T[1]$	$T[2]$	$T[3]$	$T[4]$	$T[5]$	$T[6]$	$T[7]$	$T[8]$	$T[9]$	$T[10]$

Now, we attempt to insert 33 into slot 33 mod 11 $= 0$. The slot $T[0]$ is occupied, as are slots $T[1], T[2]$ and $T[3]$. So we end up inserting 33 into slot $T[4]$:

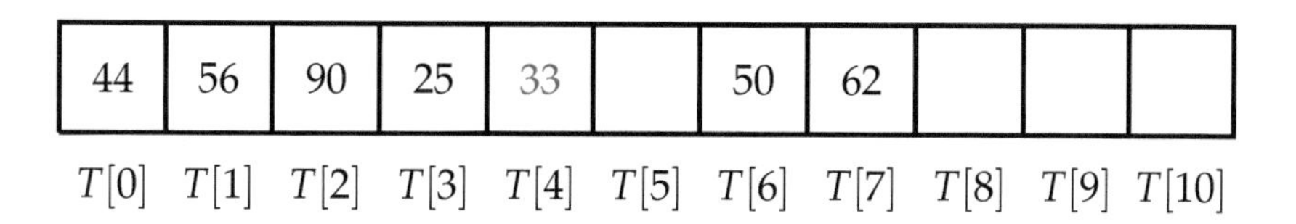

Now, suppose we delete the key 90 (from slot $T[2]$). We mark this slot as *deleted* (or *del*, for short) to indicate that an item has been deleted from this slot:

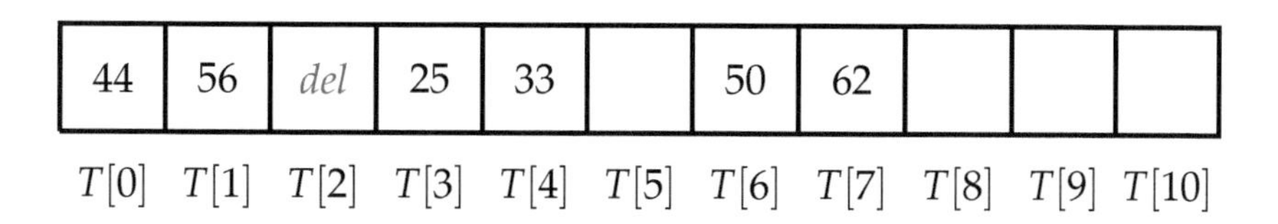

Next, suppose we want to insert the key 22 into T. We first search for 22. We begin at slot $T[0]$ because 22 mod 11 $= 0$. We have to examine successive slots until we encounter an empty slot. Therefore we check slots $T[0], T[1], \ldots, T[5]$. $T[5]$ is the first empty slot in the probe sequence, so at this point we can conclude that the key 22 is not present in the table. Thus we can insert the key 22 into T. We would place the key 22 in slot $T[2]$, because it is the first slot in the probe sequence that is marked as *deleted* or *empty*.

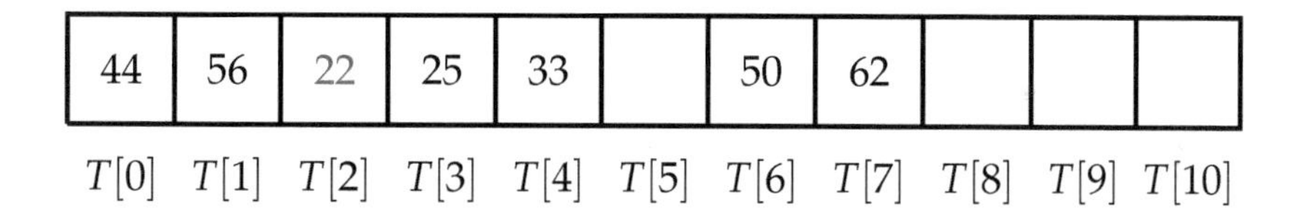

□

Algorithm 3.8 is an algorithm that searches for a key x in a hash table T of size n using linear probing. This algorithm returns two values, *success* and *index*. These are interpreted as follows:

- If x is in the table T, then *success* = **true** and *index* is the position where x is found (i.e., $T[index] = x$).
- If x is not in T, then *success* = **false** and $index \neq \infty$ is the position of the *first available* slot (hence, in particular, $T[index] =$ *deleted* or *empty*). Thus we can insert the value x into slot $T[index]$. If $index = \infty$, then the table T is full.

Knuth proved the following fundamental results in 1963 concerning the expected complexity of searching using linear probing.

Algorithm 3.8: LINEARPROBESEARCH(x)

```
global T, n
comment: T is an array of length n
external h
start ← h(x)
j ← start
searching ← true
index ← ∞
while searching
  do {
     if T[j] = x
       then { index ← j
              success ← true
              searching ← false
     else if T[j] = empty
       then { if index = ∞
                then index ← j
              success ← false
              searching ← false
       else { if T[j] = deleted and index = ∞
                then index ← j
              j ← j + 1 mod n
              if j = start
                then { success ← false
                       searching ← false
return (success, index)
```

Algorithm 3.9: LINEARPROBEINSERT(x)

```
global T, n
comment: T is an array of length n
external LINEARPROBESEARCH
(success, index) ← LINEARPROBESEARCH(x)
if not success and index ≠ ∞
  then T[index] ← x
```

Algorithm 3.10: LINEARPROBEDELETE(x)

```
global T, n
comment: T is an array of length n
external LINEARPROBESEARCH
(success, index) ← LINEARPROBESEARCH(x)
if success
  then T[index] ← deleted
```

Algorithm 3.11: HASHTABLEINITIALIZE(n)

```
comment: initialize T to be an empty array of length n
for i ← 0 to n − 1
  do T[i] ← empty
return (T)
```

THEOREM 3.3 *Suppose we have a hash table with open addressing having load factor α. Assuming there are no* DELETE *operations, the expected number of probes for a successful search using a random hash function with linear probing is approximately*

$$\frac{1}{2}\left(1+\frac{1}{1-\alpha}\right) \tag{3.4}$$

while the expected number of probes for an unsuccessful search is approximately

$$\frac{1}{2}\left(1+\frac{1}{(1-\alpha)^2}\right). \tag{3.5}$$

The proofs of these two results are rather complicated and we do not discuss them here. However, it turns out to be easier to analyze the expected number of probes for a ***random probe sequence***. This means that, for any key k, we assume that the probe sequence $H(k,0),\dots,h(k,M-1)$ is a ***random permutation*** of $\{0,1,\dots,M-1\}$. Requiring the probe sequence to be a permutation means that we will eventually consider every possible slot, without any repetitions, when hashing any given key. This is of course an idealized model. In particular, it is more general than the specific probe sequence that is used in double hashing (where the "step size" $h_2(k)$ is fixed for a given key k).

The following analysis was due to Peterson in 1957.

THEOREM 3.4 *Suppose we have a hash table with open addressing having load factor α. Assuming there are no* DELETE *operations, the expected number of probes for a successful* *search using a random hash function with a random probe sequence is approximately*

$$\frac{1}{\alpha} \ln \left(\frac{1}{1-\alpha} \right) \tag{3.6}$$

while the expected number of probes for an unsuccessful search is approximately

$$\frac{1}{1-\alpha}. \tag{3.7}$$

PROOF We first provide a simple derivation of the formula (3.7). Since the search is unsuccessful, we investigate random slots, stopping as soon as an empty slot is encountered. Recall that there are n filled slots and $M - n$ empty slots, and $\alpha = n/M \leq 1$.

Let $\mathbf{X}$ be a random variable that records the number of probes that are required. The maximum number of probes is n because there are n filled slots in the hash table. For $1 \leq i \leq n$, we observe that $\mathbf{X} \geq i$ if and only if the first $i-1$ slots that are examined are all filled.

Suppose $1 \leq i \leq n$. There are $\binom{n}{i-1}$ ways to choose $i-1$ filled slots and $\binom{M}{i-1}$ ways to choose $i-1$ arbitrary slots (filled or empty). Therefore,

$$\begin{aligned}
\mathbf{Pr}[\mathbf{X} \geq i] &= \frac{\binom{n}{i-1}}{\binom{M}{i-1}} \\
&= \frac{n(n-1)\cdots(n-i+2)}{M(M-1)\cdots(M-i+2)} \\
&= \frac{n}{M} \times \frac{n-1}{M-1} \times \cdots \times \frac{n-i+2}{M-i+2}.
\end{aligned}$$

It is easy to check that each of the $i-1$ terms in this product is at most $n/M = \alpha$, so

$$\mathbf{Pr}[\mathbf{X} \geq i] \leq \alpha^{i-1}. \tag{3.8}$$

Note that (3.8) also holds for integers $i > n$ because $\mathbf{Pr}[\mathbf{X} \geq i] = 0$ if $i > n$.

At this point, we use the formula (1.30) proven in Theorem 1.13. We have

$$\begin{aligned}
E[\mathbf{X}] &= \sum_{i=1}^{\infty} \mathbf{Pr}[\mathbf{X} \geq i] \\
&\leq \sum_{i=1}^{\infty} \alpha^{i-1} \quad \text{from (3.8)} \\
&= \frac{1}{1-\alpha},
\end{aligned} \tag{3.9}$$

as desired, since the sum is a geometric sequence.

Now we consider a successful search. Suppose there are n filled slots, and the n keys occupying these slots were inserted in the order $k_1, k_2, \ldots, k_n$. At the point in time when k_i is inserted into the table, there are $i-1$ occupied slots, so the load factor is $\alpha = (i-1)/M$. An *unsuccessful* search for k_i is performed, and k_i is inserted into the first empty slot in the probe sequence. Let $\mathbf{X}_i$ be a random variable whose value is the number of probes required to insert k_i. Using (3.9) with $\alpha = (i-1)/M$, we have

$$\begin{aligned} E[\mathbf{X}_i] &= \frac{1}{1-\frac{i-1}{M}} \\ &= \frac{M}{M-i+1}. \end{aligned} \tag{3.10}$$

For a randomly chosen key in the table, the expected number of probes is

$$\begin{aligned} E\left[\frac{1}{n}\sum_{i=1}^{n}\mathbf{X}_i\right] &= \frac{1}{n}E\left[\sum_{i=1}^{n}\mathbf{X}_i\right] \\ &= \frac{1}{n}\sum_{i=1}^{n}E\left[\mathbf{X}_i\right] \quad \text{by linearity of expectation} \\ &= \frac{1}{n}\sum_{i=1}^{n}\frac{M}{M-i+1} \quad \text{from (3.10)} \\ &= \frac{M}{n}\sum_{j=M-n+1}^{M}\frac{1}{j} \\ &= \frac{M}{n}(H_M - H_{M-n}) \\ &\approx \frac{M}{n}\ln\frac{M}{M-n} \quad \text{from (1.8)} \\ &= \frac{1}{\alpha}\ln\left(\frac{1}{1-\alpha}\right). \end{aligned}$$

∎

Figure 3.10 plots the expected number of probes for linear and random probing, in the cases of a successful and unsuccessful search. We use the formulas proven in Theorems 3.3 and 3.4. It is interesting to observe that all of these exhibit quite good behavior if the load factor $\alpha \leq 0.6$, say.

3.7 Notes and References

We have given a brief summary and review of basic data structures in this chapter. Two books on algorithm design that discuss data structures in considerably more detail are:

- *Introduction to Algorithms. Third Edition*, by Cormen, Leiserson, Rivest and Stein [16].

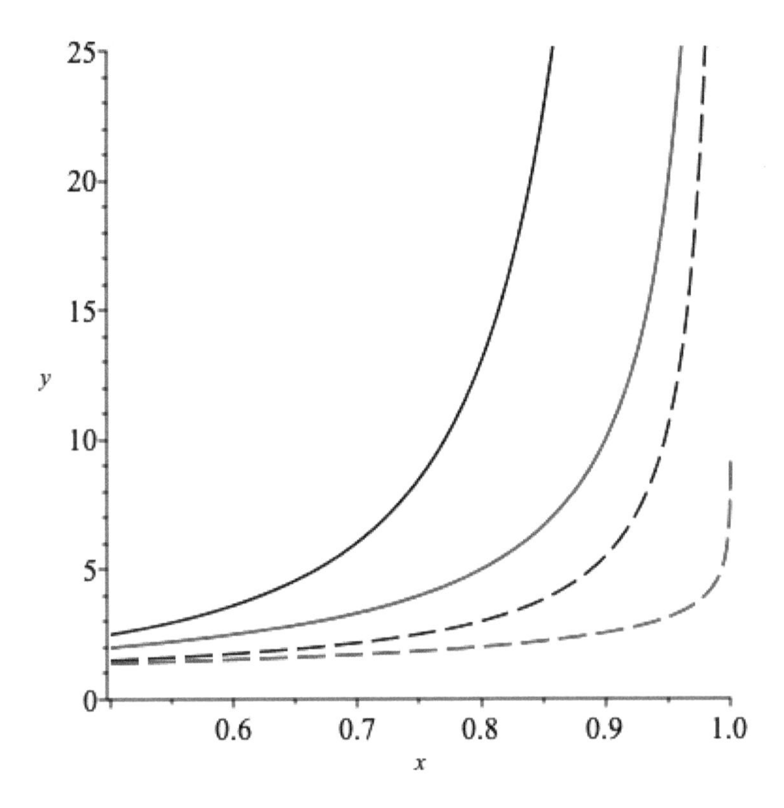

solid black	linear probing, unsuccessful search
dashed black	linear probing, successful search
solid red	random probing, unsuccessful search
dashed red	random probing, successful search

FIGURE 3.10: Expected number of probes for random hash functions.

- *Algorithm Design and Applications*, by Goodrich and Tamassia [27].

The (array-based) heap data structure was invented by Williams [62]. Floyd's algorithm for building a heap is from [20].

AVL trees were first described in [1]. Red-black trees were introduced by Bayer [3] using different terminology; the designation of nodes as "red" or "black" is first found in Guibas and Sedgewick [30].

An exact worst-case analysis of the number of comparisons required in Algorithm 3.6 can be found in Suchenek [57] or Paparrizos [51]. The maximum number of comparisons to build a heap with n items is $2n - 2\mu(n) - \sigma(n)$, where $\mu(n)$ is the number of ones in the binary representation of n, and $\sigma(n)$ is the number of trailing zeroes in the binary representation of n.

Knuth's original analysis of linear probing is given in [41]. Peterson's analysis of open addressing using random probe sequences is from [52].

Exercises

3.1 For each node N in a doubly-linked list, there are two pointers. $N.next$ points to the next node in the list and $N.prev$ points to the previous node in the list. If N is the last node in the list, then $N.next = null$, and if N is the first node in the list, then $N.prev = null$.

(a) Given a node N in a doubly-linked list, describe how pointers would be updated if N is deleted. Be sure to handle the cases where N is the first or last node in the list.

(b) Describe how to insert a new node into a doubly-linked list immediately following a node N in the list.

3.2 Suppose we store a queue in an array. When the array becomes full, we can resize it in a manner similar to what we did for stacks in Section 3.3.2. When we copy the items in the queue to the new array (which is twice the size of the old array), it could be useful to ensure that the front of the queue is at the beginning of the new array. Present pseudocode to do this.

3.3 Suppose A is a heap consisting of n integer values. Suppose we are given an integer b and we wish to output all the items in A that are $\leq b$. Give a pseudocode description of an efficient algorithm to do this. Your algorithm should have complexity $O(r)$, where r is the number of integers that the algorithm outputs.

3.4 Suppose we build a heap of size $n = 2^t - 1$ using Algorithm 3.6.

(a) Show that the minimum number of comparisons of heap elements that is required is $n - 1 = 2^t - 2$.

(b) Show that the maximum number of comparisons of heap elements that is required is $2^{t+1} - 2t - 2$.

3.5 Suppose that $\mathcal{T}$ is a given binary search tree. We will denote the successor of a node N by $\mathsf{succ}(N)$.

(a) Prove the following two facts.

i. If N has a right child, then $\mathsf{succ}(N)$ is the leftmost node in the right subtree of N.

ii. If N does not have a right child, then $\mathsf{succ}(N)$ is the first node M on the path from N to the root node of $\mathcal{T}$ such that $M.key > N.key$. (If no such node M exists, then $\mathsf{succ}(N)$ is undefined.)

(b) Given a node N in the binary search tree $\mathcal{T}$, give a pseudocode description of an algorithm that finds $\mathsf{succ}(N)$. Briefly discuss the complexity of your algorithm (the complexity should be $O(\mathsf{depth}(\mathcal{T}))$).

HINT Each node M in $\mathcal{T}$ has pointers to its left and right children, which are denoted respectively by *M.left* and *M.right*. We also have a pointer *root* to the root node of $\mathcal{T}$. However, there is no efficient way to move directly from a node to its parent. Therefore you should implement the second case of $\mathsf{succ}(N)$ in such a way that you do not move "upward" in the tree. A hint is to consider how a binary search works.

3.6 Cuckoo hashing is a type of open addressing in a hash table T where we make use of two hash functions, say h_1 and h_2. Every possible key k can be stored in one of two possible slots, namely, $T[h_1(k)]$ or $T[h_1(k)]$.

(a) Give short pseudocode descriptions of SEARCH and DELETE (note that both of these operations have complexity $O(1)$).

(b) Suppose we want to INSERT a new k into the hash table T of size n. Write out detailed pseudocode for this operation, based on the following informal description.

- If slot $T[h_1(k)]$ is empty, insert k into this slot and quit.
- Otherwise, if slot $T[h_2(k)]$ is empty, insert k into this slot and quit.
- Otherwise, choose one of the slots $T[h_1(k)]$ and $T[h_2(k)]$ and insert k, evicting the existing key k' that is in this slot. Then attempt to reinsert k' into its "alternate" slot, evicting another key if necessary. Repeat this process as often as required.

It is possible that we could encounter an infinite cycle of evictions, so we should terminate the process if the number of evictions during a given INSERT operation exceeds some prespecified threshold M. (In this case, the INSERT operation fails. However, in practice, a resizing operation would be done at this time.)

Chapter 4

Divide-and-Conquer Algorithms

In this chapter, we discuss one of the main algorithm design techniques, which is known as *divide-and-conquer*. Divide-and-conquer algorithms are recursive, so their analysis depends on solving recurrence relations. Thus, we first spend some time discussing various methods to solve recurrence relations.

4.1 Recurrence Relations

Suppose $T(1), T(2), \dots,$ is an infinite sequence of real numbers. A ***recurrence relation*** is a formula that expresses a general term $T(n)$ as a function of one or more previous terms, $T(1), \dots, T(n-1)$. A recurrence relation will also specify one or more ***initial values*** starting at $T(1)$.

Solving a recurrence relation means finding a formula for $T(n)$, as a function of n, that does *not* involve any previous terms, $T(1), \dots, T(n-1)$. There are many methods of solving recurrence relations. Two important techniques are ***guess-and-check*** and the ***recursion tree method***. We will make extensive use of the recursion tree method. However, we first take a look at the guess-and-check process.

4.1.1 Guess-and-check Method

Guess-and-check usually involves the following four steps:

step 1

Tabulate some values $T(1), T(2), \dots$ using the recurrence relation.

step 2

Guess that the solution $T(n)$ has a specific form, possibly involving undetermined constants.

step 3

Use $T(1), T(2), \dots$ to determine the actual values of the unspecified constants.

step 4

Use *induction* to prove your guess for $T(n)$ is correct.

DOI: 10.1201/9780429277412-4

Example 4.1 Suppose we have the recurrence

$$\begin{aligned} T(n) &= T(n-1) + 6n - 5 \qquad \text{if } n > 0 \\ T(0) &= 4. \end{aligned}$$

We first use the recurrence relation to compute a few values of $T(n)$ for small n:

$$\begin{aligned} T(1) &= T(0) + 6 - 5 &&= 4 + 6 - 5 &&= 5 \\ T(2) &= T(1) + 6 \times 2 - 5 &&= 5 + 12 - 5 &&= 12 \\ T(3) &= T(2) + 6 \times 3 - 5 &&= 12 + 18 - 5 &&= 25 \\ T(4) &= T(3) + 6 \times 4 - 5 &&= 25 + 24 - 5 &&= 44. \end{aligned}$$

If we are sufficiently perspicacious, we might guess that $T(n)$ is a quadratic function, e.g.,

$$T(n) = an^2 + bn + c,$$

where a, b and c are as-yet-unspecified constants.

Next, we use $T(0) = 4$, $T(1) = 5$, $T(2) = 12$ to compute a, b and c by solving three equations in three unknowns. Substituting $n = 0, 1$ and 2 into our conjectured formula for $T(n)$, we get the following system of equations:

$$\begin{aligned} c &= 4 \\ a + b + c &= 5 \\ 4a + 2b + c &= 12. \end{aligned}$$

Solving this system of equations, we get $a = 3$, $b = -2$, $c = 4$.

Finally, we can use induction to prove that $T(n) = 3n^2 - 2n + 4$ for all $n \geq 0$. We already know that the formula is valid for $n = 0, 1$ and 2. Suppose that the formula is valid for $n = m - 1$, for some $m \geq 3$. Now we prove that the formula is correct for $n = m$:

$$\begin{aligned} T(m) &= T(m-1) + 6m - 5 \\ &= 3(m-1)^2 - 2(m-1) + 4 + 6m - 5 \qquad \text{by induction} \\ &= 3m^2 - 6m + 3 - 2m + 2 + 4 + 6m - 5 \\ &= 3m^2 - 2m + 4, \end{aligned}$$

as desired. Therefore the formula is correct for all $n \geq 0$, by induction. □

REMARK In general, if the difference between $T(n)$ and $T(n-1)$ is a polynomial of degree d, then the solution to the recurrence relation will be a polynomial of degree $d + 1$. ∎

Example 4.2 Consider the recurrence

$$\begin{aligned} T(n) &= T\left(\left\lfloor \frac{n}{2} \right\rfloor\right) + T\left(\left\lfloor \frac{n}{3} \right\rfloor\right) + n \qquad \text{if } n > 2 \\ T(1) &= 1 \\ T(2) &= 2. \end{aligned}$$

Suppose we tabulate some values of $T(n)$ and then guess that

$$T(n) \leq cn$$

for all $n \geq 1$, for some unspecified constant c. We can use empirical data to guess an appropriate value for c. However, an alternative approach is to carry out the induction proof in order to determine a value of c that works. We show how this can be done now.

The base cases $n = 1, 2$ are all right if $c \geq 1$, so we will henceforth assume that $c \geq 1$. Now suppose $m \geq 3$ and $T(n) \leq cn$ for $1 \leq n \leq m - 1$. Then we have

$$\begin{aligned} T(m) &= T\left(\left\lfloor \frac{m}{2} \right\rfloor\right) + T\left(\left\lfloor \frac{m}{3} \right\rfloor\right) + m \\ &\leq c\left\lfloor \frac{m}{2} \right\rfloor + c\left\lfloor \frac{m}{3} \right\rfloor + m \qquad \text{by induction} \\ &\leq \frac{cm}{2} + \frac{cm}{3} + m \\ &= m\left(\frac{5c}{6} + 1\right). \end{aligned}$$

Therefore, we want

$$\frac{5c}{6} + 1 \leq c,$$

which is true if $c \geq 6$. Hence, we can take $c = 6$ and we have proven by induction that $T(n) \leq 6n$ for all $n \geq 1$. Thus, $T(n) \in O(n)$. It is obvious that $T(n) \in \Omega(n)$ because $T(n) \geq n$, so it follows that $T(n) \in \Theta(n)$. □

We can also use the recursion tree method (which is discussed in the next subsection) to derive the same result.

4.1.2 Linear Recurrence Relations

We now use the guess-and-check method to solve a larger class of recurrence relations, which are known as ***linear recurrence relations***. These recurrence relations have the following form:

$$\begin{aligned} T(n) &= rT(n-1) + s \qquad \text{if } n > 0 \\ T(0) &= d, \end{aligned}$$

where r, s and d are constants. We will generally assume that $r > 0$.

As usual, we use the recurrence relation to compute a few values of $T(n)$ for small n:

$$\begin{aligned}
T(0) &= d \\
T(1) &= rd + s \\
T(2) &= r^2 d + s(r+1) \\
T(3) &= r^3 d + s(r^2 + r + 1).
\end{aligned}$$

Based on the pattern evident in the values computed above, we might conjecture that

$$T(n) = r^n d + s(r^{n-1} + \cdots + 1) \tag{4.1}$$

for all $n \geq 1$. The formula (4.1) is clearly valid for $n = 1, 2$ and 3. Suppose that (4.1) is valid for $n = m - 1$, for some $m \geq 2$. Now we prove that (4.1) is correct for $n = m$:

$$\begin{aligned}
T(m) &= rT(m-1) + s \\
&= r(r^{m-1}d + s(r^{m-2} + \cdots + 1)) + s \qquad \text{by induction} \\
&= r^m d + sr(r^{m-2} + \cdots + 1) + s \\
&= r^m d + s(r^{m-1} + \cdots + r + 1),
\end{aligned}$$

as desired. Therefore the formula (4.1) is correct for all $n \geq 1$, by induction.

When $n = 0$, the formula (4.1) is also correct if we interpret the expression $r^{n-1} + \cdots + r + 1$ as having the value 0.

The sum $r^{n-1} + \cdots + r + 1$ in the formula (4.1) is just a geometric sequence. We can therefore provide a simpler formula by evaluating this sum using the formulae in (1.2). We end up with three cases, depending on whether $r > 1$, $r = 1$ or $0 < r < 1$. The resulting simplified formulae are given in the following corollary.

THEOREM 4.1 *The recurrence relation*

$$\begin{aligned}
T(n) &= rT(n-1) + s \qquad \textit{if } n > 0 \\
T(0) &= d,
\end{aligned}$$

where r, s and d are constants and $r > 0$, has the following solution:

$$T(n) = \begin{cases} r^n d + s\left(\dfrac{r^n - 1}{r - 1}\right) & \textit{if } r > 1 \\ d + sn & \textit{if } r = 1 \\ r^n d + s\left(\dfrac{1 - r^n}{1 - r}\right) & \textit{if } 0 < r < 1. \end{cases} \tag{4.2}$$

Notice that the formulae in (4.2) are valid for all $n \geq 0$.

4.1.3 Recursion Tree Method

The ***recursion tree method*** is another way of solving recurrence relations. It consists of repeatedly "expanding" the recurrence and thereby constructing a tree. Summing all the values in the tree gives the solution to the recurrence. This method has some potential advantages as compared to the guess-and-check approach. First, it does not require us to make a possibly non-obvious guess as to the form of the solution. Second, it does not require an induction proof.

We illustrate the basic method by solving the following recurrence relation, which arises in the analysis of MERGESORT (see Section 4.3.1):

$$
\begin{aligned}
T(n) &= 2T\left(\frac{n}{2}\right) + cn \qquad \text{if } n > 1 \text{ is a power of 2} \\
T(1) &= d,
\end{aligned}
$$

where c and d are positive constants.

We can solve this recurrence relation when n is a power of two. We write $n = 2^j$ and we construct a ***recursion tree*** using the following four steps:

step 1

Start with a *one-node tree*, say N, having the value $T(2^j)$.

step 2

Grow *two children* of N. These children have the value $T(2^{j-1})$, and the value of the root node is changed from $T(2^j)$ to $c2^j$.

step 3

Repeat this process recursively, terminating when we encounter nodes that receive the value $T(1) = d$.

step 4

Sum the values of all the nodes on each level of the tree, and then compute the *sum of all these sums*; the result is $T(2^j)$.

The first three steps in growing the recursion tree are shown in Figure 4.1. This diagram also can help us understand why the recursion tree method works correctly. Suppose we define the ***value of a tree*** to be the sum of the values in all the nodes in the tree. Thus, the value of the original one-node tree is $T(2^j)$. The next tree, which has three nodes, has value

$$c2^j + 2T(2^{j-1}).$$

The important observation is that these two trees have the same value, because

$$T(2^j) = c2^j + 2T(2^{j-1}).$$

This is directly obtained from the recurrence relation, by substituting $n = 2^j$.

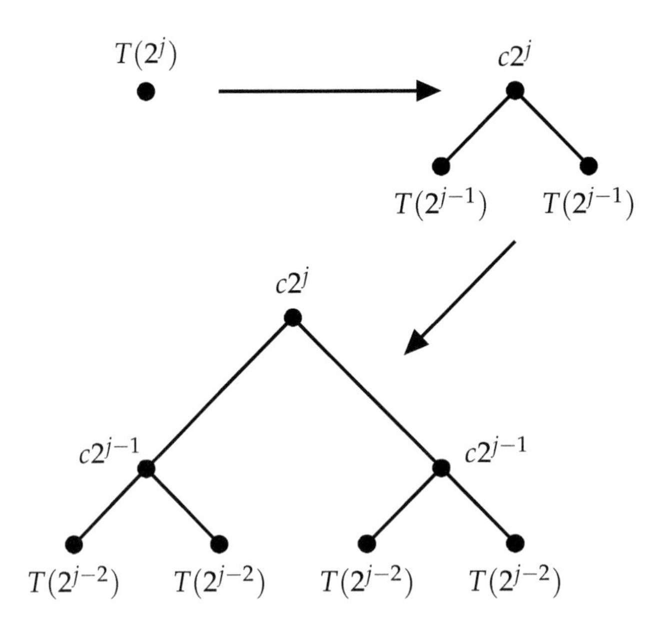

FIGURE 4.1: The recursion tree for the MERGESORT recurrence

When we proceed to the third tree, we replace each of the two nodes having value $T(2^{j-1})$ with three nodes whose values sum to

$$c2^{j-1} + 2T(2^{j-2}).$$

However, if we apply the recurrence relation with $n = 2^{j-1}$, we have

$$T(2^{j-1}) = c2^{j-1} + 2T(2^{j-2}).$$

So again the total value of the tree is not changed.

We keep "expanding" the tree until we reach base cases, which will be the leaf nodes of the recursion tree. Since the value of the tree remains constant throughout this process, the value of the final tree will be $T(2^j)$. However, the value of the final tree can be computed simply by summing the values of all the nodes. This is easily done as indicated in **step 4**.

We tabulate the number of nodes at each level, along with their values, in Table 4.1. When we sum the values at all levels of the recursion tree, we obtain

$$T(2^j) = d2^j + cj2^j.$$

This formula is valid for all $j \geq 0$.

If we want to determine the complexity of $T(n)$ as a function of n, we proceed as follows. Since $n = 2^j$, we have $j = \log_2 n$ and therefore

$$T(n) = dn + cn \log_2 n \in \Theta(n \log n).$$

TABLE 4.1: Computing the value of a recursion tree for $n = 2^j$

level	# nodes	value at each node	value of the level
0	1	$c2^j$	$c2^j$
1	2	$c2^{j-1}$	$c2^j$
2	2^2	$c2^{j-2}$	$c2^j$
$\vdots$	$\vdots$	$\vdots$	$\vdots$
$j-1$	2^{j-1}	$c2^1$	$c2^j$
j	2^j	d	$d2^j$
all			$d2^j + cj2^j$

Note that we have derived an exact formula step-by-step, so no induction proof is required. In this example, it turns out that every level has the same sum (except for the bottom level, which is level j), but that is not generally the case.

As another example, recall the recurrence

$$T(n) = T\left(\left\lfloor \frac{n}{2} \right\rfloor\right) + T\left(\left\lfloor \frac{n}{3} \right\rfloor\right) + n$$

from Example 4.2. We showed by induction that $T(n) \in O(n)$. We will now use the recursion tree method to obtain the same O-bound. The idea is to ignore all "floors" and compute the sum of the all the levels of the corresponding "infinite" tree $\mathcal{T}$. This tree $\mathcal{T}$ has the structure indicated in Figure 4.2.

The sums of the values of the nodes on the first three levels (i.e., levels 0, 1 and 2) of $\mathcal{T}$ are $n, 5n/6$ and $25n/36$, respectively. This is just a geometric series with ratio $5/6$. Assuming that $\mathcal{T}$ has an infinite number of levels, we would get the following infinite sum:

$$\sum_{i=0}^{\infty} n \left(\frac{5}{6}\right)^i.$$

Since the ratio in this geometric sequence is less than one, the infinite sum is finite, from equation (1.3):

$$\sum_{i=0}^{\infty} n \left(\frac{5}{6}\right)^i = 6n \in \Theta(n).$$

The "real" recursion tree would only have a finite number of nodes, and the value of each node would be no more than the value of the corresponding node in the tree $\mathcal{T}$. Therefore, the above analysis provides us with an upper bound on $T(n)$. Hence $T(n) \leq 6n$ and therefore $T(n)$ is $O(n)$.

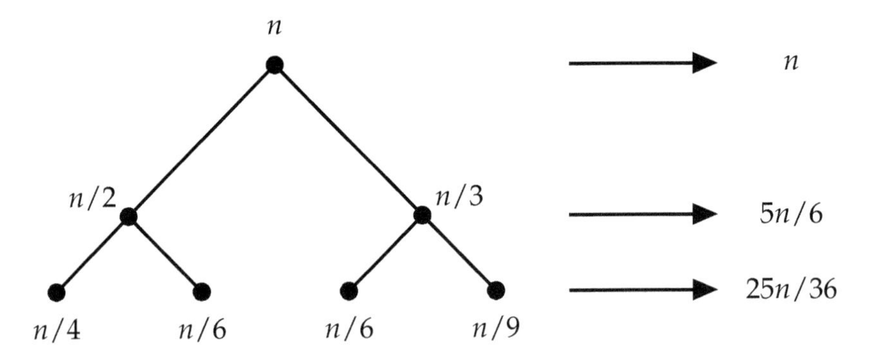

FIGURE 4.2: Another recursion tree

4.2 The Master Theorem

The ***Master Theorem*** provides a formula for the solution of many recurrence relations typically encountered in the analysis of divide-and-conquer algorithms. It was first described in 1980 by Bentley, Haken and Saxe. Theorem 4.2 is a simplified version of the Master Theorem. (We will see a more general version of this theorem a bit later.)

THEOREM 4.2 *Suppose that $a \geq 1$ and $b \geq 2$ are integers and c and y are positive constants. Consider the recurrence*

$$\begin{aligned} T(n) &= aT\left(\frac{n}{b}\right) + cn^y \qquad \text{if } n > 1 \text{ is a power of } b \\ T(1) &= d, \end{aligned} \tag{4.3}$$

where n is a power of b. Then

$$T(b^j) = da^j + cb^{jy} \sum_{i=0}^{j-1} \left(\frac{a}{b^y}\right)^i \tag{4.4}$$

for all integers $j \geq 0$.

PROOF Using the recursion tree method, we will derive an exact solution to the recurrence (4.3) when n is a power of b. Thus, we let $n = b^j$ where j is a non-negative integer. Table 4.2 lists the number of nodes in each level of the tree, along with their values. Summing the values at all levels of the recursion tree, we have that

$$T(b^j) = da^j + \sum_{i=0}^{j-1} ca^i \left(\frac{n}{b^i}\right)^y$$

TABLE 4.2: The recursion tree in the Master Theorem when $n = b^j$

level	# nodes	value at each node	value of the level
0	1	cn^y	cn^y
1	a	$c\left(\frac{n}{b}\right)^y$	$ca\left(\frac{n}{b}\right)^y$
2	a^2	$c\left(\frac{n}{b^2}\right)^y$	$ca^2\left(\frac{n}{b^2}\right)^y$
$\vdots$	$\vdots$	$\vdots$	$\vdots$
$j-1$	a^{j-1}	$c\left(\frac{n}{b^{j-1}}\right)^y$	$ca^{j-1}\left(\frac{n}{b^{j-1}}\right)^y$
j	a^j	d	da^j

$$= da^j + cn^y \sum_{i=0}^{j-1} \left(\frac{a}{b^y}\right)^i \tag{4.5}$$

$$= da^j + cb^{jy} \sum_{i=0}^{j-1} \left(\frac{a}{b^y}\right)^i,$$

which gives us the desired formula (4.4). ∎

Of course the sum in (4.4) is just a geometric sequence with ratio a/b^y. We apply the three formulae in (1.2). Note that, in the special case where $a = b^y$, we have

$$\begin{aligned} da^j + cb^{jy} \sum_{i=0}^{j-1} \left(\frac{a}{b^y}\right)^i &= da^j + cjb^{jy} \\ &= da^j + cja^j \\ &= a^j(d + cj). \end{aligned}$$

We obtain the following:

$$T(b^j) = \begin{cases} da^j + cb^{jy}\left(\frac{(a/b^y)^j - 1}{(a/b^y) - 1}\right) & \text{if } a > b^y \\ a^j(d + cj) & \text{if } a = b^y \\ da^j + cb^{jy}\left(\frac{1-(a/b^y)^j}{1-(a/b^y)}\right) & \text{if } 0 < a < b^y. \end{cases} \tag{4.6}$$

We now analyze the complexity of (4.4), which will depend on the value of the ratio a/b^y.

COROLLARY 4.3 *Suppose that $a \geq 1$ and $b > 1$. Consider the recurrence*

$$T(n) = aT\left(\frac{n}{b}\right) + \Theta(n^y),$$

where n is a power of b. Denote $x = \log_b a$. *Then*

$$T(n) \in \begin{cases} \Theta(n^x) & \text{if } y < x \\ \Theta(n^x \log n) & \text{if } y = x \\ \Theta(n^y) & \text{if } y > x. \end{cases}$$

PROOF Recall that $b^x = a$ and $n = b^j$. Hence

$$a^j = (b^x)^j = (b^j)^x = n^x.$$

As mentioned above, the formula for $T(n)$ is a geometric sequence with ratio

$$r = \frac{a}{b^y} = b^{x-y}.$$

We will use (4.5) as our starting point:

$$T(n) = dn^x + cn^y \sum_{i=0}^{j-1} r^i.$$

Let

$$S = \sum_{i=0}^{j-1} r^i;$$

then we have

$$T(n) = dn^x + cn^y S.$$

Since S is a sum of a geometric sequence, we have from (1.2) that

$$S \in \begin{cases} \Theta(r^j) & \text{if } r > 1, \text{ i.e., if } x > y \\ \Theta(j) = \Theta(\log n) & \text{if } r = 1, \text{ i.e., if } y = x \\ \Theta(1) & \text{if } r < 1, \text{ i.e., if } y > x. \end{cases}$$

case 1

Here, we have $x > y$ so $r > 1$. $S \in \Theta(r^j)$, so $T(n) \in \Theta(n^x + n^y r^j)$. However,

$$r^j = (b^{x-y})^j = (b^j)^{x-y} = n^{x-y}.$$

Therefore

$$T(n) \in \Theta(n^x + n^y n^{x-y}) = \Theta(n^x).$$

case 2

Here, we have $x = y$ so $r = 1$. $S \in \Theta(j) = \Theta(\log n)$, so

$$T(n) \in \Theta(n^x + n^y \log n) = \Theta(n^x + n^x \log n) = \Theta(n^x \log n).$$

case 3

Here, we have $x < y$ so $r < 1$. $S \in \Theta(1)$, so

$$T(n) \in \Theta(n^x + n^y) = \Theta(n^y).$$

TABLE 4.3: The three cases in the Master Theorem

case	r	y, x	complexity of $T(n)$
heavy leaves	$r > 1$	$y < x$	$T(n) \in \Theta(n^x)$
balanced	$r = 1$	$y = x$	$T(n) \in \Theta(n^x \log n)$
heavy top	$r < 1$	$y > x$	$T(n) \in \Theta(n^y)$

Observe that the complexity does not depend on the initial value d. ∎

In Table 4.3, we summarize the three cases analyzed above. They have the following interpretation:

heavy leaves

This means that the value of the recursion tree is dominated by the values of the leaf nodes.

balanced

This means that the values of the levels of the recursion tree are constant (except for the bottom-most level).

heavy top

This means that the value of the recursion tree is dominated by the value of the root node.

We now provide a few examples to illustrate the application of Theorem 4.3.

Example 4.3 Suppose

$$T(n) = 2T\left(\frac{n}{2}\right) + cn.$$

This recurrence was already discussed in Section 4.1.3. Here we have

$$a = 2, \quad b = 2, \quad y = 1, \quad \text{and} \quad x = 1.$$

We are in **case 2** (the balanced case) because $y = x$, and therefore

$$T(n) \in \Theta(n^1 \log n) = \Theta(n \log n).$$

□

Example 4.4 Suppose

$$T(n) = 3T\left(\frac{n}{2}\right) + cn.$$

Here we have

$$a = 2, \quad b = 3, \quad y = 1, \quad \text{and} \quad x = \log_2 3 \approx 1.59.$$

We are in **case 1** (heavy leaves) because $y < x$, and therefore

$$T(n) \in \Theta(n^{\log_2 3}).$$

□

Example 4.5 Suppose

$$T(n) = 4T\left(\frac{n}{2}\right) + cn.$$

Here we have

$$a = 2, \quad b = 4, \quad y = 1, \quad \text{and} \quad x = \log_2 4 = 2.$$

We are in **case 1** (heavy leaves) because $y < x$, and

$$T(n) \in \Theta(n^2).$$

□

Example 4.6 Suppose

$$T(n) = 2T\left(\frac{n}{2}\right) + cn^{3/2}.$$

Here we have

$$a = 2, \quad b = 2, \quad y = 3/2, \quad \text{and} \quad x = 1.$$

We are in **case 3** (heavy top) because $y > x$, and

$$T(n) \in \Theta(n^{3/2}).$$

□

4.2.1 Another Proof of the Master Theorem

It is also possible to prove the Master Theorem by "reducing" it to a linear recurrence relation. Suppose we begin with the recurrence (4.3):

$$\begin{aligned} T(n) &= aT\left(\frac{n}{b}\right) + cn^y \qquad \text{if } n > 1 \text{ is a power of } b \\ T(1) &= d. \end{aligned}$$

The trick is to define

$$S(j) = \frac{T(b^j)}{b^{jy}}$$

for all integers $j \geq 0$. Then

$$\begin{aligned} S(j) &= \frac{T(b^j)}{b^{jy}} \\ &= \frac{aT(b^{j-1}) + cb^{jy}}{b^{jy}} \end{aligned}$$

$$
\begin{aligned}
&= \frac{aT(b^{j-1})}{b^{jy}} + c \\
&= \frac{a}{b^y} \times \frac{T(b^{j-1})}{b^{(j-1)y}} + c \\
&= \frac{a}{b^y} S(j-1) + c.
\end{aligned}
$$

The initial condition is $S(0) = d$. Therefore we have transformed the original recurrence relation for T into a linear recurrence relation. We obtain the solution for $S(n)$ from (4.2) by taking $r = a/b^y$ and $s = c$:

$$
S(j) = \begin{cases}
(a/b^y)^j d + c\left(\frac{(a/b^y)^j - 1}{(a/b^y) - 1}\right) & \text{if } a > b^y \\
d + cj & \text{if } a = b^y \\
(a/b^y)^j d + c\left(\frac{1-(a/b^y)^j}{1-(a/b^y)}\right) & \text{if } 0 < a < b^y.
\end{cases}
$$

Since $T(b^j) = b^{jy} S(j)$, we have the following:

$$
T(b^j) = \begin{cases}
a^j d + cb^{jy}\left(\frac{(a/b^y)^j - 1}{(a/b^y) - 1}\right) & \text{if } a > b^y \\
a^j(d + cj) & \text{if } a = b^y \\
a^j d + cb^{jy}\left(\frac{1-(a/b^y)^j}{1-(a/b^y)}\right) & \text{if } 0 < a < b^y.
\end{cases}
$$

These formulae are identical to (4.6).

4.2.2 A General Version of the Master Theorem

We now state and prove a more general version of the Master Theorem.

THEOREM 4.4 *Suppose that $a \geq 1$, $b > 1$ and $d > 0$. Denote $x = \log_b a$. Suppose $T(n)$ is defined by the recurrence*

$$
\begin{aligned}
&T(n) = aT\left(\frac{n}{b}\right) + f(n) \qquad \textit{if } n > 1 \\
&T(1) = d.
\end{aligned}
$$

Then, for $n = b^j$, it holds that

$$
T(n) \in \begin{cases}
\Theta(n^x) & \textit{if } f(n) \in O(n^{x-\epsilon}) \textit{ for some } \epsilon > 0 \\
\Theta(n^x \log n) & \textit{if } f(n) \in \Theta(n^x) \\
\Theta(f(n)) & \textit{if } f(n)/n^{x+\epsilon} \textit{ is an increasing function of } n \\
& \textit{for some } \epsilon > 0.
\end{cases}
$$

PROOF The recursion tree method can be employed to obtain a formula for $T(n)$

when $n = b^j$. We omit the details, but the resulting formula is

$$
\begin{aligned}
T(n) &= a^j f(1) + \sum_{i=0}^{j-1} a^i f\left(\frac{n}{b^i}\right) \\
&= dn^x + \sum_{i=0}^{j-1} a^i f\left(\frac{n}{b^i}\right), \qquad (4.7)
\end{aligned}
$$

where we use the fact that

$$a^j = (b^x)^j = (b^j)^x = n^x$$

to derive (4.7).

case 1

Here we have $f(n) \in O(n^{x-\epsilon})$ for some $\epsilon > 0$. Denote $y = x - \epsilon$; then $f(n) \in O(n^y)$. Note that $x - y = \epsilon > 0$, so $b^{x-y} > 1$.

First, from (4.7), we have $T(n) \geq dn^x$, so $T(n) \in \Omega(n^x)$.

For the upper bound, we use the fact that $f(n) \in O(n^y)$, which implies that there is a constant $c > 0$ such that $f(n) \leq cn^y$ for all $n > 1$. Now, from (4.7), we have

$$
\begin{aligned}
T(n) &\leq dn^x + \sum_{i=0}^{j-1} ca^i \left(\frac{n}{b^i}\right)^y \\
&= dn^x + cn^y \sum_{i=0}^{j-1} \left(\frac{a}{b^y}\right)^i \\
&= dn^x + cn^y \sum_{i=0}^{j-1} (b^{x-y})^i \\
&\in O(n^x + n^y (b^{x-y})^j) \qquad \text{since } b^{x-y} > 1 \\
&= O(n^x + n^y (b^j)^{x-y}) \\
&= O(n^x + n^y n^{x-y}) \\
&= O(n^x).
\end{aligned}
$$

Because $T(n) \in \Omega(n^x)$ and $T(n) \in O(n^x)$, we have that $T(n) \in \Theta(n^x)$.

case 2

In this case, we have that $f(n) \in \Theta(n^x)$. The proof of this case is the same as the proof of the corresponding case of the simplified version of the Master Theorem.

case 3

Here, $f(n)/n^{x+\epsilon}$ is an increasing function of n. Denote $y = x + \epsilon$. Then $x - y = \epsilon < 0$, so $b^{x-y} < 1$. Note also that $n^y \in O(f(n))$.

First, from (4.7), we have $T(n) \geq f(n)$, so $T(n) \in \Omega(f(n))$.

For the upper bound, we use the fact that $f(n)/n^y$ is an increasing function of n. Therefore,

$$\frac{f(n)}{n^y} > \frac{f\left(\frac{n}{b}\right)}{\left(\frac{n}{b}\right)^y} > \frac{f\left(\frac{n}{b^2}\right)}{\left(\frac{n}{b^2}\right)^y} > \cdots,$$

which implies that

$$f(n) > b^y f\left(\frac{n}{b}\right) > b^{2y} f\left(\frac{n}{b^2}\right) > \cdots > b^{(j-1)y} f\left(\frac{n}{b^{j-1}}\right). \tag{4.8}$$

Now, from (4.7) and (4.8), we obtain

$$\begin{aligned}
T(n) &= dn^x + \sum_{i=0}^{j-1} a^i f\left(\frac{n}{b^i}\right) \\
&< dn^x + \sum_{i=0}^{j-1} \left(\frac{a}{b^y}\right)^i f(n) \\
&= dn^x + f(n) \sum_{i=0}^{j-1} \left(\frac{a}{b^y}\right)^i \\
&\in O(n^x + f(n)) \qquad \text{since } b^{x-y} < 1 \\
&= O(f(n)),
\end{aligned}$$

since $x < y$ and $n^y \in O(f(n))$.

∎

Example 4.7 Suppose

$$T(n) = 3T\left(\frac{n}{4}\right) + n \log n.$$

Here we have

$$a = 3, \quad b = 4, \quad f(n) = n \log n \quad \text{and} \quad x = \log_4 3 \approx .793.$$

We have

$$\frac{f(n)}{n^{x+\epsilon}} = \frac{n \log n}{n^{.793+\epsilon}}.$$

If we take $\epsilon = .1$, then we have

$$\frac{f(n)}{n^{x+\epsilon}} = \frac{n \log n}{n^{.893}} = n^{.107} \log n,$$

which is an increasing function of n. We are in case 3 of the modified general version of the Master Theorem and hence

$$T(n) \in \Theta(n \log n).$$

□

TABLE 4.4: A recursion tree for a more complicated example

level	# nodes	value at each node	value of the level
0	1	$j2^j$	$j2^j$
1	2	$(j-1)2^{j-1}$	$(j-1)2^j$
2	2^2	$(j-2)2^{j-2}$	$(j-2)2^j$
$\vdots$	$\vdots$	$\vdots$	$\vdots$
$j-1$	2^{j-1}	2^1	2^j
j	2^j	1	2^j

Example 4.8 Suppose

$$T(n) = 2T\left(\frac{n}{2}\right) + n\log n.$$

Here we have

$$a = 2, \quad b = 2, \quad f(n) = n\log n \quad \text{and} \quad x = 1.$$

Clearly, cases 1 and 2 do not apply. What about case 3? We compute

$$\frac{f(n)}{n^{x+\epsilon}} = \frac{n\log n}{n^{1+\epsilon}} = \frac{\log n}{n^{\epsilon}}.$$

However, $\log n \in o(n^{\epsilon})$ for any $\epsilon > 0$. So $\log n/n^{\epsilon}$ cannot be an increasing function of n. Hence the modified general version of the Master Theorem does not apply. □

We can actually solve the recurrence $T(n) = 2T(n/2) + n\log n$ (from Example 4.8) using the recursion tree method directly. Assume $T(1) = 1$ and let $n = 2^j$. Consider the values tabulated in Table 4.4.

Summing the values at all levels of the recursion tree, we have that

$$\begin{aligned} T(n) &= 2^j\left(1 + \sum_{i=1}^{j} i\right) \\ &= 2^j\left(1 + \frac{j(j+1)}{2}\right) \qquad \text{from (1.1).} \end{aligned}$$

Since $n = 2^j$, we have $j = \log_2 n$ and $T(n) \in \Theta(n(\log n)^2)$. This demonstrates that the recursion tree method is sometimes more powerful than the Master Theorem.

4.3 Divide-and-Conquer Design Strategy

Divide-and-conquer is one of the most important algorithm design techniques. Actually, a better term might be ***divide-and-conquer-and-combine***, as these three steps (rather than just "divide" and "conquer") are usually included in an algorithm of this type. We now describe how these three steps typically work.

divide

Given a problem instance I, construct one or more smaller problem instances. If we have a of these smaller problem instances, we can denote them by $I_1, \ldots, I_a$ (these are called ***subproblems***). Usually, we want the size of these subproblems to be small compared to $\mathsf{size}(I)$, e.g., half the size.

conquer

For $1 \leq j \leq a$, solve the subproblem I_j recursively. Thus we obtain a solutions to subproblems, which we will denote by $S_1, \ldots, S_a$.

combine

Given the solutions $S_1, \ldots, S_a$ to the a subproblems, use an appropriate ***combining*** function to find the solution S to the problem instance I, i.e.,

$$S \leftarrow \textsc{Combine}(S_1, \ldots, S_a),$$

where COMBINE is the combining function. The COMBINE operation is often the most complicated step of a divide-and-conquer algorithm.

Analysis of a divide-and-conquer algorithm typically involves solving a recurrence relation. Often the Master Theorem can be applied.

4.3.1 Merge Sort

MERGESORT is an efficient sorting algorithm. As before, an instance of **Sorting** consists of an array A of n integers. The required output is the same array, sorted in increasing order. The size of the problem instance is n.

MERGESORT is a divide-and-conquer sorting algorithm. We describe the three steps (divide, conquer and combine) now.

divide

Split A into two subarrays of (roughly) equal size: A_L consists of the first $\lceil \frac{n}{2} \rceil$ elements in A (the "left half") and A_R consists of the last $\lfloor \frac{n}{2} \rfloor$ elements in A (the "right half").

conquer

Run MERGESORT recursively on A_L and A_R, obtaining S_L and S_R, respectively. S_L is the left half of A, sorted, and S_R is the right half of A, sorted.

combine

After constructing S_L and S_R, use a combining function called MERGE to merge the two sorted arrays S_L and S_R into a single sorted array, S. This can be done in time $\Theta(n)$ with a single pass through S_L and S_R. We simply keep

Algorithm 4.1: MERGESORT(A, n)

```
comment: A is an array of length n
external MERGE
if n = 1
  then S ← A
        ⎧ n_L ← ⌈n/2⌉
        ⎪ n_R ← ⌊n/2⌋
        ⎪ A_L ← [A[1], ..., A[n_L]]
  else  ⎨ A_R ← [A[n_L + 1], ..., A[n]]
        ⎪ S_L ← MERGESORT(A_L, n_L)
        ⎪ S_R ← MERGESORT(A_R, n_R)
        ⎩ S ← MERGE(S_L, n_L, S_R, n_R)
return (S, n)
```

track of the "current" element of S_L and S_R, always copying the smaller one into the sorted array. Thus, S is the sorted array obtained by merging the sorted arrays S_L and S_R.

Pseudocode descriptions of MERGESORT and MERGE are presented as Algorithms 4.1 and 4.2, respectively.

In Algorithm 4.2, the first **while** loop continues as long as there are elements remaining to be processed in both S_L and S_R. When this loop terminates, there will still be elements in one of S_L and S_R. Then one of the two remaining two **while** loops will be executed, to copy the remaining elements of S_L or S_R into S.

Example 4.9 Suppose that we have obtained the following two sorted arrays and we wish to merge them:

$$S_L = [1, 5, 8, 9]$$
$$S_R = [2, 4, 6, 7].$$

Algorithm 4.2 would proceed as shown in Figure 4.3. Note that, when $i = 6$, we have reached the end of S_R. Therefore, in the last two iterations, we copy the last two elements of S_L into S. □

Now we analyze the complexity of MERGESORT. Let $T(n)$ denote the time to run MERGESORT on an array of length n.

divide

This step takes time $\Theta(n)$, because we are copying the n elements in A into A_L and A_R.

conquer

The two recursive calls take time $T\left(\lceil \frac{n}{2} \rceil\right) + T\left(\lfloor \frac{n}{2} \rfloor\right)$.

Algorithm 4.2: MERGE(S_L, n_L, S_R, n_R)

comment: merge S_L (of length n_L) and S_R (of length n_R)
$i_L \leftarrow 1$
$i_R \leftarrow 1$
$i \leftarrow 1$
while $(i_L \leq n_L)$ **and** $(i_R \leq n_R)$
do {
  if $S_L[i_L] < S_R[i_R]$
    then { $S[i] \leftarrow S_L[i_L]$; $i_L \leftarrow i_L + 1$ }
    else { $S[i] \leftarrow S_R[i_R]$; $i_R \leftarrow i_R + 1$ }
  $i \leftarrow i + 1$ }
while $i_L \leq n_L$
do { $S[i] \leftarrow S_L[i_L]$; $i_L \leftarrow i_L + 1$; $i \leftarrow i + 1$ }
while $i_R \leq n_R$
do { $S[i] \leftarrow S_R[i_R]$; $i_R \leftarrow i_R + 1$; $i \leftarrow i + 1$ }
return (S)

combine

Merging the two sorted subarrays takes time $\Theta(n)$. This is clear because we proceed once through each of the two subarrays.

As a result, we obtain the following recurrence relation:

$$T(n) = T\left(\left\lceil \frac{n}{2} \right\rceil\right) + T\left(\left\lfloor \frac{n}{2} \right\rfloor\right) + \Theta(n) \qquad \text{if } n > 1$$
$$T(1) = d.$$

We will study this recurrence in the next section.

4.4 Divide-and-Conquer Recurrence Relations

4.4.1 Sloppy and Exact Recurrence Relations

We now investigate the recurrence relation for MERGESORT in more detail. Our first simplification is to replace the $\Theta(n)$ term by cn, where c is an unspecified

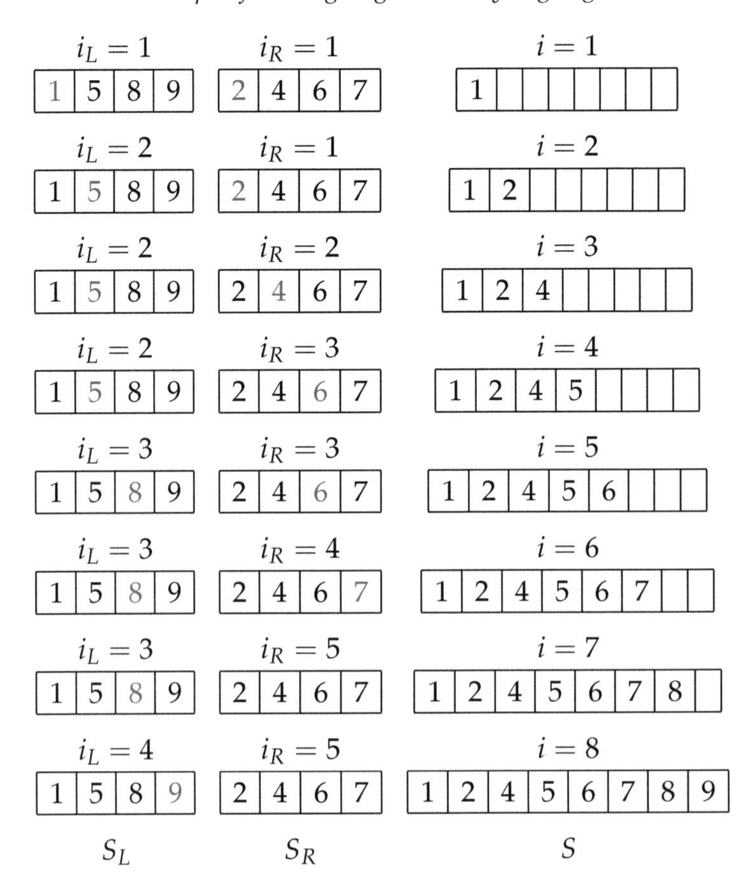

FIGURE 4.3: Merging two sorted arrays of length 4

constant. The resulting recurrence relation is called the ***exact recurrence relation***:

$$\begin{aligned} T(n) &= T\left(\left\lceil \frac{n}{2} \right\rceil\right) + T\left(\left\lfloor \frac{n}{2} \right\rfloor\right) + cn \qquad \text{if } n > 1 \\ T(1) &= d. \end{aligned} \tag{4.9}$$

If we then remove the floors and ceilings, we obtain the so-called ***sloppy recurrence relation***:

$$\begin{aligned} T(n) &= 2T\left(\frac{n}{2}\right) + cn \qquad \text{if } n > 1 \\ T(1) &= d. \end{aligned} \tag{4.10}$$

The exact and sloppy recurrences are *identical* when n is a power of two. Further, the sloppy recurrence makes sense *only* when n is a power of two. For example, if we try to determine the value of $T(5)$ using (4.10) , we obtain

$$T(5) = 2T(2.5) + 5c.$$

But this is nonsense, because $T(n)$ is only defined for integer values of n.

The Master Theorem provides the *exact* solution of the recurrence when $n = 2^j$ (it is in fact a *proof* for these values of n). We can express this solution (for powers of 2) as a function of n, using Θ-notation. It can be shown that the resulting function of n will in fact yield the *complexity* of the solution of the exact recurrence for *all values* of n. This derivation of the complexity of $T(n)$ is *not a proof*, however.

The exact recurrence is messy to analyze due to the presence of ceilings and floors. But it is still the case that $T(n) \in O(n \log n)$. There are a couple of ways to prove this. One approach is to show directly that $T(n) \leq cn \log n$ for some positive constant c by an induction proof. Another way is to first prove by induction that $T(n) \leq T(n+1)$ for all positive integers n, and then to make use of the solution we have already obtained for $T(2^j)$. Both methods are presented in detail in the next subsection. However, in general, we will not be analyzing exact recurrences in this book.

4.4.2 The Exact Recurrence for MergeSort

The exact MERGESORT recurrence was given in (4.9). To simplify it, we will let $d = 0$ (which makes sense if we are counting the number of comparisons) and therefore we consider the recurrence

$$\begin{aligned} T(n) &= T\left(\left\lfloor \frac{n}{2} \right\rfloor\right) + T\left(\left\lceil \frac{n}{2} \right\rceil\right) + cn \quad \text{if } n > 1 \\ T(1) &= 0, \end{aligned}$$

where c is a positive constant.

We want to prove that $T(n)$ is $O(n \log n)$. First, we give a complete proof from first principles. We will find constants $k, n_0 > 0$ such that

$$T(n) \leq kn \log_2 n \tag{4.11}$$

for all $n \geq n_0$.

We begin with the base case, $n = 1$. We have $T(1) = 0$ and $k \times 1 \times \log_2 1 = 0$, so any k satisfies (4.11) for the base case.

Now we make an induction assumption that (4.11) is satisfied for $1 \leq n \leq m - 1$, where $m \geq 2$. We want to prove that (4.11) is satisfied for $n = m$. Note that the constant k is unspecified so far; we will determine an appropriate value for k as we proceed.

We have

$$T(m) = T\left(\left\lfloor \frac{m}{2} \right\rfloor\right) + T\left(\left\lceil \frac{m}{2} \right\rceil\right) + cm,$$

so it follows that

$$T(m) \leq k \left\lfloor \frac{m}{2} \right\rfloor \log_2 \left\lfloor \frac{m}{2} \right\rfloor + k \left\lceil \frac{m}{2} \right\rceil \log_2 \left\lceil \frac{m}{2} \right\rceil + cm, \tag{4.12}$$

by applying the induction hypothesis for $n = \left\lfloor \frac{m}{2} \right\rfloor$ and $n = \left\lceil \frac{m}{2} \right\rceil$. Using the facts that

$$\left\lfloor \frac{m}{2} \right\rfloor \leq \frac{m}{2} \quad \text{and} \quad \left\lceil \frac{m}{2} \right\rceil \leq m,$$

we obtain the following from (4.12):

$$\begin{aligned}
T(m) &\leq k \left\lfloor \frac{m}{2} \right\rfloor \log_2 \frac{m}{2} + k \left\lceil \frac{m}{2} \right\rceil \log_2 m + cm \\
&= k \left\lfloor \frac{m}{2} \right\rfloor (\log_2 m - 1) + k \left\lceil \frac{m}{2} \right\rceil \log_2 m + cm \\
&= k \log_2 m \left(\left\lfloor \frac{m}{2} \right\rfloor + \left\lceil \frac{m}{2} \right\rceil \right) + cm - k \left\lfloor \frac{m}{2} \right\rfloor \\
&= km \log_2 m + cm - k \left\lfloor \frac{m}{2} \right\rfloor .
\end{aligned}$$

In the last line, we are using the easily verified fact that

$$\left\lfloor \frac{m}{2} \right\rfloor + \left\lceil \frac{m}{2} \right\rceil = m.$$

Recall that we are trying to prove that $T(m) \leq km \log_2 m$. Therefore we will be done provided that

$$cm - k \left\lfloor \frac{m}{2} \right\rfloor \leq 0.$$

This is equivalent to

$$k \geq \frac{cm}{\left\lfloor \frac{m}{2} \right\rfloor}.$$

Since k is required to be a constant, we must find an upper bound for the expression on the right side of this inequality. It is not hard to show that

$$\frac{m}{\left\lfloor \frac{m}{2} \right\rfloor} \leq 3$$

for all integers $m \geq 2$. Therefore we can take $k = 3c$ and we will have

$$T(m) \leq km \log_2 m,$$

as desired.

Hence, by the principle of mathematical induction, we have proven that

$$T(n) \leq 3n \log_2 n$$

for all $n \geq 1$ (so we can take $n_0 = 1$).

REMARK Suppose the recurrence instead had the form

$$\begin{aligned}
T(n) &= T\left(\left\lfloor \frac{n}{2} \right\rfloor \right) + T\left(\left\lceil \frac{n}{2} \right\rceil \right) + cn \quad \text{if } n > 1 \\
T(1) &= d,
\end{aligned}$$

where $c, d > 0$. In this situation, we could not use $n = 1$ as the base case for the induction (why?). It turns out that we would have to take $n_0 = 2$ and start with $n = 2$ and $n = 3$ as base cases for the induction. We leave the details as an exercise. ∎

Now we describe another approach that actually turns out to be a bit easier. It follows two steps: step 1 is to prove that $T(n)$ is a monotone increasing sequence, which is an "easy" induction. Step 2 is to make use of the exact solution to $T(n)$ when n is a power of two. This can be computed using the Master Theorem to be $T(2^j) = cj2^j$.

So we start by using the exact recurrence to prove (by induction) that $T(n) \leq T(n+1)$ for all positive integers n.

We have $T(1) = 0$ and $T(2) = 2$, so $T(1) \leq T(2)$. As an induction assumption, suppose that $T(n) \leq T(n+1)$ holds for $1 \leq n \leq m-1$, where $m \geq 2$. We want to prove that $T(m) \leq T(m+1)$.

It is convenient to consider the cases of even and odd m separately. First, suppose that m is even. Then

$$T(m) = 2T\left(\frac{m}{2}\right) + cm$$

and

$$T(m+1) = T\left(\frac{m}{2}\right) + T\left(\frac{m}{2}+1\right) + c(m+1).$$

Thus

$$T(m+1) - T(m) = T\left(\frac{m}{2}+1\right) - T\left(\frac{m}{2}\right) + c.$$

We have

$$T\left(\frac{m}{2}+1\right) - T\left(\frac{m}{2}\right) \geq 0$$

by induction, so it follows that $T(m) \leq T(m+1)$.

The case where m is odd is similar. We have

$$T(m) = T\left(\frac{m-1}{2}\right) + T\left(\frac{m+1}{2}\right) + cm$$

and

$$T(m+1) = 2T\left(\frac{m+1}{2}\right) + c(m+1).$$

Thus

$$T(m+1) - T(m) = T\left(\frac{m+1}{2}\right) - T\left(\frac{m-1}{2}\right) + c.$$

We have

$$T\left(\frac{m+1}{2}\right) - T\left(\frac{m-1}{2}\right) \geq 0$$

by induction, so it follows that $T(m) \leq T(m+1)$.

The second step is to consider an arbitrary positive integer n. Then

$$2^{j-1} < n \leq 2^j$$

for a unique positive integer j. Note that this implies that

$$2^j < 2n \quad \text{and} \quad j < \log_2(2n). \tag{4.13}$$

Then, because $T(n)$ is monotone increasing and $n \leq 2^j$, we have

$$\begin{aligned} T(n) &\leq T(2^j) \\ &= cj2^j \qquad \text{from the Master Theorem} \\ &\leq c(2n)\log(2n) \quad \text{from (4.13)} \\ &= 2cn(\log n + 1). \end{aligned}$$

It is clear that

$$2cn(\log n + 1) \in O(n \log n),$$

so we have shown that $T(n) \in O(n \log n)$.

4.5 Binary Search

BINARYSEARCH is one of the simplest divide-and-conquer algorithms. This technique can be applied whenever we have a *sorted* array, say $A = [A[1], \dots, A[n]]$. As usual, we will assume that A consists of distinct elements. Suppose we want to search for a specific *target value* T in the array A. The idea is to look at the "middle element" $A[mid]$, where

$$mid = \begin{cases} \dfrac{n}{2} & \text{if } n \text{ is even} \\ \dfrac{n+1}{2} & \text{if } n \text{ is odd.} \end{cases}$$

There are now three cases to consider:

1. If $T = A[mid]$, then we have found T in the array A and we are done.
2. If $T < A[mid]$, then we recursively search the left half of A.
3. If $T > A[mid]$, then we recursively search the right half of A.

Observe that we make (at most) one recursive call, to a subproblem of (roughly) half the size. This is the reason for the nomenclature "binary search."

A detailed pseudocode description of BINARYSEARCH is presented as Algorithm 4.3.

The recurrence relation for BINARYSEARCH is (roughly)

$$T(n) = T\left(\frac{n}{2}\right) + c.$$

Applying the Master Theorem, the solution is $T(n) \in O(\log n)$.

Algorithm 4.3: BINARYSEARCH(lo, hi, T)

```
global A
precondition: A is a sorted array of length n
if lo > hi
  then return ("not found")
       ⎧ mid ← ⌈(lo+hi)/2⌉
       ⎪ if A[mid] = T
       ⎪   then return (mid)
  else ⎨ else if A[mid] < T
       ⎪   then BINARYSEARCH(mid + 1, hi, T)
       ⎩   else BINARYSEARCH(lo, mid − 1, T)
```

4.6 Non-dominated Points

We discuss an interesting geometric problem next. Given two points $(x_1, y_1), (x_2, y_2)$ in the Euclidean plane, we say that (x_1, y_1) ***dominates*** (x_2, y_2) if $x_1 > x_2$ and $y_1 > y_2$. Problem 4.1 is known as **Non-dominated Points**. It asks us to find all the points in a given set of points that are not dominated by any other point in the set. This problem has a trivial $\Theta(n^2)$ algorithm to solve it, based on comparing all pairs of points in S. Can we do better?

Problem 4.1: Non-dominated Points

Instance: A set S of n points in the Euclidean plane, say

$$S = \{S[1], \ldots, S[n]\}.$$

For simplicity, we will assume that the x-co-ordinates of all these points are distinct, and the y-co-ordinates of all these points are also distinct.
Question: Find all the ***non-dominated points*** in S, i.e., all the points that are not dominated by any other point in S.

The following observation is relevant. The points dominated by a point (x_1, y_1) are all the points that are in the "southwest quadrant" with respect to (x_1, y_1). Consequently, the non-dominated points form a *staircase* and all the other points are "under" this staircase. The *treads* of the staircase are determined by the y-co-ordinates of the non-dominated points. The *risers* of the staircase are determined by the x-co-ordinates of the non-dominated points. The staircase descends from

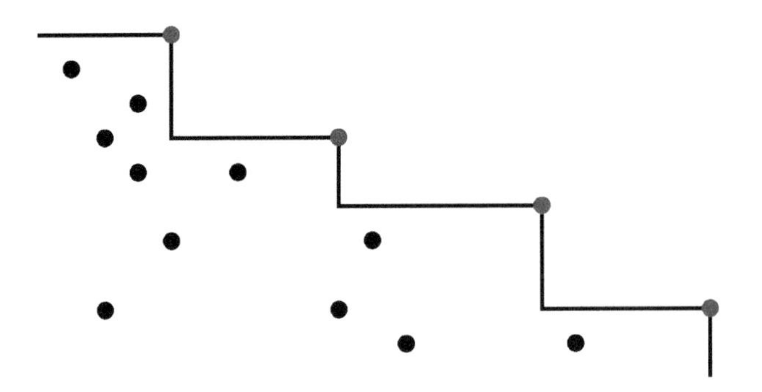

FIGURE 4.4: A staircase of non-dominated points

left to right. A typical example is shown in Figure 4.4. The red points are the non-dominated points.

In order to solve the **Non-dominated Points** problem, it turns out to be useful to pre-process the data in a suitable manner. This would be done once, before the initial invocation of the recursive algorithm. What we do is pre-sort the points in S in increasing order with respect to their x-co-ordinates. This takes time $\Theta(n \log n)$.

Now we can discuss the divide, conquer and combine operations.

divide

Let the first $n/2$ points (after they are sorted) be denoted by S_1 and let the last $n/2$ points be denoted by S_2. This divides the given problem instance into two equal-sized subproblems.

conquer

Recursively solve the subproblems defined by S_1 and S_2.

combine

Suppose that we have (recursively) determined the non-dominated points in S_1 and the non-dominated points in S_2, which we denote by $\mathsf{ND}(S_1)$ and $\mathsf{ND}(S_2)$, respectively. How do we then find the non-dominated points in S? First, we observe that *no point in S_1 dominates a point in S_2*. Therefore we only need to eliminate the points in S_1 that are dominated by a point in S_2. It turns out that this can be done in time $O(n)$.

Denote

$$S = (P_1, \dots, P_n)$$

(this is the set of all n points, such that the x-coordinates are in increasing order). Suppose

$$\mathsf{ND}(S_1) = (Q_1, \dots, Q_\ell) \quad \text{and} \quad \mathsf{ND}(S_2) = (R_1, \dots, R_m),$$

where the points in $\mathsf{ND}(S_1)$ and $\mathsf{ND}(S_2)$ are in increasing order of x-co-ordinates (hence they are also in decreasing order of y-co-ordinates).

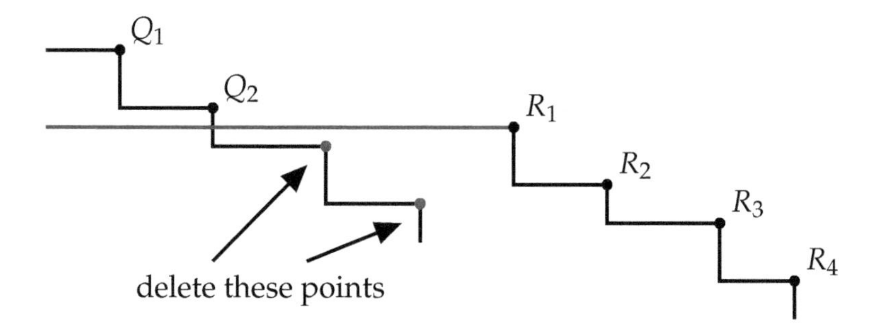

FIGURE 4.5: Combining two sets of non-dominated points

We compute k to be the maximum index i such that

$$\text{the } y\text{-co-ordinate of } Q_i \text{ is} > \text{the } y\text{-co-ordinate of } R_1.$$

This is just a linear search. (We could actually do a binary search, but the overall complexity will not be affected.) Then,

$$\text{COMBINE}(\mathsf{ND}(S_1), \mathsf{ND}(S_2)) = (Q_1, \dots, Q_k, R_1, \dots, R_m).$$

The x-co-ordinates of the points in $\text{COMBINE}(\mathsf{ND}(S_1), \mathsf{ND}(S_2))$ are in increasing order, so this can be regarded as a post-condition of the algorithm.

A small example is given in Figure 4.5, where we show how two sets of four non-dominated points would be combined. We see that $k = 2$ in this example, so

$$\text{COMBINE}(\mathsf{ND}(S_1), \mathsf{ND}(S_2)) = (Q_1, Q_2, R_1, R_2, R_3, R_4).$$

We obtain Algorithm 4.4, which is based on the ideas described above.

Algorithm 4.4: NON-DOM($S_1, \dots, S_n$)

```
comment: S_1, ..., S_n are in increasing order WRT x-co-ordinates
if n = 1 then return (S_1)
     ⎧ (Q_1, ..., Q_ℓ) ← NON-DOM(S_1, ..., S_⌊n/2⌋)
     ⎪ (R_1, ..., R_m) ← NON-DOM(S_⌊n/2⌋+1, ..., S_n)
else ⎨ i ← 1
     ⎪ while i ≤ ℓ and Q_i.y > R_1.y
     ⎩   do i ← i + 1
return (Q_1, ..., Q_{i-1}, R_1, ..., R_m)
comment: Q_1, ..., R_m are in increasing order WRT x-co-ordinates
```

The COMBINE operation takes time $O(n)$. The recursive part of the algorithm takes time $O(n \log n)$ (from the Master Theorem). The initial sort takes time $\Theta(n \log n)$, so the overall complexity is $\Theta(n \log n)$.

4.7 Stock Profits

Suppose we have an array A of n positive integers, which represents the price of a certain stock on n consecutive days. Thus, $A[i]$ is the price of the stock on day i. Suppose we want to determine the maximum profit we could have made by buying one share of the stock on day i and then selling it on day j, where $j \geq i$. This problem **Stock Profit** is defined more formally as follows:

Problem 4.2: Stock Profit

Instance: An array of n positive integers,

$$A = [A[1], \ldots, A[n]].$$

Find: The maximum profit $A[j] - A[i]$, where $1 \leq i \leq j \leq n$.

Unfortunately, we don't have any good algorithm to predict the optimal times to buy and sell stocks!

A trivial algorithm would just compute all possible values $A[j] - A[i]$ with $i \leq j$, keeping track of the maximum. This would take time $\Theta(n^2)$. Note that this maximum is non-negative, since $i = j$ is allowed.

In a divide-and conquer approach, we would divide A into two equal (or almost equal) parts, say A_L and A_R, and recursively solve these two subproblems, computing the maximum profit in each part. In the combine stage, we would need to consider the possible values of $A[j] - A[i]$ where $A[j] \in A_R$ and $A[i] \in A_L$. Evidently this quantity would be maximized by choosing the maximum value $A[j] \in A_R$ and the minimum value $A[i] \in A_L$. Finally, the maximum profit is the maximum of the three values we have computed.

Example 4.10 Suppose $n = 8$ and

$$A = [3, 8, 1, 5, 6, 7, 2, 4].$$

Here we would have

$$A_L = [3, 8, 1, 5] \quad \text{and} \quad A_R = [6, 7, 2, 4].$$

The maximum profit in A_L is $8 - 3 = 5$ and the maximum profit in A_R is $4 - 2 = 2$. The maximum profit attained by an $A[j] \in A_R$ and $A[i] \in A_L$ is computed by finding the maximum element in A_R, which is 7, and the minimum element in A_L, which is 1. Then $7 - 1 = 6$ is the maximum profit where $A[j] \in A_R$ and $A[i] \in A_L$. The maximum overall profit is

$$\max\{5, 2, 6\} = 6.$$

□

The maximum value $A[j] \in A_R$ and the minimum value $A[i] \in A_L$ could obviously be computed in time $\Theta(n)$. The resulting algorithm would satisfy the same recurrence relation as MERGESORT and thus the complexity would be $\Theta(n \log n)$.

Can we do better? The trick is to compute the desired maximum and minimum recursively, instead of doing a linear-time scan after the two subproblems have been solved. So we devise a recursive divide-and-conquer algorithm that computes three quantities:

1. the minimum element in the given subarray,
2. the maximum element in the given subarray, and
3. the maximum profit in the given subarray.

These ideas are incorporated into Algorithm 4.5.

Algorithm 4.5: BUYLOWSELLHIGH(i, j)

```
global A
comment: A is an array of length n and 1 ≤ i ≤ j ≤ n
if i = j
  then return (A[i], A[i], 0)
  else { m ← ⌈(i+j)/2⌉
         (min_L, max_L, p_L) ← MAXIMUMPROFIT(i, m − 1)
         (min_R, max_R, p_R) ← MAXIMUMPROFIT(m, j)
         min ← min{min_L, min_R}
         max ← max{max_L, max_R}
         p ← max{p_L, p_R, max_R − min_L}
         return (min, max, p)
```

Example 4.11 Suppose we revisit Example 4.10, where we considered

$$A_L = [3, 8, 1, 5] \quad \text{and} \quad A_R = [6, 7, 2, 4].$$

We would recursively compute

$$\begin{array}{llll} min_L = 1, & max_L = 8 & \text{and} & p_L = 5 \\ min_R = 2, & max_R = 7 & \text{and} & p_R = 2. \end{array}$$

Then,

$$\begin{aligned} min &= \min\{min_L, min_R\} &&= \min\{1, 2\} &&= 1 \\ max &= \max\{max_L, max_R\} &&= \max\{8, 7\} &&= 8 \\ p &= \max\{p_L, p_R, max_R - min_L\} &&= \max\{5, 2, 7 - 1\} &&= 6. \end{aligned}$$

□

The recurrence relation for Algorithm 4.5 is

$$T(n) = 2T\left(\frac{n}{2}\right) + \Theta(1),$$

so the complexity is $\Theta(n)$, from the Master Theorem. Also, we initially call BUY-LOWSELLHIGH$(1, n)$ to solve the given problem instance.

4.8 Closest Pair

We now study the problem of finding the pair of points that are closest among given n points in the Euclidean plane.

Problem 4.3: Closest Pair

Instance: A set Q of n distinct points in the Euclidean plane,

$$Q = \{Q[1], \ldots, Q[n]\}.$$

Find: Two distinct points $Q[i] = (x, y), Q[j] = (x', y')$ such that the Euclidean distance

$$\sqrt{(x' - x)^2 + (y' - y)^2}$$

is minimized.

A trivial algorithm would just compare all pairs of points, keeping track of the minimum distance between any pair of points. This would take time $\Theta(n^2)$.

The one-dimensional version of the problem, where all the points lie on a line, can be solved easily in $O(n \log n)$ time by a "sort-and-scan" approach (we leave the details for the reader to fill in). We also show how to solve the two-dimensional version of the problem in $O(n \log n)$ time using the BENTLEY-SHAMOS divide-and-conquer algorithm from 1976.

Suppose we *pre-sort* the points in Q with respect to their x-coordinates. Of course, this takes time $\Theta(n \log n)$). Then we can easily find the *vertical line* that partitions the set of points Q into two sets of size $n/2$: this line has equation $x = Q[m].x$, where $m = n/2$.

divide

We have two subproblems, consisting of the first $n/2$ points and the last $n/2$ points.

conquer

Recursively solve the two subproblems.

combine

Given that we have determined the shortest distance among the first $n/2$ points and the shortest distance among the last $n/2$ points, what additional work is required to determine the overall shortest distance? In particular, how do we handle pairs of points where one point is in the left half and one point is in the right half?

Suppose δ_L is the minimum distance in the left half and δ_R is the minimum distance in the right half. These have been computed recursively. Let $\delta = \min\{\delta_L, \delta_R\}$.

We will construct a ***critical strip*** R of width 2δ. The critical strip consists of all points whose x-coordinates are within δ of the vertical splitting line, which has equation $x = xmid$, where $xmid = Q[m].x$. (This takes time $\Theta(n)$.) If there is a pair of points having distance $< \delta$, they must be in the critical strip.

Perhaps *all* the points are in the critical strip, so it will not be efficient to check all pairs of points in the critical strip, because

$$\frac{n}{2} \times \frac{n}{2} = \frac{n^2}{4} \in \Theta(n^2).$$

The key idea is to sort the points in the critical strip in increasing order of their y-co-ordinates. This takes time $\Theta(n \log n)$. It turns out (see Lemma 4.5) that we only need to compute the distance from each point in the critical strip to the *next seven points*. This means that there are at most $7n$ pairs of points to check, which can be done in time $\Theta(n)$.

LEMMA 4.5 *Suppose the points in the critical strip are sorted in increasing order of their y-co-ordinates. Suppose that $R[j]$ and $R[k]$ are two points in the critical strip, where $j < k$, and suppose the distance between $R[j]$ and $R[k]$ is less than δ. Then $k \leq j + 7$.*

PROOF Construct a rectangle $\mathcal{R}$ having width 2δ and height δ, in which the base is the line $y = R[j].y$. Consider $\mathcal{R}$ to be partitioned into eight squares of side $\delta/2$ (see Figure 4.6). There is at most one point inside each of these eight squares. This is because the distance between two points in the same square is at most $\delta\sqrt{2}/2 < \delta$, which is a contradiction. Also, one of these eight squares contains the point $R[j]$.

Therefore, if $k \geq j + 8$, then $R[k]$ is above the rectangle $\mathcal{R}$. Then, the distance between $R[j]$ and $R[k]$ is greater than δ, because the y-co-ordinates of these two points differ by at least this much. ∎

Summarizing the discussion above, we obtain Algorithm 4.6, which solves the **Closest Pair** problem. Algorithm 4.6 incorporates a procedure SORTY as well as two other subroutines, which we refer to as SELECTCANDIDATES and CHECKSTRIP (these are Algorithms 4.7 and 4.8, respectively).

Here are a few comments about Algorithm 4.6.

- The set Q is global with respect to the recursive procedure CLOSESTPAIR1.

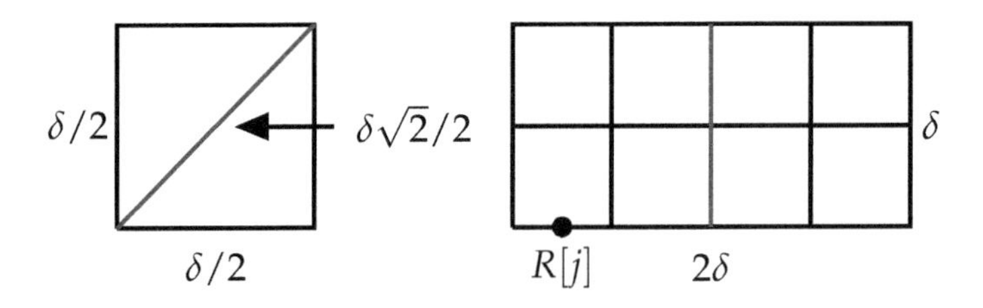

FIGURE 4.6: Subdividing a rectangle into eight squares

Algorithm 4.6: CLOSESTPAIR1(ℓ, r)

external SORTY, SELECTCANDIDATES, CHECKSTRIP
if $\ell = r$ **then** $\delta \leftarrow \infty$
else
- $m \leftarrow \lfloor(\ell + r)/2\rfloor$
- $\delta_L \leftarrow$ CLOSESTPAIR1(ℓ, m)
- $\delta_R \leftarrow$ CLOSESTPAIR1$(m+1, r)$
- $\delta \leftarrow \min\{\delta_L, \delta_R\}$
- $R \leftarrow$ SELECTCANDIDATES$(\ell, r, \delta, Q[m].x)$
- $R \leftarrow$ SORTY(R)
- $\delta \leftarrow$ CHECKSTRIP(R, δ)

return (δ)

- At any given point in the recursion, we are examining a subarray $(Q[\ell], \dots, Q[r])$, and $m = \lfloor(\ell + r)/2\rfloor$.
- The base case is a subarray of size 1, which occurs when $\ell = r$. We set $\delta = \infty$ as the solution to the base case.
- We call CLOSESTPAIR1$(1, n)$ to solve the given problem instance.

The subroutines SELECTCANDIDATES and CHECKSTRIP are straightforward. SELECTCANDIDATES just examines all the points and retains those whose x-coordinate is no more than δ away from the vertical splitting line. These are the points in the critical strip. CHECKSTRIP looks at pairs of points in the critical strip. However, as discussed above, for each point $R[j]$ in the critical strip, it is only necessary to compute the distance from $R[j]$ to the next seven points in the list.

Note that we could speed up the process of checking the vertical strip by observing we do not need to compute the distance between two points that are on the same side of the center line, as this pair of points will have already been considered in the relevant recursive call. However, this optimization would not improve the complexity of the algorithm, and therefore we do not incorporate it into Algorithm 4.6.

Algorithm 4.7: SELECTCANDIDATES($\ell, r, \delta, xmid$)

$j \leftarrow 0$
for $i \leftarrow \ell$ **to** r
  do $\begin{cases} \textbf{if } |Q[i].x - xmid| \leq \delta \\ \quad \textbf{then } \begin{cases} j \leftarrow j+1 \\ R[j] \leftarrow Q[i] \end{cases} \end{cases}$
return (R)

Algorithm 4.8: CHECKSTRIP(R, δ)

$t \leftarrow size(R)$
$\delta' \leftarrow \delta$
for $j \leftarrow 1$ **to** $t-1$
  do $\begin{cases} \textbf{for } k \leftarrow j+1 \textbf{ to } \min\{t, j+7\} \\ \quad \textbf{do } \begin{cases} x \leftarrow R[j].x \\ x' \leftarrow R[k].x \\ y \leftarrow R[j].y \\ y' \leftarrow R[k].y \\ \delta' \leftarrow \min\left\{\delta', \sqrt{(x'-x)^2 + (y'-y)^2}\right\} \end{cases} \end{cases}$
return (δ')

We now analyze the complexity of Algorithm 4.6:

- SELECTCANDIDATES takes time $\Theta(n)$,
- SORTY takes time $O(n \log n)$, and
- CHECKSTRIP takes time $O(n)$.

The recurrence relation is therefore

$$T(n) = 2T\left(\frac{n}{2}\right) + O(n \log n),$$

which has solution $T(n) \in O(n(\log n)^2)$ (we analyzed this recurrence relation at the end of Section 4.2.2). The pre-sort takes time $\Theta(n \log n)$, so the total time is $\Theta(n(\log n)^2)$. This is not bad, but we will present a variation of this algorithm with improved complexity in the next subsection.

Example 4.12 We present a small example to illustrate the recursive calls. Suppose we have eight points, ordered by x-co-ordinates:

$$(2,3), (6,4), (8,9), (10,6), (11,7), (12,2), (14,1), (16,5)$$

1. ClosestPair1$(1,8)$ generates two recursive calls:

$$\text{ClosestPair1}(1,4) \text{ and } \text{ClosestPair1}(5,8).$$

2. ClosestPair1$(1,4)$ generates two recursive calls:

$$\text{ClosestPair1}(1,2) \text{ and } \text{ClosestPair1}(3,4).$$

3. ClosestPair1$(1,2)$ generates two recursive calls:

$$\text{ClosestPair1}(1,1) \text{ and } \text{ClosestPair1}(2,2).$$

   These two recursive calls are base cases, so

$$\delta_L = \delta_R = \infty.$$

   Then we compute $\delta = \sqrt{17}$ in CheckStrip.

4. ClosestPair1$(3,4)$ generates two recursive calls:

$$\text{ClosestPair1}(3,3) \text{ and } \text{ClosestPair1}(4,4).$$

   These two recursive calls are base cases, so

$$\delta_L = \delta_R = \infty.$$

   Then we compute $\delta = \sqrt{13}$ in CheckStrip.

5. ClosestPair1$(1,4)$ receives

$$\delta_L = \sqrt{17} \quad \text{and} \quad \delta_R = \sqrt{13}.$$

   Then $\delta = \sqrt{13}$ and the value of δ is not changed in CheckStrip.

6. ClosestPair1$(5,8)$ generates two recursive calls:

$$\text{ClosestPair1}(5,6) \text{ and } \text{ClosestPair1}(7,8).$$

7. ClosestPair1$(5,6)$ generates two recursive calls:

$$\text{ClosestPair1}(5,5) \text{ and } \text{ClosestPair1}(6,6).$$

   These two recursive calls are base cases, so

$$\delta_L = \delta_R = \infty.$$

   Then we compute $\delta = \sqrt{26}$ in CheckStrip.

8. ClosestPair1$(7,8)$ generates two recursive calls:

CLOSESTPAIR1(7,7) and CLOSESTPAIR1(8,8).

These two recursive calls are base cases, so

$$\delta_L = \delta_R = \infty.$$

Then we compute $\delta = \sqrt{40}$ in CHECKSTRIP.

9. CLOSESTPAIR1(5,8) receives

$$\delta_L = \sqrt{26} \quad \text{and} \quad \delta_R = \sqrt{20}.$$

The value of δ is reduced in CHECKSTRIP to $\sqrt{5}$.

10. CLOSESTPAIR1(1,8) receives

$$\delta_L = \sqrt{13} \quad \text{and} \quad \delta_R = \sqrt{5}.$$

The value of δ is reduced in CHECKSTRIP to $\sqrt{2}$.

□

4.8.1 An Improved Algorithm

To improve the complexity of Algorithm 4.6, we eliminate the sorting of the points in critical strip with respect to their y-co-ordinates. The *precondition* for the improved algorithm, CLOSESTPAIR2, is that the relevant points in Q, namely $Q[\ell], \dots, Q[r]$, are sorted with respect to their *x-co-ordinates*. The *postcondition* for CLOSESTPAIR2 is that $Q[\ell], \dots, Q[r]$ are sorted with respect to their *y-co-ordinates*. This can be accomplished by *merging* two sublists $Q[\ell], \dots, Q[m]$ and $Q[m+1], \dots, Q[r]$ which are *recursively* sorted with respect to their y-co-ordinates. So this ends up being identical to the MERGE step in MERGESORT (see Algorithm 4.2). Incorporating this modification, we obtain Algorithm 4.9.

Note that the $O(n \log n)$ sorting step is replaced by a MERGE which takes time $O(n)$. The recurrence relation is now

$$T(n) = 2T\left(\frac{n}{2}\right) + O(n),$$

and the overall complexity (including the pre-sort) is $\Theta(n \log n)$.

An alternative approach to improving the complexity of Algorithm 4.6 involves pre-sorting (separately) with respect to both the x- and y-co-ordinates. This can also lead to an algorithm having complexity $\Theta(n \log n)$.

4.9 Multiprecision Multiplication

We now introduce the problem of multiplication of positive integers of arbitrary size; see Problem 4.4.

Algorithm 4.9: CLOSESTPAIR2(ℓ, r)

precondition: $Q[\ell], \ldots, Q[r]$ are sorted WRT their x-co-ordinates
if $\ell = r$ **then** $\delta \leftarrow \infty$
else
- $m \leftarrow \lfloor (\ell + r)/2 \rfloor$
- $Xmid \leftarrow Q[m].x$
- $\delta_L \leftarrow$ CLOSESTPAIR2(ℓ, m)
- **comment:** $Q[\ell], \ldots, Q[m]$ are now sorted with respect to their y-coordinates
- $\delta_R \leftarrow$ CLOSESTPAIR2$(m + 1, r)$
- **comment:** $Q[m + 1], \ldots, Q[r]$ are now sorted with respect to their y-coordinates
- $\delta \leftarrow \min\{\delta_L, \delta_R\}$
- MERGE(ℓ, m, r)
- $R \leftarrow$ SELECTCANDIDATES$(\ell, r, \delta, Xmid)$
- $\delta \leftarrow$ CHECKSTRIP(R, δ)

return (δ)
postcondition: $Q[\ell], \ldots, Q[r]$ are sorted WRT their y-co-ordinates

Problem 4.4: Multiprecision Multiplication

Instance: Two k-bit positive integers, X and Y, having binary representations

$$X = [X[k-1], ..., X[0]]$$

and

$$Y = [Y[k-1], ..., Y[0]].$$

Question: Compute the $2k$-bit positive integer $Z = XY$, where

$$Z = (Z[2k-1], ..., Z[0]).$$

In the unit cost model, multiplication takes time $O(1)$, but this makes sense only for *bounded* integers. Here, we are interested in the ***bit complexity*** of algorithms that solve **Multiprecision Multiplication**. This means that the complexity is expressed as a function of the number of bits in the integers that are being multiplied. Using this measure, the size of the problem instance is $2k$ bits.

The *value* of the integers X (and Y) is $\Theta(2^k)$ (at least when the most significant bit is equal to 1). On the other hand, the *size* of X, measured in bits, is $\Theta(\log X)$. We need to always remember that "size" refers to the amount of *space* required to store some data.

The *grade-school* algorithm for multiplication requires adding k (or fewer) integers, each consisting of at most $2k - 1$ bits. Adding any two $2k$-bit integers takes time $O(k)$. Since there are $O(k)$ integers to be added, the grade-school algorithm has complexity $O(k^2)$.

Example 4.13 Suppose we want to compute the product 1101×1110 (in binary). We have:

$$\begin{array}{r} 1101 \\ 1011 \\ \hline 1101 \\ 11010 \\ 1101000 \\ \hline 10001111 \end{array}$$

Here we end up adding three integers, of four, five, and seven bits in length, to obtain the 8-bit product. □

Our objective in designing a divide-and-conquer algorithm is to find an algorithm that will multiply two k-bit integers faster than the grade-school algorithm. For simplicity, let's assume k is even. Let X_L be the integer formed by the $k/2$ high-order bits of X and let X_R be the integer formed by the $k/2$ low-order bits of X. Similarly, form Y_L and Y_R from Y in the same fashion. We can express this in algebraic notation as

$$X = 2^{k/2}X_L + X_R \quad \text{and} \quad Y = 2^{k/2}Y_L + Y_R.$$

Computing the product of X and Y, we have

$$XY = 2^k X_L Y_L + 2^{k/2}(X_L Y_R + X_R Y_L) + X_R Y_R. \tag{4.14}$$

In equation (4.14), we have indicated four subproblems in red that need to be solved. We should also note that multiplication by a power of 2 is just a left shift, so it does not constitute a "real" multiplication.

A simple divide-and-conquer algorithm based on (4.14) is presented as Algorithm 4.10.

We can choose any convenient value of k for the base case. We have used $k = 1$ in Algorithm 4.10, but a more convenient choice in practice might be something like $k = 16$ or 32.

Example 4.14 Suppose we want to compute the product 1101×1011 (in binary) and we take $k = 2$ as our base case. We have

$$X_L = 11, X_R = 01, Y_L = 10 \text{ and } Y_R = 11.$$

We solve four subproblems non-recursively, because $k = 2$ is the base case, yielding

$$\begin{aligned} Z_1 &= X_L Y_L = 0110 \\ Z_2 &= X_L Y_R = 1001 \\ Z_3 &= X_R Y_L = 0010 \\ Z_4 &= X_R Y_R = 0011. \end{aligned}$$

```
Algorithm 4.10: NOTSOFASTMULTIPLY(X, Y, k)

comment: X and Y are k-bit integers
external SHIFT
if k = 1
  then Z ← X[0] × Y[0]
        ⎧ Z1 ← NOTSOFASTMULTIPLY(X_L, Y_L, k/2)
        ⎪ Z2 ← NOTSOFASTMULTIPLY(X_R, Y_R, k/2)
  else  ⎨ Z3 ← NOTSOFASTMULTIPLY(X_L, Y_R, k/2)
        ⎪ Z4 ← NOTSOFASTMULTIPLY(X_R, Y_L, k/2)
        ⎩ Z ← SHIFT(Z1, k) + Z2 + SHIFT(Z3 + Z4, k/2)
return (Z)
```

Then we compute

$$01100000 + 100100 + 001000 + 0011 = 10001111.$$

Here we end up adding four integers, of eight, six, six, and four bits in length, to obtain the 8-bit product. □

What is the complexity of Algorithm 4.10? We need to compute four products of $(n/2)$-bit integers, namely, $X_L Y_L$, $X_L Y_R$, $X_R Y_L$ and $X_R Y_R$. The recurrence relation is therefore

$$T(k) = 4T\left(\frac{k}{2}\right) + \Theta(k),$$

where k is a power of 2 (the $\Theta(k)$ term includes all additions and shifts). Therefore, the complexity of $T(k)$ is $\Theta(k^2)$ by the Master Theorem, so this is *not* an improvement over the grade-school method.

4.9.1 The Karatsuba Multiplication Algorithm

The KARATSUBA algorithm (also known as the KARATSUBA-OFMAN algorithm) was discovered in 1960. It is a divide-and-conquer algorithm that reduces the number of subproblems to compute the product XY from four to three.

Recall that

$$XY = 2^k X_L Y_L + 2^{k/2}(X_L Y_R + X_R Y_L) + X_R Y_R.$$

Suppose we compute

$$Z_1 = X_L Y_L$$

and

$$Z_2 = X_R Y_R$$

Algorithm 4.11: KARATSUBA(X, Y, k)

```
comment: we assume k is a power of two
external SHIFT
if k = 1
  then Z ← X[0] × Y[0]
  else { X_L ← k/2 high-order bits of X
         X_R ← k/2 low-order bits of X
         Y_L ← k/2 high-order bits of Y
         Y_R ← k/2 low-order bits of Y
         Z_1 ← KARATSUBA(X_L, Y_L, k/2)
         Z_2 ← KARATSUBA(X_R, Y_R, k/2)
         X_T ← X_L + X_R
         Y_T ← Y_L + Y_R
         if X_T ≥ 2^{k/2}
           then { c_T ← 1
                  X'_T ← X_T − 2^{k/2}
           else { c_T ← 0
                  X'_T ← X_T
         if Y_T ≥ 2^{k/2}
           then { d_T ← 1
                  Y'_T ← Y_T − 2^{k/2}
           else { d_T ← 0
                  Y'_T ← Y_T
         S ← KARATSUBA(X'_T, Y'_T, k/2),
         if c_T = 1
           then S ← S + SHIFT(Y'_T, k/2)
         if d_T = 1
           then S ← S + SHIFT(X'_T, k/2)'
         if c_T = 1 and d_T = 1
           then S ← S + 2^k
         Z ← SHIFT(Z_1, k) + Z_2 + SHIFT(S − Z_1 − Z_2, k/2)
return (Z)
```

recursively, as before. However, suppose we also compute

$$S = (X_L + X_R)(Y_L + Y_R).$$

Then

$$X_L Y_R + X_R Y_L = S - Z_1 - Z_2.$$

The savings are obtained because computing S only requires solving one subproblem, provided that we first compute

$$X_T = X_L + X_R \quad \text{and} \quad Y_T = Y_L + Y_R,$$

and then multiply them together.

There is one technical issue. Note that either or both of X_T and Y_T could be $(k/2+1)$-bit integers. However, computation of S can be accomplished by multiplying $(k/2)$-bit integers and accounting for *carries* by extra additions. But some modifications to the algorithm are required to handle this, of course.

Suppose we write

$$X_T = c_T 2^{k/2} + X_T', \tag{4.15}$$

where $c_T = 0$ or 1. Here c_T is the ***carry bit*** for X_T. Similarly, we express

$$Y_T = d_T 2^{k/2} + Y_T', \tag{4.16}$$

where $d_T = 0$ or 1.

For example, if $k = 8$, then $k/2 = 4$ and we want X_T to be a 4-bit integer. If X_T is a 5-bit integer, say $X_T = 26 = 11010_2$, then we would write

$$X_T = 16 + 10 = 1 \times 2^4 + 10,$$

so $c_T = 1$ and $X_T' = 10$.

Now we can express the product $S = X_T Y_T$ as follows:

$$\begin{aligned} X_T Y_T &= (c_T 2^{k/2} + X_T')(d_T 2^{k/2} + Y_T') \\ &= c_T d_T 2^k + c_T 2^{k/2} Y_T' + d_T 2^{k/2} X_T' + X_T' Y_T'. \end{aligned}$$

So we could compute $X_T Y_T$ using the following sequence of operations:

1. rewrite X_T and Y_T using (4.15) and (4.16),
2. recursively compute $S \leftarrow X_T' Y_T'$,
3. if $c_T = 1$, then $S \leftarrow S + 2^{k/2} Y_T'$,
4. if $d_T = 1$, then $S \leftarrow S + 2^{k/2} X_T'$,
5. if $c_T = 1$ and $d_T = 1$, then $S \leftarrow S + 2^k$.

Of course, as we have already noted, the multiplications by $2^{k/2}$ are just left shifts by $k/2$ bits.

Example 4.15 As an example, let's take $k = 8$, $X_T = 26$ and $Y_T = 21$. So we have

$$c_T = 1, \quad X'_T = 10, \quad d_T = 1 \quad \text{and} \quad Y'_T = 5.$$

Then the product $S = X_T Y_T$ would be computed as follows:

$$\begin{array}{lll} S \leftarrow X'_T Y'_T & = 50 & \text{(this is computed recursively)} \\ S \leftarrow S + 16 \times 5 & = 130 & \text{since } c_T = 1 \\ S \leftarrow S + 16 \times 10 & = 290 & \text{since } d_T = 1 \\ S \leftarrow S + 256 & = 546 & \text{since } c_T = d_T = 1. \end{array}$$

□

Karatsuba's algorithm is presented as Algorithm 4.11. The complexity of this algorithm can be obtained by solving the recurrence relation

$$T(k) = 3T\left(\frac{k}{2}\right) + \Theta(k),$$

where k is a power of 2. Therefore the complexity of $T(k)$ is

$$\Theta(k^{\log_2 3}) = \Theta(k^{1.59})$$

by the Master Theorem (see Example 4.4 for more detail).

Various techniques can be used to handle the case when k is not a power of two. One possible solution is to pad with zeroes on the left. So, let m be the smallest power of two that is $\geq k$. The complexity of the algorithm is then $\Theta(m^{\log_2 3})$. Since $m < 2k$, the complexity is

$$\begin{aligned} O((2k)^{\log_2 3}) &= O(3k^{\log_2 3}) \\ &= O(k^{\log_2 3}). \end{aligned}$$

There are numerous further improvements known for solving the problem of **Multiprecision Multiplication**:

- The TOOM-COOK algorithm splits X and Y into three equal parts and uses five multiplications of $(k/3)$-bit integers. The recurrence is

  $$T(k) = 5T\left(\frac{k}{3}\right) + \Theta(k),$$

  and it follows from the Master Theorem that

  $$T(k) \in \Theta(k^{\log_3 5}) = \Theta(k^{1.47}).$$

- The 1971 SCHONHAGE-STRASSEN algorithm (based on the Fast Fourier Transform) has complexity $O(n \log n \log\log n)$.

- The 2007 FURER algorithm has complexity $O(n \log n 2^{O(\log^* n)})$.

- It was announced in April 2019 that Harvey and van der Hoeven had developed an algorithm having complexity $O(n \log n)$. It is conjectured that this is the optimal complexity for any algorithm for **Multiprecision Multiplication**.

4.10 Modular Exponentiation

In the ***RSA Cryptosystem***, encryption and decryption both require a modular exponentiation to be performed. This problem is defined as follows:

Problem 4.5: Modular Exponentiation

Instance: A modulus $n > 1$, and integers x and e such that $0 < x < n$ and $0 < e < n$.
Question: Compute $y = x^e \bmod n$.

For typical RSA computations, n could be the product of two 1024-bit primes, so n would be 2048 bits in length. We are going to base our modular exponentiation algorithm on an underlying modular multiplication algorithm, MODMULT, which takes as input a modulus n and two integers x_1 and x_2 (such that $0 < x_1 < n$ and $0 < x_2 < n$) and computes their product in $\mathbb{Z}_n$. Thus, we will treat MODMULT as an oracle. For the time being, we will assume that MODMULT has complexity $\Theta(1)$, as we do in a reduction. Later, we will briefly discuss "real" algorithms that we can use to instantiate the modular multiplication oracle.

It is quite straightforward to devise a divide-and-conquer algorithm to solve **Modular Exponentiation**. The idea is to focus on the exponent e and compute $x^e \bmod n$ by solving a "subproblem" of the form $x^f \bmod n$, where $f < e$. In line with previous divide-and-conquer algorithms, we could consider taking $f = e/2$ (at least when e is even). If e is even, we could base an algorithm on the formula

$$x^e = x^{e/2} \times x^{e/2}.$$

(In the equation above, we are assuming that "$\times$" denotes modular multiplication.) Thus, we would compute $x^{e/2} \bmod n$ recursively, and then multiply it by itself modulo n using MODMULT. Using this approach, we solve one subproblem and then make use of a combine operation (namely, MODMULT) that takes time $\Theta(1)$.

What could we do if e is odd? It would not be a good idea to use the identity

$$x^e = x^{(e+1)/2} \times x^{(e-1)/2},$$

because this would require us to recursively solve two subproblems, namely $x^{(e+1)/2} \bmod n$ and $x^{(e-1)/2} \bmod n$. This approach is correct, but it turns out that we can devise an algorithm with lower complexity.

Here is a better solution. Again, we are assuming that e is odd. Consider the identity

$$x^e = x \times x^{(e-1)/2} \times x^{(e-1)/2}.$$

Here, we would perform the following operations:

1. compute $x^{(e-1)/2} \bmod n$ recursively,

2. multiply the above number by itself modulo n using MODMULT, and
3. multiply the result by x modulo n using MODMULT.

Here, we solve only one subproblem, just like we did in the case where e is even. The combine operation now consists of two invocations of MODMULT, but it still takes time $\Theta(1)$.

Summarizing the above discussion, we obtain Algorithm 4.12.

Algorithm 4.12: FASTMODEXP(n, x, e)

```
comment: we assume 0 ≤ x ≤ n − 1
external MODMULT
if e = 1
  then y ← x
  else if e is even
         then f ← e/2
              z ← FASTMODEXP(n, x, f)
              y ← MODMULT(z, z, n)
         else f ← (e − 1)/2
              z ← FASTMODEXP(n, x, f)
              z ← MODMULT(z, z, n)
              y ← MODMULT(x, z, n)
return (y)
```

As an example, it is easy to check that a call to FASTMODEXP with $e = 121$ will give rise to subproblems having $e = 60, 30, 15, 7, 3$ and 1.

Now let's analyze the complexity of Algorithm 4.12. Recall that we are temporarily treating MODMULT as a $\Theta(1)$-time algorithm. We will write down a recurrence relation that expresses the complexity of FASTMODEXP as a function of the exponent e. In order to be precise, we will consider the following recurrence, which applies to all values of $e \geq 1$ due to the use of the floor function:

$$\begin{aligned} T(e) &= T\left(\left\lfloor \frac{e}{2} \right\rfloor\right) + c \qquad \text{if } e > 1 \\ T(1) &= d, \end{aligned}$$

where c and d are positive constants. The "c" term accounts for doing either one or two modular multiplications, which we are treating (for the time being) as unit cost operations. The "d" term refers to the base case.

We could of course just appeal to the Master Theorem to find the solution of this recurrence. However, it turns out that this recurrence has the following exact solution:

$$T(e) = c\lfloor \log_2 e \rfloor + d. \tag{4.17}$$

Since $e \leq n$, it then follows immediately that $T(e) \in O(\log n)$.

The formula (4.17) can be proven by induction on e. The base case, $e = 1$, is easy because

$$c\lfloor \log_2 1 \rfloor + d = d.$$

For the general case, it is convenient to distinguish between even and odd values of e. If $e > 1$ is even, say $e = 2f$, then

$$\begin{aligned} T(2f) &= T(f) + c \\ &= c\lfloor \log_2 f \rfloor + d + c \qquad \text{by induction} \\ &= c(1 + \lfloor \log_2 f \rfloor) + d \\ &= c(\lfloor 1 + \log_2 f \rfloor) + d \\ &= c(\lfloor \log_2 2f \rfloor) + d. \end{aligned}$$

If $e > 1$ is odd, say $e = 2f + 1$, then

$$\begin{aligned} T(2f+1) &= T(f) + c \\ &= c\lfloor \log_2 f \rfloor + d + c \qquad \text{by induction} \\ &= c(\lfloor \log_2 2f \rfloor) + d, \end{aligned}$$

as in the case where e is even. However,

$$\lfloor \log_2 2f \rfloor = \lfloor \log_2 (2f+1) \rfloor$$

for all positive integers f, so

$$T(2f+1) = c\lfloor \log_2 (2f+1) \rfloor + d.$$

Thus, the formula (4.17) holds for all integers $e \geq 1$, by induction.

Finally, let's remove the (unrealistic) unit cost assumption for modular multiplication. There is much literature on algorithms for modular multiplication. It is relatively straightforward to show that multiplication in $\mathbb{Z}_n$ can be done in time $O((\log n)^2)$ (this is not the most efficient method; we are just using it for purposes of illustration). Then, our modified recurrence relation would be as follows:

$$\begin{aligned} T(e) &= T\left(\left\lfloor \frac{e}{2} \right\rfloor\right) + c(\log n)^2 \qquad \text{if } e > 1 \\ T(1) &= d. \end{aligned}$$

The solution of the modified recurrence is

$$T(e) = c(\log n)^2 \lfloor \log_2 e \rfloor + d.$$

Since $e < n$, we have $T(e) \in O((\log n)^3)$.

4.11 Matrix Multiplication

We next consider the fundamental problem of multiplying two n by n matrices.

Problem 4.6: Matrix Multiplication

Instance: Two n by n matrices, A and B, having integer entries.
Question: Compute the n by n matrix product $C = AB$.

We assume we are not doing any multiprecision multiplications, so we are working in the unit cost model and each underlying multiplication of integers can be assumed to take time $\Theta(1)$. Suppose $A = (a_{ij})$, $B = (b_{ij})$ and $C = (c_{ij}) = AB$. Then

$$c_{ij} = \sum_{k=1}^{n} a_{ik} b_{kj}.$$

There are n^2 terms c_{ij}, each of which requires n multiplications (and additions) to compute. So the overall complexity of this "naive" algorithm is $\Theta(n^2 \times n) = \Theta(n^3)$.

Now we look at divide-and-conquer algorithms for matrix multiplication. We first observe that an obvious divide-and-conquer algorithm does not improve the complexity as compared to the naive algorithm. Let

$$A = \begin{pmatrix} a & b \\ c & d \end{pmatrix}, \quad B = \begin{pmatrix} e & f \\ g & h \end{pmatrix} \quad \text{and} \quad C = AB = \begin{pmatrix} r & s \\ t & u \end{pmatrix}.$$

If A, B are n by n matrices, then $a, b, ..., h, r, s, t, u$ are $\frac{n}{2}$ by $\frac{n}{2}$ matrices, where

$$\begin{aligned} r &= ae + bg & \qquad s &= af + bh \\ t &= ce + dg & \qquad u &= cf + dh. \end{aligned}$$

We require eight multiplications of $\frac{n}{2}$ by $\frac{n}{2}$ matrices in order to compute $C = AB$.

The resulting recurrence is

$$T(k) = 8T\left(\frac{k}{2}\right) + \Theta(k^2),$$

so $T(k) \in \Theta(k^3)$ by the Master Theorem. The $\Theta(k^2)$ term accounts for four additions of $\frac{k}{2} \times \frac{k}{2}$ matrices, each of which has $\Theta(k^2)$ entries.

4.11.1 Strassen Matrix Multiplication

The STRASSEN algorithm for matrix multiplication was published in 1969. It uses the approach described in the previous section; however, it incorporates an ingenious way of reducing the number of subproblems from eight to seven. Suppose we define

$$\begin{aligned} P_1 &= a(f-h) & P_2 &= (a+b)h \\ P_3 &= (c+d)e & P_4 &= d(g-e) \\ P_5 &= (a+d)(e+h) & P_6 &= (b-d)(g+h) \\ P_7 &= (a-c)(e+f). \end{aligned}$$

Then, we compute

$$\begin{aligned} r &= P_5 + P_4 - P_2 + P_6 & s &= P_1 + P_2 \\ t &= P_3 + P_4 & u &= P_5 + P_1 - P_3 - P_7. \end{aligned}$$

We now require only seven multiplications of $\frac{n}{2}$ by $\frac{n}{2}$ matrices (namely, to compute $P_1, \ldots, P_7$) in order to obtain the product $C = AB$.

The verifications are somewhat tedious but straightforward. For example,

$$\begin{aligned} & P_5 + P_4 - P_2 + P_6 \\ &= (a+d)(e+h) + d(g-e) - (a+b)h + (b-d)(g+h) \\ &= ae + ah + de + dh + dg - de - ah - bh + bg + bh - dg - dh \\ &= ae + bg, \end{aligned}$$

which agrees with the formula.

What is the complexity of the resulting divide-and-conquer algorithm? The recurrence is

$$T(k) = 7T\left(\frac{k}{2}\right) + \Theta(k^2).$$

Here we have $a = 7$, $b = 2$, so $x = \log_2 7 = 2.81$. We also have $y = 2$, so $x > y$ and we are in case 1. Therefore,

$$T(n) \in \Theta(n^x) = \Theta(n^{2.81}).$$

The STRASSEN algorithm was improved in 1990 by Coppersmith and Winograd. Their algorithm has complexity $O(n^{2.376})$. Some slight improvements have been found more recently. The current record is $O(n^{2.3728639})$, by Le Gall in 2014.

4.12 QuickSort

QUICKSORT, which was invented in 1959 by Tony Hoare, is regarded as the fastest general-purpose sorting algorithm used in practice. In QUICKSORT, we are given an array A of n items. First, we choose an element as the ***pivot*** and we restructure the array so all elements preceding the pivot are less than the pivot, and all elements following the pivot are greater than the pivot. Then we recursively sort the subarray preceding the pivot and the subarray following the pivot.

Algorithm 4.13 gives the high-level structure of QUICKSORT. The values p and r are indices in the array A that is being sorted. After the subarray $[A[p], \ldots, A[r]]$ is restructured, $A[q]$ is the position in which the pivot ends up. Thus, there are two recursive calls, to sort the subarrays $[A[p], \ldots, A[q = 1]]$ and $[A[q+1], \ldots, A[r]]$. The initial call that is made is $\text{QUICKSORT}(A, 1, n)$.

Algorithm 4.13: $\text{QUICKSORT}(A, p, r)$

comment: A is an array of length n and $1 \leq p \leq r \leq n$
external PARTITION
if $p < r$
  then $\begin{cases} q \leftarrow \text{PARTITION}(A, p, r) \\ \text{QUICKSORT}(A, p, q-1) \\ \text{QUICKSORT}(A, q+1, r) \end{cases}$

Algorithm 4.14 describes how the restructuring is done. The last element in the relevant subarray, namely $A[r]$, is chosen to be the pivot. Then a sequence of EXCHANGE operations is done to restructure the array appropriately. Each EXCHANGE operation just swaps two array elements that are specified by their indices: $\text{EXCHANGE}(A, i, j)$ would swap $A[i]$ and $A[j]$. After the array is restructured, the position of the pivot is returned to the calling program. The complexity of RESTRUCTURE is $\Theta(r - p + 1)$, because there is a single **for** loop that iterates from p to $r - 1$.

Algorithm 4.14: $\text{RESTRUCTURE}(A, p, r)$

external EXCHANGE
$x \leftarrow A[r]$
$i \leftarrow p - 1$
for $j \leftarrow p$ **to** $r - 1$
  do $\begin{cases} \textbf{if } A[j] \leq x \\ \quad \textbf{then } \begin{cases} i \leftarrow i + 1 \\ \text{EXCHANGE}(A, i, j) \end{cases} \end{cases}$
$\text{EXCHANGE}(A, i+1, r)$
return $(i + 1)$

Example 4.16 Suppose A is the following array of length 10:

$$A = [1, 10, 5, 8, 6, 7, 12, 2, 11, 9].$$

TABLE 4.5: An example of restructuring

$p \leftarrow 1$	$r \leftarrow 10$	$x \leftarrow 9$	$i \leftarrow 0$
$j \leftarrow 1$	$1 < 9$	$i \leftarrow 1$	EXCHANGE$(A,1,1)$
	$A = [1,10,5,8,6,7,12,2,11,9]$		
$j \leftarrow 2$	$10 > 9$		
$j \leftarrow 3$	$5 < 9$	$i \leftarrow 2$	EXCHANGE$(A,2,3)$
	$A = [1,5,10,8,6,7,12,2,11,9]$		
$j \leftarrow 4$	$8 < 9$	$i \leftarrow 3$	EXCHANGE$(A,3,4)$
	$A = [1,5,8,10,6,7,12,2,11,9]$		
$j \leftarrow 5$	$6 < 9$	$i \leftarrow 4$	EXCHANGE$(A,4,5)$
	$A = [1,5,8,6,10,7,12,2,11,9]$		
$j \leftarrow 6$	$7 < 9$	$i \leftarrow 5$	EXCHANGE$(A,5,6)$
	$A = [1,5,8,6,7,10,12,2,11,9]$		
$j \leftarrow 7$	$12 > 7$		
$j \leftarrow 8$	$2 < 9$	$i \leftarrow 6$	EXCHANGE$(A,6,8)$
	$A = [1,5,8,6,7,2,12,10,11,9]$		
$j \leftarrow 9$	$11 > 9$		
	EXCHANGE$(A,7,10)$		
	$A = [1,5,8,6,7,2,9,10,11,12]$		

The initial restructuring of A using Algorithm 4.14 would proceed as shown in Table 4.5.

So, the pivot, 9, ends up in position $A[7]$. After restructuring, the first six elements of A are less than 9 and the last three elements are greater than 9. □

4.12.1 Complexity Analysis

We now analyze the complexity of QUICKSORT. For simplicity of analysis, assume that all n elements are distinct.

best-case complexity

If the pivot always splits the array into two equal parts, then the running time $T(n)$ satisfies the following recurrence:

$$T(n) = T\left(\left\lfloor \frac{n-1}{2} \right\rfloor\right) + T\left(\left\lceil \frac{n-1}{2} \right\rceil\right) + \Theta(n).$$

This is essentially the MERGESORT recurrence; its solution is $T(n) \in \Theta(n \log n)$.

worst-case complexity

If the pivot is always the maximum (or minimum) element in a subarray, then a subarray of size ℓ is split into subarrays of size 0 and $\ell - 1$. In this case, we have the recurrence

$$T(n) = T(0) + T(n-1) + \Theta(n) = T(n-1) + \Theta(n).$$

The reader can easily verify that $T(n) \in \Theta(n^2)$, so the complexity is quadratic.

Since the best-case and worst-case complexities are so different, it is interesting to consider the *average-case complexity*. Recall that we already did some average-case analyses in Sections 2.4.2 and 2.4.3. Here, we will estimate the complexity by counting the number of comparisons of two array elements that are performed in the QUICKSORT algorithm.

The complexity of QUICKSORT depends only on how the n items are ordered initially (the actual values of the items are immaterial). Therefore we can fix a set of n items (say $1, \ldots, n$) and consider all $n!$ orderings (i.e., permutations) of them. We assume that each ordering is equally likely. For each permutation of the n items, we count the number of comparisons. Then we sum all these numbers and divide by $n!$ to compute the average. This will be an exact analysis.

Let c_n denote the average number of comparisons done in QUICKSORT for an array of n distinct integers. RESTRUCTURE requires $n - 1$ comparisons. Then the array is partitioned into two pieces, a "left" subarray of size i and a "right" subarray of size $n - 1 - i$.

For $i = 0, 1, \ldots, n-1$, it is not hard to see that

$$\mathbf{Pr}[\text{the left subarray has size } i] = \frac{1}{n}.$$

Hence we have the following recurrence, including initial conditions:

$$\begin{aligned} c_n &= n - 1 + \frac{1}{n} \sum_{i=0}^{n-1} (c_i + c_{n-1-i}) \qquad \text{if } n > 1 \qquad (4.18)\\ c_1 &= 0\\ c_0 &= 0. \end{aligned}$$

This can be simplified as follows:

$$\begin{aligned} c_n &= n - 1 + \frac{2}{n} \sum_{i=0}^{n-1} c_i\\ nc_n &= n^2 - n + 2 \sum_{i=0}^{n-1} c_i \qquad (4.19) \end{aligned}$$

If we replace n by $n+1$ in (4.19), we obtain

$$(n+1)c_{n+1} = n^2 + n + 2\sum_{i=0}^{n} c_i. \tag{4.20}$$

Computing (4.20) − (4.19), we have

$$\begin{aligned}
(n+1)c_{n+1} - nc_n &= 2n + 2c_n \\
(n+1)c_{n+1} - (n+2)c_n &= 2n \\
\frac{c_{n+1}}{n+2} - \frac{c_n}{n+1} &= \frac{2n}{(n+1)(n+2)} \\
\frac{c_{n+1}}{n+2} - \frac{c_n}{n+1} &= \frac{-2}{n+1} + \frac{4}{n+4}.
\end{aligned} \tag{4.21}$$

Now define

$$d_n = \frac{c_n}{n+1}.$$

Then, from (4.21), we have

$$d_{n+1} - d_n = \frac{-2}{n+1} + \frac{4}{n+2}.$$

Replacing n by $n-1, n-2$, etc., we obtain

$$\begin{aligned}
d_{n+1} - d_n &= \frac{-2}{n+1} + \frac{4}{n+2} \\
d_n - d_{n-1} &= \frac{-2}{n} + \frac{4}{n+1} \\
d_{n-1} - d_{n-2} &= \frac{-2}{n-1} + \frac{4}{n} \\
&\vdots \\
d_2 - d_1 &= \frac{-2}{2} + \frac{4}{3} \\
d_1 &= 0.
\end{aligned}$$

Summing all these equations, we have

$$d_{n+1} = \frac{4}{n+2} + 2\sum_{i=3}^{n+1} \frac{1}{i} - 1.$$

Replacing $n+1$ by n, we get

$$d_n = \frac{4}{n+1} + 2\sum_{i=3}^{n} \frac{1}{i} - 1.$$

Finally, substituting $c_n = (n+1)d_n$ and simplifying, we have

$$\begin{aligned} c_n &= 2(n+1)\sum_{i=3}^{n}\frac{1}{i} - n + 3. \\ &= 2(n+1)\sum_{i=1}^{n}\frac{1}{i} - 2(n+1)\left(1+\frac{1}{2}\right) - n + 3. \\ &= 2(n+1)\sum_{i=1}^{n}\frac{1}{i} - 3n - 3 - n + 3. \\ &= 2(n+1)H_n - 4n, \end{aligned}$$

where H_n is the nth harmonic number. Therefore, we have proven the following.

THEOREM 4.6 *The average number of comparisons of array elements done by Algorithm 4.13, computed over all $n!$ permutations of the elements of an array of length n, is*

$$2(n+1)H_n - 4n. \tag{4.22}$$

Since $H_n \in \Theta(\log n)$, it is clear that the expression in (4.22) is $\Theta(n \log n)$, so we have proven that the average-case complexity of QUICKSORT is $\Theta(n \log n)$.

A ***randomized algorithm*** is an algorithm that uses random numbers in its execution. Let M be a program implementing a randomized algorithm. Then for a fixed problem instance I, the running time $T_M(I)$ may vary, depending on the random numbers used during the execution of the algorithm.

Suppose we randomly permute the n entries in an array A before running QUICKSORT on A. We call the resulting algorithm RANDOMIZEDQUICKSORT. Then the average-case complexity of RANDOMIZEDQUICKSORT for any fixed array A is the same as the average-case complexity of QUICKSORT over all $n!$ orderings of the n items in A.

In the randomized algorithm, we start with a fixed problem instance (i.e., a fixed ordering of n items), and the average is computed over all the possible executions of the algorithm on the given problem instance. On the other hand, in the average-case analysis of "regular" QUICKSORT, the average is computed over all $n!$ problem instances (i.e., the $n!$ orderings of n elements.).

The advantage of the randomized algorithm is that there are no longer any "bad" problem instances: all instances I have the same average-case behavior. In contrast, for regular QUICKSORT, some instances are "good" and some are "bad." The randomization has the effect of "smoothing" the complexity of the algorithm so it is the same for all inputs.

Generating a random permutation of an array of n items can be done easily in time $\Theta(n)$, so the extra overhead in the randomized algorithm does not affect the complexity. In the algorithm RANDOMPERMUTE, we assume the existence of a subroutine RANDOM(i, j) that generates a random integer k such that $i \leq k \leq j$.

It is not hard to prove that RANDOMPERMUTE generates a truly random permutation of A, i.e., every possible permutation of $A[1], \ldots, A[n]$ occurs with the

Algorithm 4.15: RANDOMPERMUTE(A)

comment: A is an array of length n
external EXCHANGE, RANDOM
for $i \leftarrow 1$ **to** n
  do EXCHANGE($A[i], A[$RANDOM$(i, n)]$)

same probability $1/n!$, provided that RANDOM is a perfect random number generator.

An alternative, but equivalent, approach is to not run RANDOMPERMUTE, but instead choose the pivot randomly during the RESTRUCTURE operation.

4.13 Selection and Median

In this section, we explore algorithms to find the kth smallest element in an array of size n.

Problem 4.7: Selection

Instance: An array $A[1], \ldots, A[n]$ of distinct integer values, and an integer k, where $1 \leq k \leq n$.
Find: The kth smallest integer in the array A.

The problem **Median** is the special case of **Selection** where $k = \lceil \frac{n}{2} \rceil$. The median is the "middle" element in the array.

For purposes of comparison, here are some "simple" algorithms based on elementary data structures.

sorting-based solution

Suppose we first SORT the array A and then return the value $A[k]$. Then the complexity is $\Theta(n \log n)$, independent of the value of k.

finding k successive minima

Suppose we use a modified SELECTIONSORT, based on finding k successive minima (recall that we described SELECTIONSORT in Section 3.4). For example, we could first find the minimum element in the array and then delete it. If this process is repeated k times, the last element that we delete is the kth smallest element in the array. Then the complexity would be computed as

$$\begin{aligned}\sum_{i=0}^{k-1} \Theta(n-i) &= \Theta\left(\sum_{i=0}^{k-1}(n-i)\right)\\ &= \Theta\left(nk - \frac{k(k-1)}{2}\right) \quad \text{from (1.1)}\\ &= \Theta\left(k\left(n - \frac{k-1}{2}\right)\right)\\ &= \Theta(nk),\end{aligned}$$

since $1 \leq k \leq n$. If $k = n/2$ (i.e., we are looking for the median), then the complexity is $\Theta(n^2)$. On the other hand, if k is a constant (e.g., $k = 10$), then the complexity is $\Theta(n)$.

priority queue-based solution

We can base a solution on a priority queue (see Section 3.4). We can use a modification of HEAPSORT, consisting of BUILDHEAP followed by k DELETEMIN operations. The complexity of BUILDHEAP is $\Theta(n)$ and the complexity of each DELETEMIN operation is $O(\log n)$. Hence the complexity is $\Theta(n + k \log n)$. If $k = n/2$, then the complexity is $\Theta(n \log n)$. On the other hand, if k is a constant (e.g., $k = 10$), then the complexity is $\Theta(n)$.

Note that none of the three approaches described above allow us to find the median in $O(n)$ time.

4.13.1 QuickSelect

We describe an algorithm, published in 1961, that is commonly known as QUICKSELECT. It is very similar to QUICKSORT, but with one recursive call instead of two. Like QUICKSORT, it was also discovered by Tony Hoare. Suppose we choose a ***pivot*** element y in the array A, and we restructure A so that all elements less than y precede y in A, and all elements greater than y occur after y in A. (This is exactly what is done in Algorithm 4.14, and it takes linear time.)

Algorithm 4.14 will return the index *posn*, which is the correct position of the pivot. Thus, $A[posn] = y$ after restructuring. Let A_L be the subarray consisting of the $posn - 1$ elements

$$A[1], \ldots, A[posn-1]$$

and let A_R be the subarray (of size $n - posn$) consisting of elements

$$A[posn+1], \ldots, A[n].$$

Then the kth smallest element of A is

$$\begin{array}{ll} y & \text{if } k = posn\\ \text{the } k\text{th smallest element of } A_L & \text{if } k < posn\\ \text{the } (k-posn)\text{th smallest element of } A_R & \text{if } k > posn.\end{array}$$

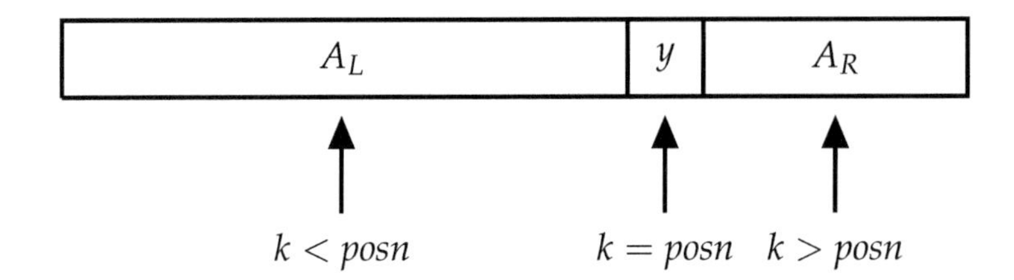

FIGURE 4.7: The three cases in QUICKSELECT

If it happens by chance that $k = posn$, then no recursive call is necessary. If $k \neq posn$, then we make a single recursive call (to one of the subarrays A_L or A_R). Therefore, we make (at most) one recursive call at each level of the recursion, as indicated in Figure 4.7.

Algorithm 4.16: QUICKSELECT(k, n, A)

> **comment:** A is an array of length n and $1 \leq k \leq n$
> **external** RESTRUCTURE2
> $(A_L, A_R, posn) \leftarrow$ RESTRUCTURE2(A, n)
> **if** $k = posn$
>   **then return** (y)
> **else if** $k < posn$
>   **then return** (QUICKSELECT$(k, posn - 1, A_L)$)
>   **else return** (QUICKSELECT$(k - posn, n - posn, A_R)$)

QUICKSELECT is presented as Algorithm 4.16. Observe that RESTRUCTURE2 is a very slightly modified restructuring algorithm; see Algorithm 4.17. Here, we return the left and right subarrays after the restructuring takes place, along with the final position of the pivot. As before, the last element is used as the pivot.

This is another situation where it is interesting to do an average-case complexity analysis. As in the case of QUICKSORT, by "average-case" we mean the average over the $n!$ possible permutations (i.e., orderings) of the n elements in the array A.

Suppose we define a pivot to be ***good*** if $posn$ is in the middle half of A, i.e., $n/4 \leq posn \leq 3n/4$. The probability that a pivot is good is $1/2$. Therefore, on average, after two iterations, we will encounter a good pivot (this follows from Theorem 1.12).

If a pivot is good, then $|A_L| \leq 3n/4$ and $|A_R| \leq 3n/4$. Thus, with an *expected* linear amount of work, the size of the subproblem is reduced by at least 25%.

Let's consider the *average-case* recurrence relation:

$$T(n) = T\left(\frac{3n}{4}\right) + \Theta(n).$$

```
Algorithm 4.17: RESTRUCTURE2(A, n)

  comment: A is an array of length n
  external EXCHANGE
  x ← A[n]
  i ← p − 1
  for j ← 1 to n − 1
         ⎧ if A[j] ≤ x
    do   ⎨          ⎧ i ← i + 1
         ⎩   then   ⎨
                    ⎩ EXCHANGE(A, i, j)
  EXCHANGE(A, i + 1, n)
  A_L ← [A[1], ..., A[i]]
  A_R ← [A[i + 2], ..., A[n]]
  return (A_L, A_R, i + 1)
```

We can apply the Master Theorem with $a = 1$, $b = 4/3$ and $y = 1$. Here

$$x = \log_{4/3} 1 = 0 < 1 = y,$$

so we are in case 3. This yields $T(n) \in \Theta(n)$ *on average*.

Here is a more rigorous proof of the average-case complexity, which uses some notions from probability theory such as random variables (see Section 1.5). We say the algorithm is in ***phase*** j if the current subarray has size s, where

$$n\left(\frac{3}{4}\right)^{j+1} < s \leq n\left(\frac{3}{4}\right)^{j}.$$

Let $\mathbf{X_j}$ be a random variable that denotes the amount of computation time occurring in phase j. If the pivot is in the middle half of the current subarray, then we transition from phase j to phase $j+1$. This occurs with probability $1/2$, so the expected number of recursive calls in any given phase is 2. The computing time for each recursive call is linear in the size of the current subarray, so

$$E[\mathbf{X_j}] \leq 2cn\left(\frac{3}{4}\right)^{j}$$

(where, as usual, $E[\cdot]$ denotes the expectation of a random variable). The total time of the algorithm is given by the random variable

$$\mathbf{X} = \sum_{j \geq 0} \mathbf{X_j}.$$

Therefore, by linearity of expectation (Theorem 1.17), we have

$$\begin{aligned}
E[\mathbf{X}] &= \sum_{j\geq 0} E[\mathbf{X_j}] \\
&\leq 2cn \sum_{j\geq 0} \left(\frac{3}{4}\right)^j \\
&= 8cn \\
&\in O(n).
\end{aligned}$$

In the derivation above, we are using the fact that an infinite geometric sequence with ratio r has a finite sum if $0 < r < 1$ (see equation (1.3)).

4.13.2 A Linear-time Selection Algorithm

We now describe how to achieve $O(n)$ worst-case complexity to solve the **Selection** problem. The algorithm is based on a clever strategy for choosing the pivot which was published in 1973 by Blum, Floyd, Pratt, Rivest and Tarjan. More precisely, we choose the pivot to be a certain ***median-of-medians***:

step 1

Given $n \geq 15$, write

$$n = 10r + 5 + \theta, \tag{4.23}$$

where $r \geq 1$ and $0 \leq \theta \leq 9$ (note that r and θ are defined uniquely).

step 2

Divide A into $2r+1$ disjoint subarrays of 5 elements. Denote these subarrays by $B_1, \dots, B_{2r+1}$.

step 3

For $1 \leq i \leq 2r+1$, find the median of B_i *non-recursively* (i.e., by brute force), and denote it by m_i.

step 4

Define M to be the array consisting of the $2r+1$ elements $m_1, \dots, m_{2r+1}$.

step 5

Find the median y of the array M *recursively*.

step 6

Use the element y as the pivot for A.

The strategy for the worst-case linear time algorithm is to divide the array into (roughly) $n/5$ subsets of size 5, find the median of each subset of five elements, and then recursively find the median of the resulting $n/5$ medians (i.e., the *median-of-medians*). This element y is then chosen to be the pivot in the basic QUICKSELECT algorithm. Observe that, as a result of this process, there will be an *odd* number of groups of five elements of A, and at most nine "extra" elements.

Example 4.17 Suppose A is the following array of length 15:

$$A = [1, 10, 5, 8, 21, 34, 6, 7, 12, 23, 2, 4, 30, 11, 25].$$

We divide A into three groups of size 5. We then (non-recursively) compute the median of each group, as follows:

						median
1	10	5	8	21	$\rightarrow$	8
34	6	7	12	23	$\rightarrow$	12
2	4	30	11	25	$\rightarrow$	11

Then we (recursively) find the median of the three medians, which is 11. Using 11 as the pivot to restructure the array A, we have

$$\begin{aligned} A_L &= [1, 10, 5, 8, 6, 7, 2, 4] \\ A_R &= [21, 34, 12, 23, 30, 25]. \end{aligned}$$

The pivot, 11, now occupies position $A[9]$. Then, if $k \neq 9$, we will make an appropriate recursive call to either A_L or A_R to find the kth smallest element of A. To illustrate, suppose that $k = 12$. Then the recursive call would be QUICKSELECT$(3, 6, A_R)$. □

We refer to the resulting algorithm as MOM-QS (this is an abbreviation of MEDIAN-OF-MEDIANS-QUICKSELECT). A high-level description of the algorithm is presented as Algorithm 4.18.

Algorithm 4.18: MOM-QS(k, n, A)

comment: A is an array of length n, and $1 \leq k \leq n$
external RESTRUCTURE2

1. **if** $n \leq 14$ **then** $C \leftarrow$ SORT(A) and **return** $(C[k])$
2. write $n = 10r + 5 + \theta$, where $0 \leq \theta \leq 9$
3. construct $B_1, \dots, B_{2r+1}$ ($2r + 1$ subarrays of A, each of size 5)
4. find medians $m_1, \dots, m_{2r+1}$ of $B_1, \dots, B_{2r+1}$ non-recursively
5. $M \leftarrow [m_1, \dots, m_{2r+1}]$
6. $y \leftarrow$ MOM-QS$(r + 1, 2r + 1, M)$
7. find y in the array A and exchange it with $A[n]$
8. $(A_L, A_R, posn) \leftarrow$ RESTRUCTURE2(A)
9. **if** $k = posn$ **then return** (y)
10.   **else if** $k < posn$ **then return** (MOM-QS$(k, posn - 1, A_L)$)
11.   **else return** (MOM-QS$(k - posn, n - posn, A_R)$)

Except for the method of choosing the pivot, Algorithm 4.18 is pretty much the same as the basic QUICKSELECT algorithm. We have already discussed steps 1–6 of this modified algorithm. Step 7 ensures that the desired pivot is placed in the last position of A before the call is made to RESTRUCTURE2. Steps 8–11 are the same as the corresponding parts of Algorithm 4.16.

We can analyze the complexity of the various steps in Algorithm 4.18:

step 1 $\Theta(1)$ (base case)

step 2 $\Theta(1)$

step 3 $\Theta(n)$

step 4 $\Theta(n)$

step 5 $\Theta(n)$

step 6 $T(n/5)$

step 7 $\Theta(n)$

step 8 $\Theta(n)$ (as in QUICKSORT)

steps 9–11 $\leq T(7n/10)$ (see the following discussion for the proof).

We need to justify the claimed complexity of steps 9–11. To this end, we first provide an imprecise estimate to show that the number of elements $> y$ is (roughly) at most $7n/10$. Similarly, we claim that the number of elements $< y$ is also at most $7n/10$.

Consider $n/5$ groups of five numbers, $B_1, \ldots, B_{n/5}$, and let m_i be the median of B_i, for $1 \leq i \leq n/5$. Let y be the median of the m_i's. There are $n/10$ j's such that $m_j < y$. For each such j, there are three elements in B_j that are less than y (namely, m_j and two other elements of B_j). So there are *at least* $3n/10$ elements that are less than y and hence there are *at most* $7n/10$ elements that are greater than y. Similarly, there are at most $7n/10$ elements that are less than y.

The diagram in Figure 4.8 can be considered a *conceptual aid*; it is not constructed by the algorithm. Suppose we write out every B_i in increasing order, so the middle element of each B_i is the median. These will be the rows in the diagram. Then permute the rows so the medians are in increasing order (from top to bottom). Consider the middle element in the diagram (the median-of-medians); denote it by y. The (roughly) $3n/10$ elements in the lower right quadrant are all $\geq y$. Therefore there are (roughly) at most $7n/10$ elements that are $\leq y$.

Here is a more careful analysis. Recall from (4.23) that

$$n = 10r + 5 + \theta,$$

where $0 \leq \theta \leq 9$. The red rectangle in Figure 4.8 contains $3(r+1)$ elements, all of which are $\geq y$. Therefore, the number of elements $< y$ is at most

$$n - 3(r+1) = 10r + 5 + \theta - 3r - 3$$

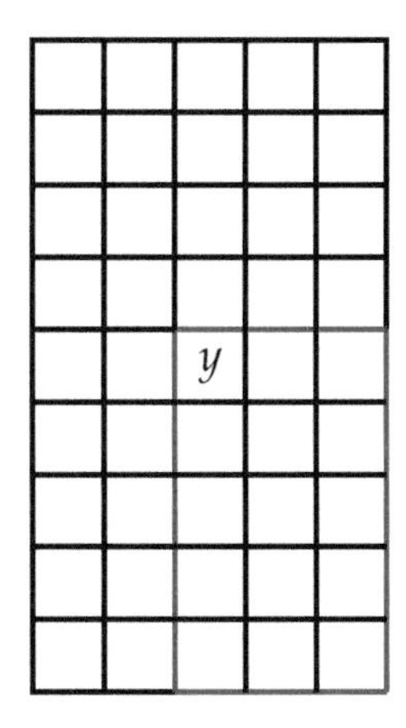

FIGURE 4.8: Recurrence relation for median-of-medians

$$\begin{aligned}
&= 7r + \theta + 2 \\
&= \frac{7(n - \theta - 5)}{10} + \theta + 2 \qquad \text{from (4.23)} \\
&= \frac{7n + 3\theta - 15}{10} \\
&\leq \frac{7n + 12}{10}, \qquad \text{since } \theta \leq 9.
\end{aligned}$$

Similarly, the number of elements $> y$ is at most $(7n + 12)/10$.

It follows that the worst-case complexity $T(n)$ of this algorithm satisfies the following recurrence:

$$T(n) \leq \begin{cases} T\left(\left\lfloor \frac{n}{5} \right\rfloor\right) + T\left(\left\lfloor \frac{7n+12}{10} \right\rfloor\right) + \Theta(n) & \text{if } n \geq 15 \\ \Theta(1) & \text{if } n \leq 14. \end{cases}$$

How do we prove that $T(n)$ is $O(n)$? The key fact is that

$$\frac{1}{5} + \frac{7}{10} = \frac{19}{20} < 1.$$

The analysis is similar to the recurrence relation considered in Example 4.2. It is possible to prove that $T(n) \in \Theta(n)$ formally using guess-and-check (induction) or informally using the recursion tree method.

If we use the recursion tree method and ignore the "+12" term, the structure of the recursion tree is as depicted in Figure 4.9. The sums of the levels of the recursion tree form a geometric sequence with ratio 19/20. Therefore the infinite geometric sequence has a finite sum (from (1.3)) and the complexity of the algorithm is linear:

$$\sum_{i=0}^{\infty} n \left(\frac{9}{10}\right)^i = 10n \in \Theta(n).$$

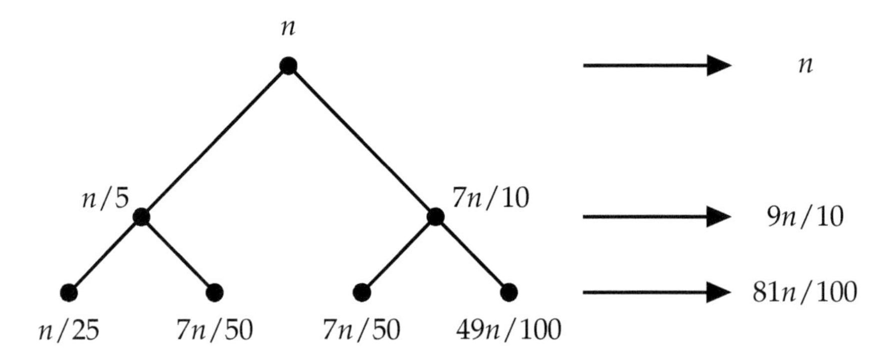

FIGURE 4.9: QUICKSELECT recursion tree

4.14 Notes and References

The Master Theorem was first described in Bentley, Haken and Saxe in [5], but the term "Master Theorem" was popularized by Cormen, Leiserson and Rivest in [15]. The general version of the Master Theorem (Theorem 4.4) is a simplification, due to Timothy Chan, of a theorem given in [15]. The second case is the same as in Theorem 4.3, but the first and third cases are more general in that $f(n)$ is no longer required to be a polynomial function of n. The condition in the third case has been replaced by a simpler condition than the one used in [15].

Non-dominated points are also called the *maxima of a point set*. Algorithms to find non-dominated points were found by Kung, Luccio and Preparata in [45], where algorithms having complexity $O(n \log n)$ are described for arbitrary dimensions $d \geq 2$. Problem 4.1 is just the two-dimensional version of the problem.

The $O(n \log n)$ divide-and-conquer algorithm to solve **Closest Pair** is first described in Bentley and Shamos [6].

The KARATSUBA-OFMAN algorithm is from [37]. At the time this book was written, the $O(n \log n)$ algorithm for **Multiprecision Multiplication** developed by Harvey and van der Hoeven had not been published; however, their paper describing the algorithm can be found in the HAL archives [32].

The STRASSEN algorithm for **Matrix Multiplication** was published in [56]. The algorithm due to Le Gall was published in [46].

QUICKSORT was first published by Hoare in [33]. QUICKSELECT was published at the same time in [34]. MEDIAN-OF-MEDIANS-QUICKSELECT is from Blum, Floyd, Pratt, Rivest and Tarjan [9].

Exercises

4.1 Consider the following pseudocode:

Algorithm 4.19: STRANGEDC(A, ℓ, r)

```
comment: A is an array of length n and 1 ≤ ℓ ≤ r ≤ n
if r − ℓ ≤ 5
  then return ()
m ← (r − ℓ)/2
STRANGEDC(A, ℓ, ℓ + m)
STRANGEDC(A, ℓ + m, r)
for i ← ℓ to r
  do { for j ← r to ℓ
       do output (A[j])
STRANGEDC(A, ℓ + (m/2), r − (m/2))
```

Analyze the time complexity of STRANGEDC as a function of $n = r - \ell$ by giving a recurrence $T(n)$. Then solve the recurrence with the method of your choice, obtaining a Θ-bound. For convenience, you can assume $r - \ell$ is a power of 2.

4.2 Find the exact solution to the following recurrence relation when n is a power of 2 by using the recursion-tree method.

$$T(n) = 2T\left(\frac{n}{2}\right) + n - 1 \quad \text{if } n > 1$$
$$T(1) = 3.$$

4.3 Consider the following recurrence:

$$T(n) = 9T\left(\left\lfloor n^{1/3} \right\rfloor\right) + \log^2 n \quad \text{if } n \geq 64$$
$$T(n) = 1 \quad \text{if } n < 64.$$

Use induction to prove the upper bound $T(n) \in O(\log^2 n \log\log n)$.

4.4 Consider the following recurrence:

$$T(n) = T\left(\left\lfloor \frac{n}{2} \right\rfloor\right) + T\left(\left\lfloor \frac{n}{4} \right\rfloor\right) + 1 \quad \text{if } n \geq 4$$
$$T(n) = 0 \quad \text{if } n < 4.$$

Use induction to prove the upper bound $T(n) \in O(n^{0.7})$.

HINT use $T(n) \leq cn^{0.7} - c'$ as your "guess," where c and c' are suitable constants that you will specify when you do the induction proof.

4.5 Use the general version of the Master Theorem (Theorem 4.4) to solve the following recurrence relation (you just need to determine the complexity as a function of n):

$$T(n) = 2T\left(\frac{n}{3}\right) + n^{0.6}\log_2 n \quad \text{if } n \geq 2$$
$$T(1) = 1.$$

4.6 In this question, we analyze the average number of nodes examined in a binary search of a "random" binary search tree on n nodes.

(a) Given a binary search tree T having n nodes, let $C(T)$ denote the average number of nodes that are examined in a binary search for a key in T. Show that

$$C(T) = \frac{1}{n}\sum_{N \in T}(\mathsf{depth}(N) + 1).$$

(b) Suppose that π is any permutation of $\{1, \dots, n\}$ and we create a binary tree by successively inserting $\pi[1], \dots, \pi[n]$, as described in Section 3.5. In this way we obtain $n!$ binary search trees having the n keys $1, \dots, n$. Define $C(n)$ to be the average value of $C(T(\pi))$ over the $n!$ trees $T(\pi)$ and define $D(n) = nC(n)$. Prove that $D(n)$ satisfies the following recurrence relation:

$$D(n) = \frac{1}{n}\sum_{i=1}^{n}(D(i-1) + D(n-i) + n) \qquad \text{if } n \geq 2$$
$$D(1) = 1.$$

(c) Solve the recurrence relation in the previous part of this question.

HINT Try to use techniques similar to those employed in Section 4.12.1.

4.7 Consider an array A of n integers that is sorted in increasing order. A contains the $n-1$ even integers $2, 4, \dots 2n-2$ along with one odd integer. For example, if $n = 6$, then A could look like

$$2, 4, 6, 7, 8, 10 \quad \text{or} \quad 2, 3, 4, 6, 8, 10$$

(various other configurations are possible). Describe and analyze a divide-and-conquer algorithm that will find the unique odd integer in A in $O(\log n)$ time.

4.8 In both parts of this question, you are required to use the STRASSEN algorithm for matrix multiplication as a building block. (The details of the STRASSEN algorithm are not important; the only relevant fact is that it computes the product of two $n \times n$ matrices in $\Theta(n^{\log_2 7})$ time.)

(a) Suppose that A is an $n \times kn$ matrix and B is a $kn \times n$ matrix. Describe how to compute the matrix product AB. Determine the complexity as a function of k and n.

(b) Suppose that A is an $n \times kn$ matrix and B is a $kn \times n$ matrix. Describe how to compute the matrix product BA. Determine the complexity as a function of k and n.

4.9 It can be proven that the Fibonacci numbers (see Section 6.1) satisfy the following recurrence relation:

$$\begin{aligned} f_0 &= 0 \\ f_1 &= 1 \\ f_{2n} &= f_n(2f_{n+1} - f_n) \quad \text{if } n \geq 1 \\ f_{2n+1} &= {f_{n+1}}^2 + {f_n}^2 \quad \text{if } n \geq 1. \end{aligned}$$

(a) Use this recurrence relation to design an efficient divide-and-conquer algorithm to compute a given Fibonacci number f_n. The idea is to compute pairs of consecutive Fibonacci numbers using the recurrence. It will be helpful to make use of the fact that $f_{k+1} = f_k + f_{k-1}$ for all $k \geq 1$.

(b) Analyze the complexity of your algorithm, under the (unrealistic) assumption that all addition, subtraction and multiplication operations take time $O(1)$.

4.10 Consider a card game involving two parties: yourself and a dealer. In this game there are n cards, denoted $C_1, \dots, C_n$, which are face down on a table. Each card C_i has a symbol S_i on the front (non-visible side) of the card. You cannot look at any of the cards, but you can ask the dealer if two specified cards have the same symbol (the dealer's response will be "*yes*" if they are the same, and "*no*" if they are different). Thus the dealer can be regarded as an oracle, say SAME, where

$$\text{SAME}(i,j) = \begin{cases} \textit{yes} & \text{if } S_i = S_j \\ \textit{no} & \text{if } S_i \neq S_j. \end{cases}$$

Your task is to determine if there are more than $n/2$ cards having the same symbol. Furthermore, if the answer is "*yes*," you should also output an index i of some card C_i whose symbol appears more than $n/2$ times (of course, i is not unique).

Design a divide-and-conquer algorithm to accomplish this task. Your algorithm should the call the oracle $O(n \log n)$ times.

4.11 Suppose two people rank a list of n items (say movies), denoted $M_1, \dots, M_n$. A *conflict* is a pair of movies $\{M_i, M_j\}$ such that $M_i > M_j$ in one ranking and $M_j > M_i$ in the other ranking. The number of conflicts between two rankings is a measure of how different the rankings are.

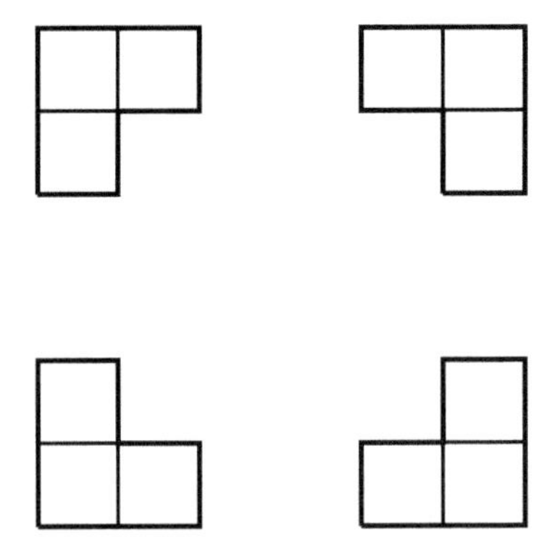

FIGURE 4.10: The four orientations of an L-tile

For example, consider the following two rankings:

$$M_1 > M_2 > M_3 > M_4 \quad \text{and} \quad M_2 > M_4 > M_1 > M_3.$$

The number of conflicts is three: $\{M_1, M_2\}$, $\{M_1, M_4\}$, and $\{M_3, M_4\}$ are the conflicting pairs.

Give a pseudocode description of an efficient divide-and-conquer algorithm to compute the number of conflicts between two rankings of n items. Briefly justify its correctness and analyze the complexity using a recurrence relation.

You can assume n is a power of two, for simplicity. To simplify the notation in the algorithm, you can assume that the first ranking is

$$M_1 > M_2 > \cdots > M_n.$$

HINT Think of a MERGESORT-like algorithm that solves two subproblems of size $n/2$. After solving the subproblems, you will also need to compute the number of conflicts between the two sublists during the MERGE step.

4.12 An *L-tile* consists of three square 1×1 cells that form a letter "L." There a four possible *orientations* of an L-tile, as shown in Figure 4.10.

The purpose of this question is to find a divide-and-conquer algorithm to tile an $n \times n$ square grid with $(n^2 - 1)/3$ L-tiles in such a way that only *one corner cell* is not covered by an L-tile. This can be done whenever $n \geq 2$ is a power of two.

Give a pseudocode description of a divide-and-conquer algorithm to solve this problem, briefly justify its correctness, and analyze the complexity using a recurrence relation. Remember that we are assuming n is a power of two.

HINT The basic idea is to split the $n \times n$ grid into four $n/2 \times n/2$ subgrids. This defines four subproblems that can be solved recursively. Then you have to combine the solutions to the four subproblems to solve the original problem instance.

4.13 The matrices $H_0, H_1, \ldots$ are defined as follows:

$$H_0 = (1),$$

and

$$H_k = \begin{pmatrix} H_{k-1} & H_{k-1} \\ H_{k-1} & -H_{k-1} \end{pmatrix}$$

for $k \geq 1$. Thus,

$$H_1 = \begin{pmatrix} 1 & 1 \\ 1 & -1 \end{pmatrix},$$

$$H_2 = \begin{pmatrix} 1 & 1 & 1 & 1 \\ 1 & -1 & 1 & -1 \\ 1 & 1 & -1 & -1 \\ 1 & -1 & -1 & 1 \end{pmatrix},$$

etc.

Given an integer vector $\mathbf{v}$ of length 2^k, we want to compute the vector-matrix product $\mathbf{v}H_k$. For example, if $k = 2$, then

$$(2, 3, 4, -1)H_2 = (8, 4, 2, -6).$$

(a) Design a divide-and-conquer algorithm to compute a matrix-vector product $\mathbf{v}H_k$. You can assume that all arithmetic operations take $\Theta(1)$ time. Your algorithm will take two inputs, namely $\mathbf{v}$ and k, where $\mathbf{v}$ has length 2^k. Note that you should *not* explicitly construct the matrix H_k.

(b) Write down a recurrence relation for the complexity of your algorithm and solve it using the Master Theorem. You should aim for a complexity of $\Theta(n \log n)$, where $n = 2^k$.

4.14 Suppose we are given $4n$ points in the Euclidean plane, say

$$(x_1, y_1), \ldots, (x_{4n}, y_{4n}).$$

You can assume that all the x_i's are distinct, and all the y_i's are distinct. Describe an algorithm to create four disjoint rectangular regions, each of which contains exactly n of the $4n$ given points. The sides of these rectangles should be parallel to the x- and y-axes. The complexity of the algorithm should be $O(n)$.

For each of the four rectangles, your output would be four numbers $(x_\ell, x_h, y_\ell, y_h)$ where each point in the rectangle, say (x, y), would satisfy $x_\ell \leq x \leq x_h$ and $y_\ell \leq y \leq y_h$.

Chapter 5

Greedy Algorithms

Greedy algorithms follow a particularly simple design strategy and can be implemented very efficiently. They are well suited as a method to solve certain optimization problems. Unfortunately, for many problems, the greedy approach does result in a correct algorithm. However, for certain problems, the greedy approach can be proven to yield a valid algorithm.

5.1 Optimization Problems

Many of the most important practical problems are optimization problems. We will study various methods for solving optimization problems in this and subsequent chapters. We begin by defining some terminology relating to optimization problems.

optimization problem

In an optimization problem, given a problem instance, we want to find a feasible solution that maximizes (or minimizes) a specified objective function. (The terms "feasible solution" and "objective function" are defined below.)

problem instance

As always, a problem instance is the ***input*** for the specified problem.

problem constraints

Constraints are necessary conditions (i.e., ***requirements***) that must be satisfied.

feasible solution

If all the specified constraints are satisfied, then we have a ***feasible solution***. For any problem instance I, $\mathsf{feasible}(I)$ denotes the set of all outputs (i.e., solutions) for the instance I that satisfy the given constraints (i.e., all the feasible solutions). Note that it may the case for certain problems and certain instances I that there are no feasible solutions.

objective function

An ***objective function*** assigns a non-negative real number to any feasible

DOI: 10.1201/9780429277412-5

solution, so we can regard it as a function

$$f : \mathsf{feasible}(I) \rightarrow \mathbb{R}^+ \cup \{0\}.$$

We often think of an objective function f as specifying a ***profit*** or a ***cost***. Thus, for a feasible solution $X \in \mathsf{feasible}(I)$, $f(X)$ denotes the cost or profit of X.

optimal solution

An ***optimal solution*** is a feasible solution $X \in \mathsf{feasible}(I)$ such that the profit $f(X)$ is maximized (or the cost $f(X)$ is minimized). Typically, we want to *maximize* a profit or *minimize* a cost.

5.2 Greedy Design Strategy

The greedy design strategy is often used to design greedy algorithms for optimization problems. The following items describe a typical setting in which a greedy algorithm can be considered.

partial solutions

Given a problem instance I, it should be possible to write a feasible solution X as a tuple $[x_1, x_2, \ldots, x_n]$ for some integer n. A tuple $[x_1, \ldots, x_i]$ where $i < n$ is a ***partial solution*** if no constraints are violated. Note that it is possible that a given partial solution cannot be extended to a feasible solution.

choice set

For a partial solution $X = [x_1, \ldots, x_i]$ where $i < n$, we define the ***choice set***

$$\mathsf{choice}(X) = \{y : [x_1, \ldots, x_i, y] \text{ is a partial solution}\}.$$

Thus, the choice set determines all possible ways to extend a partial solution (an i-tuple) to an $(i+1)$-tuple.

local evaluation criterion

A ***local evaluation criterion*** is a function g such that, for any partial solution $X = [x_1, \ldots, x_i]$ and any $y \in \mathsf{choice}(X)$, $g(x_1, \ldots, x_i, y)$ measures the cost or profit of extending the partial solution X to include y.

extension

Given a partial solution $X = [x_1, \ldots, x_i]$ where $i < n$, choose $y \in \mathsf{choice}(X)$ so that $g(y)$ is as small (or large) as possible. Update X to be the $(i+1)$-tuple $[x_1, \ldots, x_i, y]$.

greedy algorithm

Starting with the "empty" partial solution, repeatedly extend it until a feasible solution X is constructed. This feasible solution may or may not be optimal. Thus, when we extend a partial solution (an i-tuple) to an $(i+1)$-tuple, we choose the extension that yields the *maximum local improvement*. This may or may not turn out to be a good strategy, however.

Here are a few basic observations about greedy algorithms.

- Greedy algorithms do no ***looking ahead*** and do not incorporate any ***backtracking*** (however, we note that backtracking algorithms are discussed in detail in Chapter 8).
- Greedy algorithms can usually be implemented efficiently. Often they consist of a ***preprocessing step*** based on the function g, followed by a ***single pass*** through the data. For a greedy algorithm to be efficient, we need a fast way to find the "best" extension of any given partial solution X.
- In a greedy algorithm, only ***one feasible solution*** is constructed.
- The execution of a greedy algorithm is based on ***local criteria*** (i.e., the values of the function g).
- *Correctness*: For certain greedy algorithms, it is possible to prove that they always yield optimal solutions. However, these proofs can be tricky and complicated! (Recall that for any algorithm to be correct, it has to find the optimal solution for *every* problem instance.)

5.3 Interval Selection

We begin by considering a problem known as **Interval Selection** (it is also often called the **Activity Selection** problem); see Problem 5.1.

Here are three possible greedy strategies for **Interval Selection**. We will see that one of these three strategies gives rise to a correct greedy algorithm.

strategy 1: earliest starting time

Sort the intervals in increasing order of *starting times*. At any stage, choose the ***earliest starting*** interval that is disjoint from all previously chosen intervals (i.e., the local evaluation criterion is the value s_i).

strategy 2: minimum duration

Sort the intervals in increasing order of *duration*. At any stage, choose the interval of ***minimum duration*** that is disjoint from all previously chosen intervals (i.e., the local evaluation criterion is the value $f_i - s_i$).

Problem 5.1: Interval Selection

Instance: A set $\mathcal{A} = \{A_1, \ldots, A_n\}$ of ***intervals***. For $1 \leq i \leq n$, the interval A_i has the form $A_i = [s_i, f_i)$, where s_i is the ***start time*** of interval A_i and f_i is the ***finish time*** of A_i. The s_i's and f_i's are assumed to be positive integers, and we assume that $s_i < f_i$ for all i.
Note that an interval, by definition, includes its start time and excludes its finish time.

Feasible solution: A subset $\mathcal{B} \subseteq \mathcal{A}$ of ***pairwise disjoint intervals***.

Find: A feasible solution of maximum size (i.e., one that maximizes $|\mathcal{B}|$).

strategy 3: earliest finishing time

Sort the intervals in increasing order of *finishing times*. At any stage, choose the ***earliest finishing*** interval that is disjoint from all previously chosen intervals (i.e., the local evaluation criterion is the value f_i).

To show that a greedy algorithm is not correct, it suffices to provide a single counterexample. It is easy to see that the first two strategies are not always correct.

The following counterexample shows that strategy 1 does not always yield the optimal solution:

$$[0, 10), [1, 3), [5, 7).$$

Here we choose $[0, 10)$ and we're done. But the other two intervals comprise an optimal solution.

The following counterexample shows that strategy 2 does not always yield the optimal solution:

$$[0, 5), [6, 10), [4, 7).$$

Here we choose $[4, 7)$ and we're done. But the other two intervals comprise an optimal solution.

However, will show that strategy 3 (where we sort the intervals in increasing order of finishing times) always yields the optimal solution. For future reference, we present the greedy algorithm based on this strategy as Algorithm. 5.1.

The complexity of Algorithm 5.1 is easily seen to be

$$\Theta(n \log n) + \Theta(n) = \Theta(n \log n).$$

The $\Theta(n \log n)$ term accounts for sorting the intervals, while the $\Theta(n)$ term is the time to "process" the intervals one at a time.

First we give an induction proof that Algorithm 5.1 always finds the optimal solution. Let $\mathcal{B}$ be the *greedy solution*,

$$\mathcal{B} = (A_{i_1}, \ldots, A_{i_k}),$$

Algorithm 5.1: GREEDYINTERVALSELECTION($\mathcal{A}$)

```
comment: A is a set of n intervals
rename the intervals, by sorting if necessary, so that f_1 ≤ ··· ≤ f_n
B ← {A_1}
prev ← 1
comment: prev is the index of the last selected interval
for i ← 2 to n
  do { if s_i ≥ f_prev
       then { B ← B ∪ {A_i}
              prev ← i
return (B)
```

where $i_1 < \cdots < i_k$. Let $\mathcal{O}$ be any *optimal solution,*

$$\mathcal{O} = (A_{j_1}, \ldots, A_{j_\ell}),$$

where $j_1 < \cdots < j_\ell$. Observe that $\ell \geq k$ because $\mathcal{O}$ is optimal; we want to prove that $\ell = k$.

We will assume $\ell > k$ and obtain a contradiction. The correctness proof depends on the following lemma, which can be descriptively named *greedy stays ahead*. This phrase indicates that the greedy solution is in fact the "best possible" partial solution at every stage of the algorithm.

LEMMA 5.1 (Greedy stays ahead) *For all $m \geq 1$, it holds that*

$$f_{i_m} \leq f_{j_m}.$$

PROOF The initial case is $m = 1$. We have $f_{i_1} \leq f_{j_1}$ because the greedy algorithm begins by choosing $i_1 = 1$ (A_1 has the earliest finishing time).

Our induction assumption is that

$$f_{i_{m-1}} \leq f_{j_{m-1}}.$$

Now consider A_{i_m} and A_{j_m}. We have

$$s_{j_m} \geq f_{j_{m-1}} \geq f_{i_{m-1}}.$$

A_{i_m} has the earliest finishing time of any interval that starts after $f_{i_{m-1}}$ finishes. Therefore $f_{i_m} \leq f_{j_m}$. (See Figure 5.1 for an illustration.) ∎

Now we complete the correctness proof for Algorithm 5.1. From Lemma 5.1, we have that $f_{i_k} \leq f_{j_k}$. Suppose that $\ell > k$. $A_{j_{k+1}}$ starts after A_{j_k} finishes, and $f_{i_k} \leq f_{j_k}$. So $A_{j_{k+1}}$ is a feasible interval with respect to the greedy algorithm, and

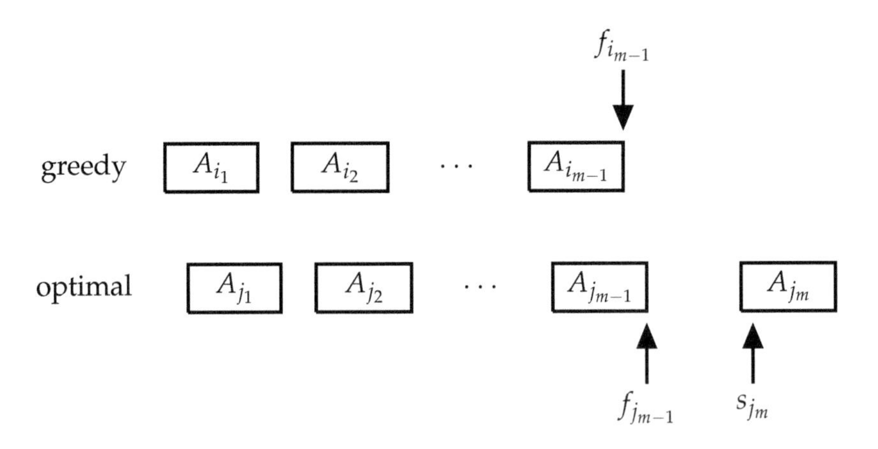

FIGURE 5.1: Greedy stays ahead.

therefore the greedy solution would not have terminated with A_{i_k}. This contradiction shows that $\ell = k$. See Figure 5.2 for an illustration.

Induction is a standard way to prove correctness of greedy algorithms; however, sometimes shorter "slick" proofs are possible. We present a slick proof of correctness of Algorithm 5.1. Recall that $\mathcal{B}$ is the greedy solution and $\mathcal{O}$ is the optimal solution. Let $F = \{f_{i_1}, \ldots, f_{i_k}\}$ be the finishing times of the intervals in $\mathcal{B}$. The key observation is that there is no interval in $\mathcal{O}$ that is "between" $f_{i_{m-1}}$ and f_{i_m} for any $m \geq 2$. This is easily seen because A_{i_m} would not be chosen by the greedy algorithm in this case. As well, there is no interval in $\mathcal{O}$ that finishes before f_{i_1}, or one that starts after f_{i_k}.

Therefore, every interval in $\mathcal{O}$ contains a point in F (or has a point in F as a finishing time). No two intervals in $\mathcal{O}$ contain the same point in F because the intervals are disjoint. Hence, there is an injective (i.e., one-to-one) mapping from $\mathcal{O}$ to F and therefore $|\mathcal{O}| \leq |F|$. Then we have

$$\ell = |\mathcal{O}| \leq |F| = |\mathcal{B}| = k.$$

But $\ell \geq k$, so $\ell = k$. Figure 5.3 illustrates the main idea of the proof.

5.4 Interval Coloring

The next problem we study is the **Interval Coloring** problem; see Problem 5.2.

As usual, a greedy algorithm will consider the intervals one at a time. At a given point in time, suppose we have colored the first $i < n$ intervals using d colors. We will color the $(i+1)$st interval with ***any permissible color***. If it cannot

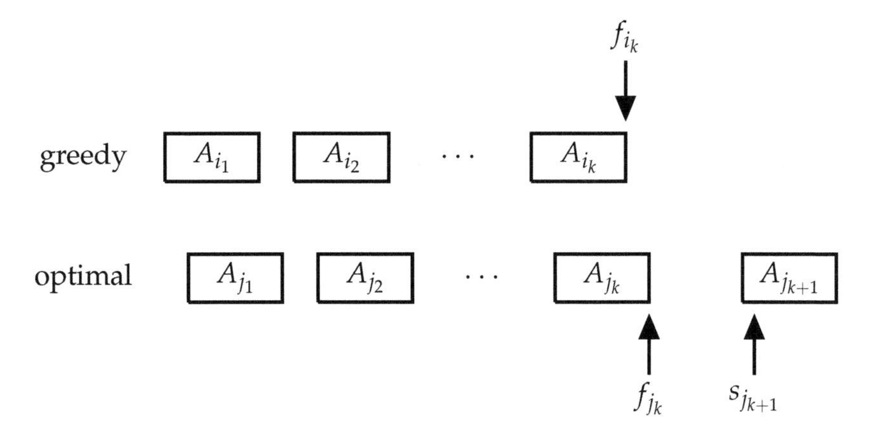

FIGURE 5.2: Completing the proof.

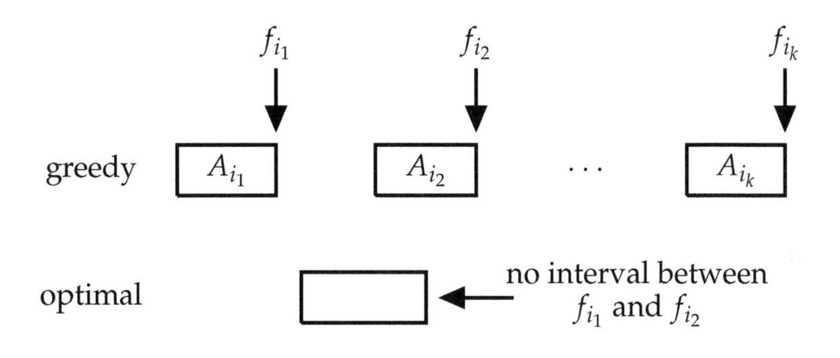

FIGURE 5.3: The slick proof.

be colored using any of the existing d colors, then we introduce a ***new color*** and d is increased by 1. The main question is to decide the order in which we should consider the intervals. The order matters, as can be seen in the next example.

Example 5.1 Suppose we have the following set of intervals.

$$A_1 = [0,3) \quad A_2 = [8,11) \quad A_3 = [14,20) \quad A_4 = [4,9)$$
$$A_5 = [16,20) \quad A_6 = [6,13) \quad A_7 = [10,15) \quad A_8 = [0,7)$$
$$A_9 = [12,20) \quad A_{10} = [0,5)$$

If we consider the intervals in the given order, $A_1, A_2, \ldots,$ then they will receive colors

$$1,1,1,2,2,3,2,4,4,3$$

Problem 5.2: Interval Coloring

Instance: A set $\mathcal{A} = \{A_1, \dots, A_n\}$ of ***intervals***. For $1 \leq i \leq n$, $A_i = [s_i, f_i)$, where s_i is the ***start time*** of interval A_i and f_i is the ***finish time*** of A_i.

Feasible solution: A ***c-coloring*** is a mapping $col : \mathcal{A} \rightarrow \{1, \dots, c\}$ that assigns each interval a ***color*** such that two intervals receiving the same color are always disjoint. Equivalently, overlapping intervals must receive different colors.

Find: A c-coloring of $\mathcal{A}$ with the minimum number of colors. Note that we can always assign a different color to every interval, so $c \leq n$.

respectively. But this does not yield an optimal solution—it is possible to color the given set of intervals with three colors.

The correct strategy is to consider the intervals in increasing order of *start times* (ties are broken arbitrarily). In the above example, we could consider the intervals in the order

$$A_1, A_8, A_{10}, A_4, A_6, A_2, A_7, A_9, A_3, A_5.$$

When we greedily color the intervals in this order, we see that three colors suffice. The intervals receive colors

$$1, 2, 3, 1, 3, 2, 1, 2, 3, 1$$

respectively. □

Algorithm 5.2 implements the above-described greedy strategy. Note that, at any point in time, *finish*$[c]$ denotes the finishing time of the *last interval* that has received color c. Therefore, a new interval A_i can be assigned color c if $s_i \geq$ *finish*$[c]$. If *flag* is false after the **while** loop, it means that no existing color can be used. In this case, we have to introduce a new color.

Excluding the sort, the complexity of the algorithm is $O(nD)$, where D is the value of d returned by the algorithm. We don't know the exact value of D ahead of time; all we know is that $1 \leq D \leq n$. If it turns out that $D \in \Omega(n)$, then the best we can say is that the complexity is $O(n^2)$.

For each interval, suppose we search the d existing colors to determine if one of them is suitable. This is a *linear search*. However, perhaps we can use a suitable data structure that would allow a more efficient algorithm to be designed. A possible modification is to use the color of the interval having the *earliest finishing time* among the most recently chosen intervals of each color. We can use a *priority queue* to keep track of these finishing times (priority queues and the complexities of operations that are performed on them were discussed in Section 3.4).

Whenever we color interval A_i with color c, we INSERT(f_i, c) into the priority queue (here f_i is the "key"). When we want to color the next interval A_i, we look at the minimum key f in the priority queue. If $f \leq s_i$, then we do a DELETEMIN

Algorithm 5.2: GREEDYINTERVALCOLORING($\mathcal{A}$)

```
comment: A is a set of n intervals
sort the intervals so that s_1 ≤ ··· ≤ s_n
d ← 1
color[1] ← 1
finish[1] ← f_1
for i ← 2 to n
  do { flag ← false
       c ← 1
       while c ≤ d and not flag
         do { if finish[c] ≤ s_i
                then { color[i] ← c
                       finish[c] ← f_i
                       flag ← true
                else c ← c + 1
       if not flag
         then { d ← d + 1
                color[i] ← d
                finish[d] ← f_i
return (d, color)
```

operation, yielding the pair (f, c) and we use color c for interval A_i. If $f > s_i$, we introduce a new color. Note that each interval is inserted once and deleted once from the priority queue. The priority queue containing D items can be implemented using a heap, in which case all the operations have complexity $O(\log D)$. Therefore, the complexity of this approach is $O(n \log D)$. Since $D \leq n$, it is $O(n \log n)$. (The initial sort is also $O(n \log n)$.)

The correctness of Algorithm 5.2 can be proven inductively as well as by a "slick" method. Here, we give the "slick" proof. In the following discussion, the reader can refer to Figure 5.4, which illustrates our claims.

Let D denote the number of colors used by the algorithm. Suppose $A_\ell = [s_\ell, f_\ell)$ is the *first* interval to receive the *last* color, D. For every color $c < D$, there is an interval $A_c = [s_c, f_c)$ such that $s_c \leq s_\ell < f_c$. That is, for every color $c < D$, the interval A_c overlaps the interval A_ℓ (if A_c did not overlap A_ℓ, then A_ℓ could have received color c.) Therefore we have D intervals, all of which contain the point s_ℓ. Clearly, these D intervals must all receive different colors, so there is no coloring with fewer than D colors.

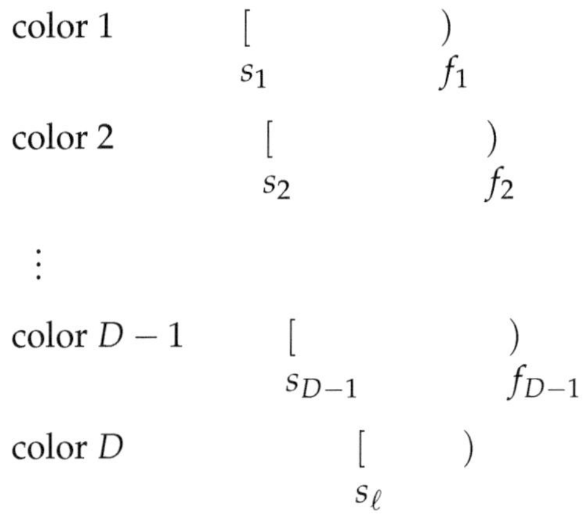

color 1 $[$ s_1 $)$ f_1

color 2 $[$ s_2 $)$ f_2

$\vdots$

color $D-1$ $[$ s_{D-1} $)$ f_{D-1}

color D $[$ s_ℓ $)$

FIGURE 5.4: Greedy interval coloring.

5.5 Wireless Access Points

Problem 5.3: Wireless Access Points

Instance: A list

$$X = [x_1, x_2, \ldots, x_n]$$

of distinct integers, and a positive integer D. The values $x_1, x_2, \ldots, x_n$ represent the locations of n stores on a straight road (x_i is the distance of the ith store from a specified origin).

Feasible solution: A set of *wireless access points*, located along this road, such that every store has distance $\leq D$ from the nearest wireless access point. That is, a feasible solution is a list

$$Y = [y_1, \ldots y_k],$$

where the y_j's are integers, such that, for every $i \in \{1, \ldots, n\}$, there exists a $j \in \{1, \ldots, k\}$ such that $|y_j - x_i| \leq D$.

Find: The feasible solution Y having the smallest possible value of k.

Problem 5.3 asks us to find an optimal way to locate various wireless access points. We design a greedy algorithm to solve this problem. First, we pre-process the data by sorting the x_i's so they are in increasing order. Then we successively choose wireless access points $y_1, \ldots, y_k$ in increasing order, in a greedy fashion. Each y_i is assigned the *largest* value such that its distance from the first "uncovered" store is no more than D. More precisely, we end up with Algorithm 5.3.

Algorithm 5.3: WIRELESSACCESSPOINTS(X)

comment: $X = [x_1, x_2, \ldots, x_n]$ is a list of n distinct integers
sort the x_i's so that $x_1 < \cdots < x_n$
$y_0 \leftarrow -\infty$
$m \leftarrow 0$
for $i \leftarrow 1$ **to** n
do { **if** $x_i > y_m + D$
  then { $y_{m+1} \leftarrow x_i + D$
    $m \leftarrow m + 1$ }}
return $(Y = [y_1, \ldots, y_m])$

Example 5.2 Suppose $D = 3$ and suppose the x_i's, in increasing order, are

$$-10, -8, -5, -2, -1, 1, 2, 4, 5, 8, 11, 12, 14, 16, 19.$$

We first put a wireless access point at $y_1 = -10 + 3 = -7$. The next uncovered store is at -2, so we put the next wireless access point at 1. Continuing in this fashion, we get the solution $Y = [-7, 1, 8, 15, 22]$. ▯

To establish the correctness of Algorithm 5.3, we first prove a useful "greedy stays ahead" lemma.

LEMMA 5.2 (Greedy Stays Ahead) *Suppose that $x_1 < \cdots < x_n$. Let*

$$\mathcal{O} = [y_1^*, \ldots, y_\ell^*]$$

be an optimal solution, where $y_1^* < \cdots < y_\ell^*$, and let

$$\mathcal{G} = [y_1, \ldots, y_m]$$

be the greedy solution, where $y_1 < \cdots < y_m$. Then $y_i^* \leq y_i$ for $1 \leq i \leq \ell$.

PROOF Note that $\ell \leq m$ because $\mathcal{O}$ is optimal. Therefore $y_1, \ldots, y_\ell$ are all defined.

We will prove that $y_i^* \leq y_i$ for $1 \leq i \leq \ell$ by induction on j. For $j = 1$ (the base case), because the distance from x_1 to y_1^* is at most D, we must have

$$y_1^* \leq x_1 + D = y_1.$$

(Note that the greedy algorithm assigns $y_1 = x_1 + D$.)

Now, as an induction assumption, suppose $y_j^* \leq y_j$ for some j with $1 \leq j < \ell$. We want to prove that $y_{j+1}^* \leq y_{j+1}$. Observe that

$$y_{j+1} = x_i + D$$

for some i, where $x_i > y_j + D$ (this is how the greedy algorithm assigns the y-values). We have $y_j^* \leq y_j$ by induction, so

$$x_i > y_j^* + D.$$

Therefore we must have

$$y_{j+1}^* \leq x_i + D = y_{j+1},$$

as desired. By induction on j, the proof is complete. ∎

Now, using Lemma 5.2, we can prove that the greedy solution is optimal.

THEOREM 5.3 *Suppose that $x_1 < \cdots < x_n$. Let*

$$\mathcal{O} = [y_1^*, \ldots, y_\ell^*]$$

be an optimal solution, where $y_1^* < \cdots < y_\ell^*$, and let

$$\mathcal{G} = [y_1, \ldots, y_m]$$

be the greedy solution, where $y_1 < \cdots < y_m$. Then $\ell = m$.

PROOF Note that $\ell \leq m$ because $\mathcal{O}$ is optimal. We will assume that $\ell < m$ and derive a contradiction.

The greedy algorithm chooses a wireless access point $y_{\ell+1}$, which means that

$$y_{j+1} = x_i + D$$

for some i, where

$$x_i > y_\ell + D.$$

By Lemma 5.2, we have

$$y_\ell^* \leq y_\ell.$$

Therefore,

$$x_i > y_\ell^* + D.$$

This means that the distance from x_i to any y_j is greater than D, so $\mathcal{O}$ is not a feasible solution. This contradiction proves that $\ell = m$. ∎

There is also a slick correctness proof for Algorithm 5.3, which is outlined in the Exercises.

5.6 A House Construction Problem

We consider a certain scheduling problem in this section, namely, Problem 5.4. We note that this problem is not an optimization problem. Rather, we just want to determine if it is possible to design a feasible schedule satisfying certain constraints.

Problem 5.4: House Construction

Instance: Two lists of n positive integers,

$$D = [d_1, \ldots, d_n] \quad \text{and} \quad L = [\ell_1, \ldots, \ell_n].$$

These lists are interpreted as follows: A construction company has n customers, each of whom wants a house to be built. Each house H_i has a *deadline* d_i by which it is supposed to be completed. The *construction time* of house H_i is the length of time required to build H_i; this is denoted by ℓ_i.

Feasible solution: The builder can only construct one house at a time and any house must be completed before the next one is started. Is it possible to build all n houses, starting at month 0, so each house is completed by its deadline? If so, the set of houses $\mathcal{H} = \{H_1, \ldots, H_n\}$ is *feasible*.

Example 5.3 Suppose that there are $n = 3$ houses, H_1, H_2, H_3, with deadlines $D = [4, 6, 1]$ and construction times $L = [2, 3, 1]$. This is a feasible set, as can be seen by building the houses in the order H_3, H_1, H_2. Thus, H_3 is started at time 0 and it is completed at time 1; H_1 is started at time 1 and it is completed at time 3; and H_2 is started at time 3 and it is completed at time 6. □

We want to devise a correct greedy algorithm to solve Problem 5.4. This involves examining one particular schedule or building the n houses. We will prove that $\mathcal{H}$ is feasible if and only if no constraints are violated when the houses in $\mathcal{H}$ are scheduled in *increasing order of deadline*. Here is an informal description of the algorithm.

step 1

Suppose $\mathcal{H} = \{H_1, \ldots, H_n\}$. Sort the houses in increasing order by deadline and rename them $H_1, \ldots, H_n$ with corresponding deadlines $d_1 \leq \cdots \leq d_n$ and construction times $\ell_1, \ldots, \ell_n$.

step 2

Now schedule the houses in the order $H_1, \ldots, H_n$ and check to see if this specific schedule $\mathcal{S}$ is feasible. Return the result "*true*" or "*false*" based on the feasibility of the schedule $\mathcal{S}$. This amounts to checking that $\ell_1 \leq d_1$, $\ell_1 + \ell_2 \leq d_2$, etc. These conditions can be written as

$$\sum_{i=1}^{j} \ell_i \leq d_j,$$

for $j = 1, \ldots, n$.

The complexity of step 1 is $\Theta(n \log n)$ and the complexity of step 2 is $\Theta(n)$.

To prove the correctness of this algorithm, we show that if there exists *any* feasible schedule $\mathcal{T}$ for $H_1, \ldots, H_n$, then the *specific* schedule $\mathcal{S}$ (described above) is feasible for $H_1, \ldots, H_n$. Suppose that $\mathcal{T}$ schedules H_j immediately before $H_{j'}$, for some pair of indices j, j' such that $j > j'$. (If no such pair (j, j') exists, then $\mathcal{T} = \mathcal{S}$ and we're done.) Then $d_j \geq d_{j'}$.

Now create a new schedule $\mathcal{T}'$ by interchanging H_j and $H_{j'}$ in $\mathcal{T}$. We claim that $\mathcal{T}'$ is feasible if and only if $\mathcal{T}$ is feasible. Suppose that house $H_{j'}$ finishes at time t in $\mathcal{T}$. Then house H_j finishes at time $t - \ell_{j'}$ in $\mathcal{T}$. Since $\mathcal{T}$ is feasible, we have

$$t \leq d_{j'} \quad \text{and} \quad t - \ell_{j'} \leq d_j.$$

Now, in the modified schedule $\mathcal{T}'$, house H_j finishes at time t and house $H_{j'}$ finishes at time $t - \ell_j$. We just need to verify that

$$t \leq d_j \quad \text{and} \quad t - \ell_j \leq d_{j'}.$$

Since $t \leq d_{j'}$ and $d_{j'} \leq d_j$, we have $t \leq d_j$. Also, because $t \leq d_{j'}$, we have $t - \ell_j \leq d_{j'}$.

By a succession of transformations of this type, we can transform $\mathcal{T}$ to $\mathcal{S}$, maintaining feasibility at each step.

This type of proof is sometimes called an ***exchange argument***. The general idea of an exchange argument is to transform the optimal solution into the greedy solution, maintaining optimality at each step of the process.

REMARK We could also consider finding pairs of houses that are out of order but not adjacent in the schedule $\mathcal{T}$. The process of modifying the schedule $\mathcal{T}$ while maintaining feasibility is a bit more complicated in this case, however. ▮

5.7 Knapsack

We begin by defining two versions of the **Knapsack** problem in Problem 5.5. In both versions of the problem, we have a set of objects, each of which has a given profit and weight. We also have a ***knapsack*** with a specified capacity. We desire to place objects into the knapsack, without exceeding the capacity, so as to maximize the profit attained. In the "rational" version of the problem, we are allowed to place fractions of an object into the knapsack. In the "0-1" version, each object is indivisible, so we either have to include it in the knapsack or omit it.

The **0-1 Knapsack** problem is **NP-hard**, so it almost surely does not have a polynomial-time algorithm to solve it (see Chapter 9). However, we can solve the **Rational Knapsack** problem in polynomial time by means of a suitable greedy algorithm.

We will examine three greedy strategies for knapsack problems. It turns out that one of these strategies yields a *correct* greedy algorithm for the **Rational Knapsack** problem.

Problem 5.5: Knapsack

Instance: An instance consists of

profits	$\mathcal{P} = [p_1, \ldots, p_n]$
weights	$\mathcal{W} = [w_1, \ldots, w_n]$ and
capacity	M.

These values are all positive integers.

Feasible solution: An n-tuple $X = [x_1, \ldots, x_n]$ where

$$\sum_{i=1}^{n} w_i x_i \leq M.$$

In the **0-1 Knapsack** problem (often denoted just as **Knapsack**), we require that $x_i \in \{0, 1\}, 1 \leq i \leq n$.

In the **Rational Knapsack** problem, we require that $x_i \in \mathbb{Q}$ and $0 \leq x_i \leq 1$, $1 \leq i \leq n$.

Find: A feasible solution X that maximizes

$$\sum_{i=1}^{n} p_i x_i.$$

strategy 1

Consider the items in decreasing order of ***profit*** (i.e., the local evaluation criterion is p_i).

strategy 2

Consider the items in increasing order of ***weight*** (i.e., the local evaluation criterion is w_i).

strategy 3

Consider the items in decreasing order of ***profit divided by weight*** (i.e., the local evaluation criterion is p_i/w_i).

Example 5.4 Consider the following instance of the **Rational Knapsack** problem:

object	1	2	3
profits	50	90	20
weights	50	60	10
capacity	100		

Suppose we consider the objects in order of *decreasing profit*, i.e., in the order $1, 2, 3$. Then we obtain the feasible solution $(4/5, 1, 0)$, yielding profit 130.

Suppose we consider the objects in order of *increasing weight*, i.e., in the order $3, 1, 2$. Then we obtain the feasible solution $(1, 2/3, 1)$, yielding profit 130.

Suppose we consider the objects in order of *decreasing profit / weight ratio*. These ratios are 1 for object 1, 1.5 for object 2, and 2 for object 3. So the order here is $3, 2, 1$. We obtain the feasible solution $(3/5, 1, 1)$, yielding profit 140.

Therefore, at least for this instance, the third strategy is better than the other two strategies. We can therefore conclude that the first two strategies are not optimal. □

Algorithm 5.4 is a straightforward greedy algorithm for the **Rational Knapsack** problem, based on considering the items in decreasing order of profit / weight. The variable *CurW* denotes the *current weight* of the items in the knapsack. Basically, we continue to add objects to the knapsack as long as $CurW < M$. When the knapsack is "almost full," we might perhaps include a fraction of the next item to fill the knapsack completely. Thus, it is not hard to see that the greedy solution always has the form

$$(1, 1, \ldots, 1, 0, \ldots, 0)$$

or

$$(1, 1, \ldots, 1, x_i, 0, \ldots, 0),$$

where $0 < x_i < 1$.

The complexity of Algorithm 5.4 is

$$O(n \log n) + O(n) = O(n \log n).$$

The "$n \log n$" term is the SORT and the "n" term is the iterative part of the algorithm. For convenience, we are assuming the unit cost model here (so arithmetic operations are assumed to take $O(1)$ time).

THEOREM 5.4 *Algorithm 5.4 always finds the optimal solution to the* ***Rational Knapsack*** *problem.*

PROOF For simplicity, let's assume that the profit / weight ratios are all distinct, so

$$\frac{p_1}{w_1} > \frac{p_2}{w_2} > \cdots > \frac{p_n}{w_n}.$$

Suppose we denote the greedy solution by

$$X = (x_1, \ldots, x_n)$$

and the optimal solution by

$$Y = (y_1, \ldots, y_n).$$

We will prove that $X = Y$, i.e., $x_j = y_j$ for $j = 1, \ldots, n$. Thus, there is a unique optimal solution and it is equal to the greedy solution.

Algorithm 5.4: GREEDYRATIONALKNAPSACK($\mathcal{P}, \mathcal{W}, M$)

```
comment: P and W are lists of length n
SORT the items so that p_1/w_1 ≥ ··· ≥ p_n/w_n
X ← [0,...,0]
i ← 1
CurW ← 0
while (CurW < M) and (i ≤ n)
  do { if CurW + w_i ≤ M
         then { x_i ← 1
                CurW ← CurW + w_i
                i ← i + 1
         else { x_i ← (M − CurW)/w_i
                CurW := M
return (X)
```

The proof is by contradiction. So, we begin by assuming that $X \neq Y$. Let j be the smallest index such that $x_j \neq y_j$.

It is impossible that $x_j < y_j$, given that $x_i = y_i$ for $1 \leq i < j$. This is because Algorithm 5.4 always chooses the maximum possible value for each x_i in turn. Therefore, we have $x_j > y_j$.

Now, there must exist an index $k > j$ such that $y_k > 0$ (otherwise Y would not be optimal). Let

$$\delta = \min\{w_k y_k, w_j(x_j - y_j)\}. \tag{5.1}$$

We observe that $\delta > 0$, because $x_j > y_j$. Define

$$y_j' = y_j + \frac{\delta}{w_j} \quad \text{and} \quad y_k' = y_k - \frac{\delta}{w_k}.$$

Then let Y' be obtained from Y by replacing y_j and y_k with y_j' and y_k', respectively.

The relation between X and Y is depicted in Figure 5.5. Note that y_j and y_k are the two coordinates in Y that are modified.

To complete the proof, we will show that

1. Y' is feasible, and

2. $\mathsf{profit}(Y') > \mathsf{profit}(Y)$.

This contradicts the optimality of Y and therefore it proves that $X = Y$.

To show Y' is feasible, we must prove that $y_k' \geq 0$, $y_j' \leq 1$ and $\mathsf{weight}(Y') \leq M$.

X		Y		
x_1	$=$	y_1		
x_2	$=$	y_2		
$\vdots$		$\vdots$		
x_{j-1}	$=$	y_{j-1}		
x_j	$>$	y_j		
$\vdots$		$\vdots$		
		y_k	$>$	0

FIGURE 5.5: The relation between X and Y.

First, we have

$$\begin{aligned} y'_k &= y_k - \frac{\delta}{w_k} \\ &\geq y_k - \frac{w_k y_k}{w_k} \qquad \text{from (5.1)} \\ &= y_k - y_k \\ &= 0. \end{aligned}$$

Second,

$$\begin{aligned} y'_j &= y_j + \frac{\delta}{w_j} \\ &\leq y_j + \frac{w_j(x_j - y_j)}{w_j} \qquad \text{from (5.1)} \\ &= y_j + x_j - y_j \\ &= x_j \\ &\leq 1. \end{aligned}$$

Third,

$$\begin{aligned} \mathsf{weight}(Y') &= \mathsf{weight}(Y) + \frac{\delta}{w_j} w_j - \frac{\delta}{w_k} w_k \\ &= \mathsf{weight}(Y) \\ &\leq M. \end{aligned}$$

Finally, we compute

$$\begin{aligned} \mathsf{profit}(Y') &= \mathsf{profit}(Y) + \frac{\delta p_j}{w_j} - \frac{\delta p_k}{w_k} \\ &= \mathsf{profit}(Y) + \delta \left(\frac{p_j}{w_j} - \frac{p_k}{w_k} \right) \\ &> \mathsf{profit}(Y), \end{aligned}$$

since $\delta > 0$ and $p_j/w_j > p_k/w_k$.

∎

We note that the above proof is an exchange argument (see Section 5.6). Also, in the proof of Algorithm 5.4, we were able to show directly that any solution that differed from the greedy solution could not be optimal, so the optimal solution is in fact unique. If we considered instances where profit / weight ratios were not necessarily distinct, then a slightly more complicated exchange argument would have been required. In this case, there could be more than one optimal solution.

5.8 Coin Changing

We define a coin changing problem in Problem 5.6. In this problem, a_i denotes the number of coins of denomination d_i that are used, for $i = 1, \dots, n$. The total value of all the chosen coins must be exactly equal to T. Subject to this constraint, we want to *minimize* the number of coins used, which is denoted by N.

Problem 5.6: Coin Changing

Instance: A list of ***coin denominations***, $[d_1, d_2, \dots, d_n]$, and a positive integer T, which is called the ***target sum***.

Feasible solution: An n-tuple of non-negative integers, say $A = [a_1, \dots, a_n]$, such that

$$T = \sum_{i=1}^{n} a_i d_i.$$

Find: A feasible solution such that

$$N = \sum_{i=1}^{n} a_i$$

is minimized.

We will generally assume that there is a coin with value 1 (i.e., a ***penny***, even though the penny is obsolete in Canada). This assumption allows us to make exact change for any positive integer T.

A greedy algorithm would look at the coins in *decreasing* order of denomination, always choosing the maximum number of coins of each denomination in turn. See Algorithm 5.5.

For some specific fixed sets of coins, the greedy algorithm always gives the optimal solution for any target value T. We provide some examples.

Algorithm 5.5: GREEDYCOINCHANGING(D, T)

```
comment: D = [d_1, ..., d_n] is the list of coin values
sort the coins so that d_1 > ··· > d_n
N ← 0
for i ← 1 to n
       ⎧ a_i ← ⌊T / d_i⌋
   do  ⎨ T ← T − a_i d_i
       ⎩ N ← N + a_i
if T > 0
  then return (fail)
  else return ([a_1, ..., a_n], N)
```

Example 5.5 If $d_j \mid d_{j-1}$ for $2 \leq j \leq n$, then the greedy solution is always optimal. (See the Exercises.) □

Example 5.6 If $D = [200, 100, 50, 25, 10, 5, 1]$ (this is the Canadian coin system, including now-obsolete pennies), then the greedy solution is always optimal. □

Example 5.7 Suppose $D = [30, 24, 12, 6, 3, 1]$ (this is an old English coin system). Then the greedy solution is not always optimal. For example, suppose $T = 48$. The greedy algorithm yields

$$A = [1, 0, 1, 1, 0, 0]$$

(using three coins), whereas the optimal solution is

$$A = [0, 2, 0, 0, 0, 0]$$

(using two coins). □

Example 5.8 Suppose $D = [8, 3]$ and $T = 28$. The greedy algorithm would compute $a_1 = 3$, $a_2 = 1$, which is not a feasible solution since

$$3 \times 8 + 3 = 27.$$

However, there is a (non-greedy) feasible solution: $a_1 = 2$, $a_2 = 4$. □

We prove optimality of the greedy algorithm for a particular set of coins.

THEOREM 5.5 *Algorithm 5.5 always finds the optimal solution to Problem 5.6 if* $D = [100, 25, 10, 5, 1]$.

PROOF In this proof, we refer to coins as follows: a one-cent coin is a *penny*, a five-cent coin is a *nickel*, a ten-cent coin is a *dime*, a 25-cent coin is a *quarter* and a 100-cent coin is a *loonie*. We will make use of the following properties of any optimal solution:

property 1

There are at most four pennies in an optimal solution (if there are at least five pennies, then we can replace five pennies by a nickel and reduce the number of coins in the solution).

property 2

There is at most one nickel in an optimal solution (if there are at least two nickels, then we can replace two nickels by a dime and reduce the number of coins in the solution).

property 3

There are at most three quarters in an optimal solution (if there are at least four quarters, then we can replace four quarters by a loonie and reduce the number of coins in the solution).

property 4

In any optimal solution, the number of nickels + the number of dimes is at most two (three dimes can be replaced by a quarter and a nickel; two dimes and a nickel can be replaced by a quarter; and, from above, the number of nickels is at most one).

The proof of correctness is by induction on T. As (trivial) base cases, we can take $T = 1, 2, 3, 4$. The remainder of the proof is divided into various cases.

case 1

Suppose $5 \leq T < 10$. First, assume there is no nickel in the optimal solution. Then the optimal solution consists only of pennies, so $T \leq 4$ from property 1; this is a contradiction. Therefore the optimal solution contains at least one nickel. Clearly the greedy solution contains at least one nickel. By induction, the greedy solution for $T - 5$ is optimal. Therefore the greedy solution for T is also optimal.

case 2

Suppose $10 \leq T < 25$. First, assume there is no dime in the optimal solution. Then the optimal solution contains only nickels and pennies, so $T \leq 5 + 4 = 9$, from properties 1 and 2; this is a contradiction. Therefore the optimal solution contains at least one dime. Clearly the greedy solution contains at least one dime. By induction, the greedy solution for $T - 10$ is optimal. Therefore the greedy solution for T is also optimal.

case 3

Suppose $25 \leq T < 100$. First, assume there is no quarter in the optimal solution. Then the optimal solution contains only dimes, nickels and pennies, so $T \leq 2 \times 10 + 4 = 24$, from properties 1 and 4; this is a contradiction. Therefore the optimal solution contains at least one quarter. Clearly the greedy solution contains at least one quarter. By induction, the greedy solution for $T - 25$ is optimal. Therefore the greedy solution for T is also optimal.

case 4

Suppose $100 \leq T$. First, assume there is no loonie in the optimal solution. Then the optimal solution contains only quarters, dimes, nickels and pennies, so $T \leq 75 + 2 \times 10 + 4 = 99$, from properties 1, 3 and 4; this is a contradiction. Therefore the optimal solution contains at least one loonie. Clearly the greedy solution contains at least one loonie. By induction, the greedy solution for $T - 100$ is optimal. Therefore the greedy solution for T is also optimal.

∎

5.9 Multiprocessor Scheduling

In this section, we investigate a scheduling problem, where we have n tasks, of possibly varying lengths, that need to be executed on m identical processors. The problem is defined formally as Problem 5.7.

Problem 5.7: Multiprocessor Scheduling

Instance: A list of n positive integers,

$$T = [t_1, \ldots, t_n]$$

and a positive integer m. For $1 \leq i \leq n$, t_i is the length of the ith ***task***, and m is the number of ***processors***.

Feasible solution: A schedule for the n tasks, such that each processor executes only one task at a time.

Find: A feasible solution with the minimum ***finishing time***, where the finishing time is the time at which the last task finishes executing.

Evidently, there is no need for any processor to have any "slack time" during which it is not executing a task. Also, the tasks that are executed on any given processor can be executed in any order. Thus, we can define a ***schedule*** to be any assignment of the n tasks to the m processors. We denote the processor that will execute task i by $\rho[i]$, for $1 \leq i \leq n$. The values $\rho[i]$ are integers such that $1 \leq \rho[i] \leq m$ for all i. Given a schedule $\rho = [\rho[1], \ldots, \rho[n]]$, the ***finishing time*** of processor j is

$$F[j] = \sum_{\{i:\rho[i]=j\}} t_i,$$

for $1 \leq j \leq m$. The ***finishing time*** of the schedule ρ is

$$F_{max} = \max\{F[j] : 1 \leq j \leq m\}.$$

This is the quantity that we wish to minimize.

The **Multiprocessor Scheduling** problem is in fact **NP-hard**, so we would not expect to find a polynomial-time algorithm to solve it (see Chapter 9). However, it is not difficult to design a greedy algorithm for this problem that performs quite well. More precisely, a fairly obvious greedy algorithm always finds a feasible solution whose finishing time is at most

$$\frac{4\,F^*_{max}}{3},$$

where F^*_{max} is the finishing time of the optimal schedule.

Let's first devise a greedy algorithm for **Multiprocessor Scheduling**. As usual, we will consider the tasks in some particular order, and then schedule them one at a time. If we keep track of the current finishing time for each processor, then it seems reasonable to assign each task to the first available processor.

So the question is what the best order is to consider the tasks. Intuitively, it seems that it would be wise to examine the tasks in decreasing order of their lengths. Then, when we come to the "last" task, it will not affect any current finishing time too much. This is a common strategy in various scheduling algorithms and it is often referred to as ***LPT***, since it schedules the tasks with the ***longest processing time*** first. So we will analyze the ***LPT greedy algorithm***. See Algorithm 5.6 for a pseudocode description.

Algorithm 5.6 also keeps track of the starting and finishing times of each task, in the arrays ST and FIN. We perhaps do not require these values, but they are useful in the analysis of the algorithm that we will be carrying out.

We are most interested in how close Algorithm 5.6 comes to finding an optimal solution, so we do not worry too much about its complexity. However, we note that Algorithm 5.6 would be more efficient if we made use of a priority queue to store the "current" finishing times as the tasks are scheduled. This approach would allow us to schedule each task without searching through all m values $F[j]$, $1 \leq j \leq m$. This is similar to the use of a priority queue in Algorithm 5.2 in Section 5.4. We leave the details for the reader to verify.

Here is a small example to illustrate Algorithm 5.6.

Example 5.9 Suppose

$$T = [5,5,4,4,3,3,3]$$

and $m = 3$. Note that the tasks are already sorted in nonincreasing order of length. Algorithm 5.6 would create the following schedule:

$$\begin{array}{lll}
\rho[1] = 1 & ST[1] = 0 & FIN[1] = 5 \\
\rho[2] = 2 & ST[2] = 0 & FIN[2] = 5 \\
\rho[3] = 3 & ST[3] = 0 & FIN[3] = 4 \\
\rho[4] = 3 & ST[4] = 4 & FIN[4] = 8 \\
\rho[5] = 1 & ST[5] = 5 & FIN[5] = 8 \\
\rho[6] = 2 & ST[6] = 5 & FIN[6] = 8 \\
\rho[7] = 1 & ST[7] = 8 & FIN[7] = 11.
\end{array}$$

Algorithm 5.6: GREEDYSCHEDULING(T, m)

```
comment: T = [t_1, ..., t_n] is the list of task lengths
sort the tasks so that t_1 ≥ ··· ≥ t_n
for j ← 1 to m
  do F[j] ← 0
for i ← 1 to n
       ⎧ min ← 1
       ⎪ for j ← 2 to m
       ⎪    do ⎧ if F[j] < F[min]
       ⎪       ⎩   then min ← j
  do   ⎨ ρ[i] ← min
       ⎪ ST[i] ← F[min]
       ⎪ FIN[i] ← F[min] + t_i
       ⎩ F[min] ← F[min] + t_i
max ← 1
for j ← 2 to m
  do ⎧ if F[j] > F[max]
     ⎩   then max ← j
F_max ← F[max]
return (ρ, ST, FIN, F_max)
```

The finishing times of the three processors are

$$\begin{aligned} F[1] &= 5+3+3 = 11 \\ F[2] &= 5+3 \quad\;\;\; = 8 \qquad \textit{and} \\ F[3] &= 4+4 \quad\;\;\; = 8, \end{aligned}$$

so $F_{max} = 11$, as shown in Figure 5.6. □

We measure the "goodness" of a feasible schedule in terms of the relative error, which is defined as follows. Suppose a feasible schedule for a given instance I has finishing time F_{max} and the optimal schedule for the instance I has finishing time F^*_{max}. Then the ***relative error*** of the given feasible schedule is

$$\frac{F_{max}}{F^*_{max}} - 1.$$

In order to compute the exact value of the relative error, we need to know the optimal finishing time, F^*_{max}. The exact value of F^*_{max} is, in general, hard to compute; however, for small instances we can perhaps determine the value of F^*_{max}. In general, we will make use of a fairly obvious lower bound on F^*_{max} that will be discussed in Lemma 5.6.

We return to the instance considered in Example 5.9 to illustrate.

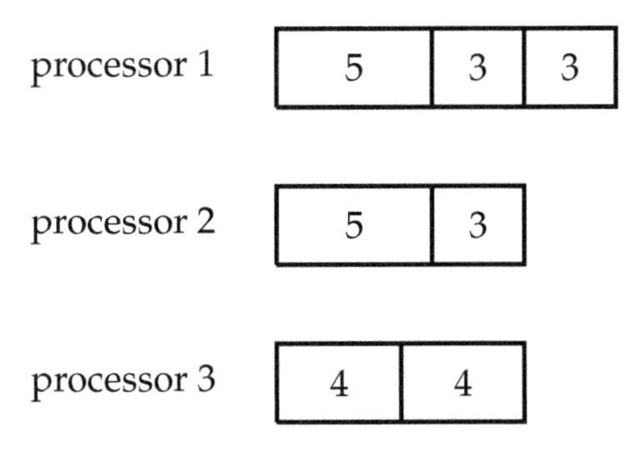

FIGURE 5.6: The greedy schedule.

Example 5.10 In Example 5.9, we looked at the instance with $T = [5, 5, 4, 4, 3, 3, 3]$ and $m = 3$. Algorithm 5.6 produced a schedule with $F_{max} = 11$.

However, it is not hard to use trial and error to find the following improved schedule:

$$\begin{array}{ll} \rho'[1] = 3 & \rho'[2] = 2 \\ \rho'[3] = 3 & \rho'[4] = 2 \\ \rho'[5] = 1 & \rho'[6] = 1 \\ \rho'[7] = 1. & \end{array}$$

The finishing times of the three processors are

$$\begin{aligned} F'[1] &= 3 + 3 + 3 = 9 \\ F'[2] &= 5 + 4 \quad\quad = 9 \qquad \textit{and} \\ F'[3] &= 5 + 4 \quad\quad = 9, \end{aligned}$$

so $F'_{max} = 9$. This is depicted in Figure 5.7.

Since the finishing times of the three processors are all equal, it is clear that this

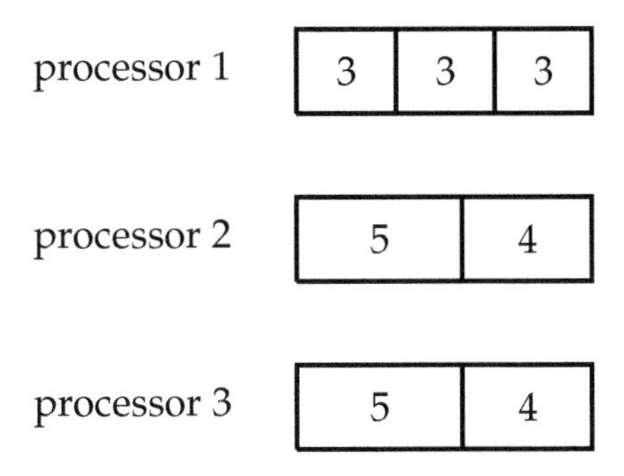

FIGURE 5.7: The optimal schedule.

schedule is optimal. Hence, $F^*_{max} = 9$ and the relative error of the greedy schedule is

$$\frac{11}{9} - 1 = \frac{2}{9}.$$

□

5.9.1 Relative Error of the Greedy Algorithm

In this section, we prove a theorem, due to Graham, concerning the relative error of the greedy schedule. First, we prove a simple lower bound on F^*_{max}.

LEMMA 5.6 *Suppose that $T = [t_1, \ldots, t_n]$ and m define an instance of the* ***Multiprocessor Scheduling*** *problem. Then F^*_{max}, the finishing time of the optimal schedule, satisfies the following inequality:*

$$F^*_{max} \geq \frac{\sum_{i=1}^{n} t_i}{m}. \tag{5.2}$$

PROOF For any feasible schedule, the quantity $(\sum_{i=1}^{n} t_i)/m$ is the average finishing time of the m processors. Therefore the maximum finishing time cannot be less than this value. ∎

The following technical lemma will also be useful in the proof of our main theorem. This lemma is intuitively obvious, but the proof is quite finicky.

LEMMA 5.7 *Suppose that $T = [t_1, \ldots, t_n]$ and m form an instance of the* ***Multiprocessor Scheduling*** *problem where $t_1 \geq \cdots \geq t_n$. Suppose that there exists an optimal schedule in which no processor executes more than two tasks. Then $n \leq 2m$ and the following schedule ρ^* is optimal:*

$$\begin{aligned} \rho^*[i] &= i && \text{for } 1 \leq i \leq m \\ \rho^*[i] &= 2m + 1 - i && \text{for } m + 1 \leq i \leq n. \end{aligned}$$

PROOF It is clear that $n \leq 2m$. Hence, there are $2m - n$ processors that execute one task and $n - m$ processors that execute one task. Without loss of generality, the first $2m - n$ processors each execute one task and the last $n - m$ processors each execute two tasks. In the rest of the proof, we refer to the first $2m - n$ tasks as "long tasks," the next $n - m$ tasks as "medium tasks" and the remaining $n - m$ tasks as "short tasks."

The proof takes place in three steps.

step 1

First, we show that the long tasks can be assigned to the first $2m - n$ processors in an optimal schedule. This is proven using a type of "exchange" argument. If we have an optimal schedule in which one of the first $2m - n$ processors executes a medium or short task, then we can exchange that task

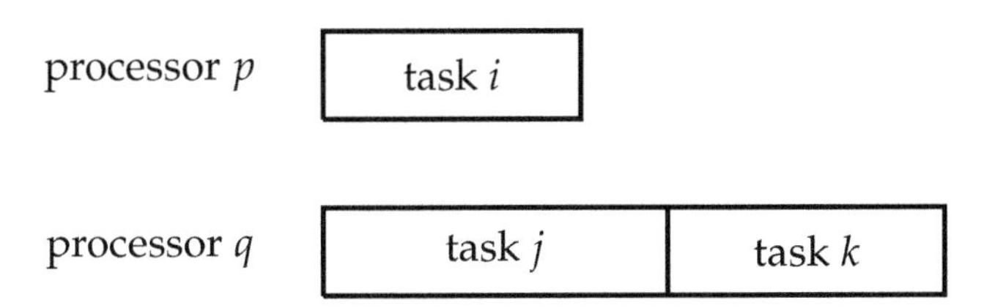

FIGURE 5.8: Properties of the optimal schedule.

with a long task that is executed by one of the last $n - m$ processors. This cannot increase the finishing time of the schedule, so we still have an optimal schedule.

More precisely, suppose that processor $p \leq 2m - n$ executes one medium or short task, say task i, and processor $q > 2m - n$ executes one long task, say task j, as well as some other task, say task k. Then processor p has finishing time t_i and processor q has finishing time $t_j + t_k$.

So we have the situation shown in Figure 5.8.

After we switch tasks i and j, processor p has finishing time t_j and processor q has finishing time $t_i + t_k$.

In order to show that the overall finishing time does not increase, we need to prove that

$$\max\{t_j, t_i + t_k\} \leq \max\{t_i, t_j + t_k\}.$$

Since task i is medium or short and task j is long, we have $t_i \leq t_j$ and hence

$$t_i + t_k \leq t_j + t_k.$$

Also, it is clear that

$$t_j \leq t_j + t_k.$$

Therefore,

$$\begin{aligned}\max\{t_j, t_i + t_k\} &\leq t_j + t_k \\ &\leq \max\{t_i, t_j + t_k\},\end{aligned}$$

as desired.

By a sequence of switches of this type, we can transform an arbitrary optimal schedule into an optimal schedule in which the $2m - n$ long tasks are executed on the first $2m - n$ processors.

step 2

The last $n - m$ processors must execute the $n - m$ medium tasks and the $n - m$ short tasks. In step 2 of the proof, we prove that there is an optimal schedule in which each of these processors executes one medium and one

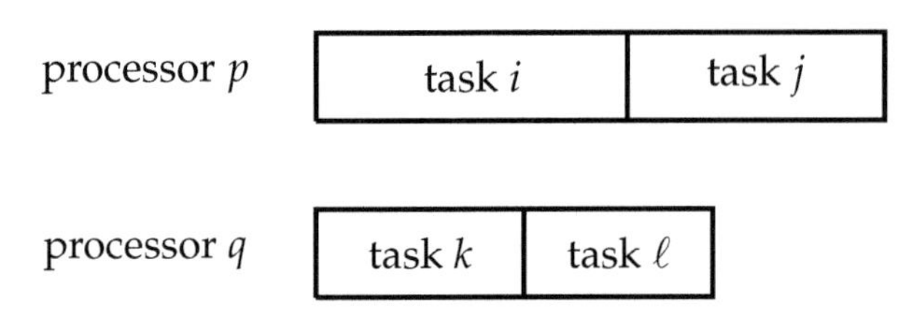

FIGURE 5.9: More properties of the optimal schedule.

short task. This can be shown by a similar type of exchange argument. Suppose we have an optimal schedule in which there is at least one of the last $n - m$ processors that does not execute one medium and one short task. This processor therefore executes two medium or two short tasks. If it executes two medium tasks, then these is another processor that executes two short tasks, and vice versa.

Thus, we suppose that processor p executes two medium tasks, say tasks i and j, and processor q executes two short tasks, say tasks k and ℓ. This is depicted in Figure 5.9.

Suppose we swap tasks i and k. Before the swap, processor p has finishing time $t_i + t_j$ and processor q has finishing time $t_k + t_\ell$. After the swap, processor p has finishing time $t_k + t_j$ and processor q has finishing time $t_i + t_\ell$.

In order to show that the overall finishing time does not increase, we need to prove that

$$\max\{t_k + t_j, t_i + t_\ell]\} \leq \max\{t_i + t_j, t_k + t_\ell\}.$$

Since task i is medium and task k is short, we have $t_k \leq t_i$ and hence

$$t_k + t_j \leq t_i + t_j.$$

Similarly, since task j is medium and task ℓ is short, we have $t_\ell \leq t_j$ and hence

$$t_i + t_\ell \leq t_i + t_j.$$

Therefore,

$$\begin{aligned}\max\{t_k + t_j, t_i + t_\ell\} &\leq t_i + t_j \\ &\leq \max\{t_i + t_j, t_k + t_\ell\},\end{aligned}$$

as desired.

By a sequence of switches of this type, we can ensure that each of the last $n - m$ processors executes one medium task and one short task.

step 3

As a result of step 2, we can assume that, for $2m - n + 1 \leq i \leq m$, processor

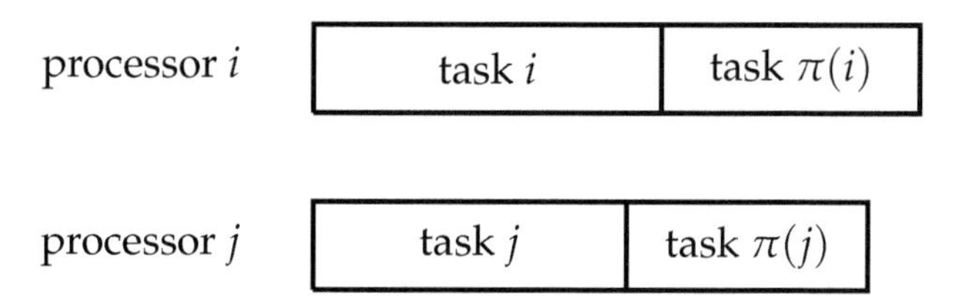

FIGURE 5.10: Still more properties of the optimal schedule.

i executes task i, which is a medium task. The proof will be completed by showing that we obtain an optimal schedule if, for $2m - n + 1 \leq i \leq m$, processor i also executes (the short) task $2m + 1 - i$. (For example, if $m = 5$ and $n = 8$, then processor 3 would execute tasks 3 and 8, processor 4 would execute tasks 4 and 7, and processor 5 would execute tasks 5 and 6.) This can also be proven by an exchange argument.

For $2m - n + 1 \leq i \leq m$, we can suppose that processor i executes tasks i and $\pi(i)$, where task i is medium and task $\pi(i)$ is short. Observe that

$$\{\pi(i) : 2m - n + 1 \leq i \leq m\} = \{m + 1, \ldots, n\}$$

since all the short tasks have to be executed on the $n - m$ given processors.

If $\pi(i) > \pi(j)$ whenever $i < j$, then it must be the case that

$$\pi(i) = m + 1 - i$$

for $2m - n + 1 \leq i \leq m$. That is, we would have

$$\begin{aligned} \pi(2m - n + 1) &= n \\ \pi(2m - n + 2) &= n - 1 \\ &\vdots \\ \pi(m) &= m + 1. \end{aligned}$$

This is precisely what we want to end up with!

So, let's suppose that $\pi(i) < \pi(j)$ for some $i < j$. We have the situation shown in Figure 5.10.

Then we swap tasks $\pi(i)$ and $\pi(j)$. By an argument similar to step 2, the modified schedule is still optimal (we leave the details for the reader to check). After carrying out a sequence of swaps of this type, we will end up with the desired optimal schedule. This completes the proof.

∎

Now we prove our main theorem.

THEOREM 5.8 *For any instance of the* **Multiprocessor Scheduling** *problem on m processors, the relative error of the greedy schedule is at most*

$$\frac{1}{3} - \frac{1}{3m}. \tag{5.3}$$

PROOF If $m = 1$, then the greedy schedule is clearly optimal and the relative error of the greedy schedule is 0. This agrees with (5.3) when $m = 1$. Thus we assume henceforth that $m \geq 2$.

Suppose that there is an instance of the **Multiprocessor Scheduling** problem with $m \geq 2$ such that

$$\frac{F_{max}}{F^*_{max}} - 1 > \frac{1}{3} - \frac{1}{3m}. \tag{5.4}$$

Among all such counterexamples with m processors, choose an instance I with the minimum possible value of n. Assume that

$$t_1 \geq \cdots \geq t_n.$$

Let F^*_{max} denote the finishing time of the optimal schedule and let F^G_{max} denote the finishing time of the greedy schedule.

Suppose that, in the greedy schedule for instance I, task n finishes before time F^G_{max}, i.e., $FIN[n] < F^G_{max}$. Construct an instance I' from I by deleting task n. The greedy schedule for I' also has finishing time F^G_{max}, so I' is a counterexample with a smaller value of n. This is a contradiction, so we conclude that $FIN[n] = F^G_{max}$ in the greedy schedule for the instance I.

Therefore, in the greedy schedule, we have

$$ST[n] = F^G_{max} - t_n.$$

In the greedy algorithm, each task is assigned to the first available processor. If we ignore the last task, the average finishing time of the m processors is

$$\frac{\sum_{i=1}^{n-1} t_i}{m}.$$

Therefore,

$$F^G_{max} - t_n \leq \frac{\sum_{i=1}^{n-1} t_i}{m},$$

or

$$F^G_{max} \leq \frac{\sum_{i=1}^{n} t_i}{m} + \frac{t_n(m-1)}{m}.$$

Applying (5.2), we obtain

$$F^G_{max} \leq F^*_{max} + \frac{t_n(m-1)}{m}. \tag{5.5}$$

From (5.4), we are assuming that

$$\frac{F^G_{max}}{F^*_{max}} - 1 > \frac{1}{3} - \frac{1}{3m},$$

which is equivalent to

$$F^G_{max} > \left(\frac{4}{3} - \frac{1}{3m}\right) F^*_{max}.$$

Substituting this into (5.5), we obtain

$$\left(\frac{4}{3} - \frac{1}{3m}\right) F^*_{max} < F^*_{max} + \frac{t_n(m-1)}{m}.$$

After some simplification, we have

$$F^*_{max} < 3t_n.$$

Since the nth task is the shortest, there cannot be more than two tasks assigned to any processor in the optimal schedule. Hence, from Lemma 5.7, we have that $n \leq 2m$ and the following schedule ρ^* is an optimal schedule:

$$\begin{aligned} \rho^*[i] &= i && \text{for } 1 \leq i \leq m \\ \rho^*[i] &= 2m + 1 - i && \text{for } m + 1 \leq i \leq n. \end{aligned}$$

We will complete the proof by showing that the greedy schedule, ρ_G, is optimal under the assumption that any optimal schedule assigns at most two tasks to any processor. Assume that $\rho_G \neq \rho^*$ and suppose that task ℓ is the first task such that $\rho_G[\ell] \neq \rho^*[\ell]$. Clearly,

$$\rho_G[i] = \rho^*[i] \qquad \text{for } 1 \leq i \leq m + 1.$$

Hence, $\ell \geq m + 2$. It is also easy to see that $\rho_G[\ell] < \rho^*[\ell]$ and processor $\rho_G[\ell]$ has (at least) three tasks assigned to it.

Suppose we define a new schedule ρ' as follows:

$$\begin{aligned} \rho'[i] &= \rho^*[i] && \text{for } i \neq \ell \\ \rho'[\ell] &= \rho_G[\ell]. \end{aligned}$$

Thus, ρ' is constructed from ρ^* by moving one task to a different processor.

For example, we could have a situation arising in the greedy schedule similar to the one depicted in Figure 5.11. In this example, task $m + 3$ is assigned to processor $m - 1$, whereas this task would be assigned to processor $m - 2$ in the optimal schedule.

In constructing ρ', we have moved one task in the optimal schedule to another processor in such a way that this task now finishes earlier. Everything else is unchanged, so it is clear that the finishing time F'_{max} for the schedule ρ' is no later than the finishing time F^*_{max} for ρ^*. Hence, ρ' is also an optimal schedule. But ρ' assigns at least three tasks to processor $\rho'[\ell]$, which is impossible for an optimal schedule. This contradiction completes the proof. ∎

processor $m-2$	task $m-2$		
processor $m-1$	task $m-1$	task $m+2$	task $m+3$
processor m	task m	task $m+1$	

FIGURE 5.11: A hypothetical greedy schedule.

Example 5.9 has $m = 3$ processors and the relative error of the greedy schedule is $2/9$. Since

$$\frac{1}{3} - \frac{1}{3 \times 3} = \frac{2}{9},$$

Example 5.9 has the maximum possible error, by Theorem 5.8. More generally, for every possible value of m, there is an instance of **Multiprocessor Scheduling** in which the relative error of the greedy schedule meets the bound of Theorem 5.8 with equality.

Example 5.11 We define an instance of **Multiprocessor Scheduling** with $n = 2m + 1$ tasks and m processors. The task lengths are

$$\begin{aligned} t_1 &= 2m - 1 \\ t_2 &= 2m - 1 \\ t_3 &= 2m - 2 \\ t_4 &= 2m - 2 \\ &\vdots \\ t_{n-4} &= m + 1 \\ t_{n-3} &= m + 1 \\ t_{n-2} &= m \\ t_{n-1} &= m \\ t_n &= m. \end{aligned}$$

It can be verified that $F^G_{max} = 4m - 1$ for this instance. Also, it is not hard to find a schedule with finishing time $3m$ by trial and error. Furthermore, from a straightforward computation, we have

$$\sum_{i=1}^{n} t_i = 3m^2.$$

It follows from Lemma 5.6 that $F^*_{max} = 3m$. Hence, the relative error of the greedy schedule is

$$\begin{aligned}\frac{F_{max}}{F^*_{max}} - 1 &= \frac{4m-1}{3m} - 1\\ &= \frac{m-1}{3m}\\ &= \frac{1}{3} - \frac{1}{3m}.\end{aligned}$$

Thus, the bound of Theorem 5.8 is met with equality and the relative error of the greedy schedule is as large as possible for this instance. □

5.10 Stable Matching and the Gale-Shapley Algorithm

We begin by defining the **Stable Matching** problem as Problem 5.8. The description given in Problem 5.8 may seem a bit complicated. However, the basic idea is to find a matching of two sets of equal size that achieves a desirable "stability" condition.

Problem 5.8: Stable Matching

Instance: Two sets of size n say

$$X = \{x_1, \dots, x_n\} \quad \text{and} \quad Y = \{y_1, \dots, y_n\}.$$

Each x_i has a ***preference ranking*** of the elements in Y, and each y_i has a preference ranking of the elements in X.

Find: A ***matching*** of the sets X and Y such that there *does not exist* a pair $\{x_i, y_j\}$ which is not in the matching, but where x_i and y_j prefer each other to their existing matches. A matching with this property is called a ***stable matching***.

In more detail, we are trying to avoid the following situation. Suppose x_i is matched with y_j, x_k is matched with y_ℓ, x_i prefers y_ℓ to y_j, and y_ℓ prefers x_i to x_k. Then this is an ***instability***; see Figure 5.12. The reason that instabilities are undesirable is that participants acting solely in their own interests might be inclined not to respect the given matching. For instance, in this example, x_i and y_ℓ might decide to ignore the specified matching and instead form a match themselves.

The **Stable Matching** problem is also known historically as the **Stable Marriage** problem. The 2012 Nobel Prize in economics was awarded to Roth and Shapley for their work in the "theory of stable allocation and the practice of market design."

There are several-world examples, dating from the 1950s, of practical applications of stable matching. These include matching medical interns to hospitals and matching organs to patients requiring transplants.

We should note that the **Stable Matching** problem is not really an optimization problem, as we are not trying to maximize or minimize a specific objective function. Rather, we are just trying to find *any* stable matching.

We will study the classic GALE-SHAPLEY algorithm to solve the **Stable Matching** problem. However, before giving the pseudocode description, we highlight several features of the algorithm and define some terminology that we will use in our discussion of the algorithm.

- Elements of X *propose* to elements of Y.
- If y_j accepts a proposal from x_i, then the pair $\{x_i, y_j\}$ is a ***matched pair***, and the elements x_i and y_j are each said to be ***matched***.
- An unmatched y_j must ***accept*** any proposal from any x_i.
- If $\{x_i, y_j\}$ is a matched pair, and y_j subsequently receives a proposal from x_k, where y_j prefers x_k to x_i, then y_j ***accepts*** and the pair $\{x_i, y_j\}$ is replaced by $\{x_k, y_j\}$. As a result, x_k is now matched and x_i is unmatched.
- If $\{x_i, y_j\}$ is a matched pair, and y_j subsequently receives a proposal from x_k, where y_j prefers x_i to x_k, then y_j ***rejects*** and nothing changes.
- A matched y_j never becomes unmatched.
- An x_i might make a number of proposals (up to n); the order of the proposals is determined by x_i's preference list.

We now present the GALE-SHAPLEY algorithm as Algorithm 5.7. The algorithm basically follows the points enumerated above.

REMARK The GALE-SHAPLEY algorithm, as it was described originally, proceeds in "rounds," during which several proposals may take place simultaneously. We are presenting a sequential variation of the algorithm, due to McVitie and Wilson,

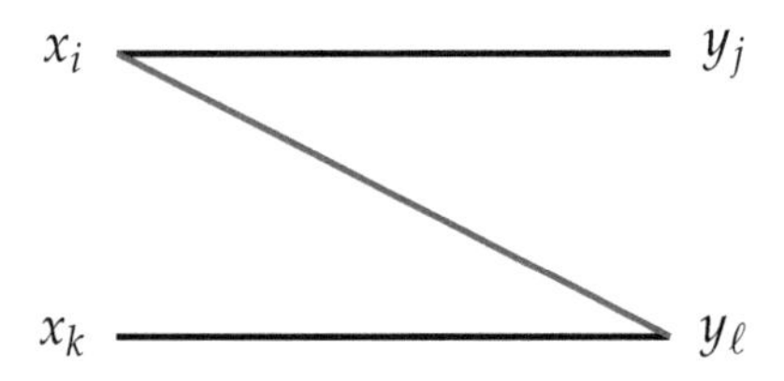

FIGURE 5.12: An instability in a matching.

Algorithm 5.7: GALE-SHAPLEY(X, Y)

```
comment: |X| = |Y| = n
Match ← ∅
comment: Match keeps track of all the current matches
while there exists an unmatched x_i
  do { let y_j be the next element in x_i's preference list
       if y_j is not matched
         then Match ← Match ∪ {x_i, y_j}
         else { suppose {x_k, y_j} ∈ Match
                if y_j prefers x_i to x_k
                  then { Match ← Match \ {x_k, y_j} ∪ {x_i, y_j}
                         comment: x_k is now unmatched
return (Match)
```

which is perhaps a bit easier to analyze. The two versions of the algorithm yield identical outcomes, however. ∎

As we already mentioned, there is no explicit objective function that we are attempting to optimize in the **Stable Matching** problem, so the GALE-SHAPLEY algorithm is not a typical greedy algorithm. However, the GALE-SHAPLEY algorithm can perhaps be thought of as "greedy" in the way that elements of X make their proposals, i.e., in order of their preferences.

We look at a couple of examples now.

Example 5.12 Suppose we have the following preference lists:

$$x_1 : y_2 > y_3 > y_1 \qquad y_1 : x_1 > x_2 > x_3$$
$$x_2 : y_1 > y_3 > y_2 \qquad y_2 : x_2 > x_3 > x_1$$
$$x_3 : y_1 > y_2 > y_3 \qquad y_3 : x_3 > x_2 > x_1$$

The GALE-SHAPLEY algorithm could be executed as follows:

proposal	result	*Match*
x_1 proposes to y_2	y_2 accepts	$\{x_1, y_2\}$
x_2 proposes to y_1	y_1 accepts	$\{x_1, y_2\}, \{x_2, y_1\}$
x_3 proposes to y_1	y_1 rejects	
x_3 proposes to y_2	y_2 accepts	$\{x_3, y_2\}, \{x_2, y_1\}$
x_1 proposes to y_3	y_3 accepts	$\{x_3, y_2\}, \{x_2, y_1\}, \{x_1, y_3\}$

□

TABLE 5.1: An example of the GALE-SHAPLEY algorithm.

proposal	result	*Match*
x_1 proposes to y_1	y_1 accepts	$\{x_1, y_1\}$
x_2 proposes to y_2	y_2 accepts	$\{x_1, y_1\}, \{x_2, y_2\}$
x_3 proposes to y_3	y_3 accepts	$\{x_1, y_1\}, \{x_2, y_2\}, \{x_3, y_3\}$
x_4 proposes to y_1	y_1 accepts	$\{x_4, y_1\}, \{x_2, y_2\}, \{x_3, y_3\}$
x_1 proposes to y_2	y_2 accepts	$\{x_4, y_1\}, \{x_1, y_2\}, \{x_3, y_3\}$
x_2 proposes to y_3	y_3 accepts	$\{x_4, y_1\}, \{x_1, y_2\}, \{x_2, y_3\}$
x_3 proposes to y_1	y_1 accepts	$\{x_3, y_1\}, \{x_1, y_2\}, \{x_2, y_3\}$
x_4 proposes to y_2	y_2 accepts	$\{x_3, y_1\}, \{x_4, y_2\}, \{x_2, y_3\}$
x_1 proposes to y_3	y_3 accepts	$\{x_3, y_1\}, \{x_4, y_2\}, \{x_1, y_3\}$
x_2 proposes to y_1	y_1 accepts	$\{x_2, y_1\}, \{x_4, y_2\}, \{x_1, y_3\}$
x_3 proposes to y_2	y_2 accepts	$\{x_2, y_1\}, \{x_3, y_2\}, \{x_1, y_3\}$
x_4 proposes to y_3	y_3 accepts	$\{x_2, y_1\}, \{x_3, y_2\}, \{x_4, y_3\}$
x_1 proposes to y_4	y_4 accepts	$\{x_2, y_1\}, \{x_3, y_2\}, \{x_4, y_3\}, \{x_1, y_4\}$

Example 5.13 Suppose we have the following preference lists:

$$x_1 : y_1 > y_2 > y_3 > y_4 \qquad y_1 : x_2 > x_3 > x_4 > x_1$$
$$x_2 : y_2 > y_3 > y_1 > y_4 \qquad y_2 : x_3 > x_4 > x_1 > x_2$$
$$x_3 : y_3 > y_1 > y_2 > y_4 \qquad y_3 : x_4 > x_1 > x_2 > x_3$$
$$x_4 : y_1 > y_2 > y_3 > y_4 \qquad y_4 : x_1 > x_2 > x_3 > x_4$$

The GALE-SHAPLEY algorithm could be executed as shown in Table 5.1: In this example, we have 13 iterations. □

5.10.1 Correctness

We now prove that the GALE-SHAPLEY algorithm always produces a stable matching. First we need to show that the algorithm always terminates. This is equivalent to showing that is impossible that an unmatched x_i has proposed to every y_j.

We have already commented that, once an element of $y_j \in Y$ is matched, y_j never becomes unmatched at a later time (even though they can change who they are matched with). Thus, if x_i has proposed to every y_j, then every y_j is matched. But if every element of Y is matched, then every element of X is matched, which is a contradiction.

We now prove that the GALE-SHAPLEY algorithm terminates with a stable matching. Suppose there is an instability, say

$$x_i \text{ is matched with } y_j,$$
$$x_k \text{ is matched with } y_\ell,$$
$$x_i \text{ prefers } y_\ell \text{ to } y_j, \text{ and}$$
$$y_\ell \text{ prefers } x_i \text{ to } x_k,$$

which can be depicted as in Figure 5.12. Observe that x_i must have proposed to y_ℓ before proposing to y_j, because x_i prefers y_ℓ to y_j.

There are three cases to consider.

case 1

y_ℓ rejected x_i's proposal. This could happen only if y_ℓ was already matched with someone they preferred to x_i. But y_ℓ ended up matched with someone they liked less than x_i. We conclude that y_ℓ did not reject a proposal by x_i.

case 2

y_ℓ accepted x_i's proposal, but later accepted another proposal. This could happen only if y_ℓ later received a proposal from someone they preferred to x_i. But y_ℓ ended up matched with someone they liked less than x_i, so this also did not happen.

case 3

y_ℓ accepted x_i's proposal, and did not accept any subsequent proposal. In this case, y_ℓ would have ended up matched to x_i, which did not happen.

All three cases have been shown to be impossible, and hence we conclude that the resulting matching is stable.

5.10.2 Complexity

First, we consider the number of iterations (i.e., the number of proposals made) in the GALE-SHAPLEY algorithm. It is obvious that the number of iterations is at most n^2, because every x_i proposes at most once to each y_j. Thus, the number of iterations is $O(n^2)$.

However, it is not difficult to prove the stronger result that the maximum number of iterations is $n^2 - n + 1$. We have already noted that a matched y_j never becomes unmatched. At some point during the execution of the algorithm, there are $n - 1$ matched elements in Y. Suppose that y_j is the remaining, unmatched element in Y. Then, as soon as y_j receives a proposal, y_j must accept the proposal and the algorithm terminates, because all the elements are matched. The first $n - 1$ elements in Y to be matched might each receive n proposals in the worst case, but the last element in Y to be matched always receives only one proposal. Hence, the maximum number of proposals (or iterations) is

$$(n-1) \times n + 1 \times 1 = n^2 - n + 1.$$

Example 5.13 required $13 = 4^2 - 4 + 1$ iterations, so it has the maximum possible number of iterations. It is not difficult to construct an example, for every n, of an instance that requires $n^2 - n + 1$ iterations, by generalizing the construction used in Example 5.13.

Now, in order to analyze the complexity of the algorithm, we need to consider how long it takes to perform each iteration. So we need to consider how to perform the following tasks efficiently.

1. Is there an efficient way to identify an unmatched x_i at any point in the algorithm?
2. How does an unmatched x_i identify the "next" y_j in their preference list?
3. If this y_j is currently matched, how does y_j determine efficiently if they prefer x_i to their current match?
4. In the case that y_j prefers x_i to their current match, how do we efficiently update the matching?

We will show, by using appropriate data structures, that all required operations 1–4 can be carried out in $O(1)$ time.

To implement operation 1, we maintain a queue Q that contains all of the unmatched x_i's at any given point in time (recall that queues were discussed in Section 3.3). Initially the queue Q contains all elements of X. When x_i is matched with a y_j, we DEQUEUE x_i. On the other hand, a matched x_i is added to the queue Q if x_i becomes unmatched, i.e., if its current match y_j accepts a proposal from some $x_k \neq x_i$. This is done using an ENQUEUE operation. Also, note that choosing an unmatched x_i is accomplished by simply looking at the first element of Q; the DEQUEUE operation would be performed when a proposal by x_i is accepted.

To carry out operation 2 efficiently, we could, for each x_i, initially create a (static) linked list containing x_i's preference list (or, alternatively, we could use an array). Each of these preference lists would be traversed, in order, during the execution of the algorithm.

Operation 3 could be done efficiently if, for each y_j, we initially create a "ranking array," R, such that

$$R[y_j, x_i] = k \text{ if } x_i \text{ is } y_j\text{'s } k\text{th favorite element of } X.$$

Then, if y_j receives a proposal from x_i and y_j is currently matched with x_k, the proposal is accepted if $R[y_j, x_i] < R[y_j, x_k]$. For example, the preference list for y_1, which is

$$x_3 > x_1 > x_2 > x_4,$$

would be recorded as the following rankings:

$$\begin{aligned} R[y_1, x_1] &= 2 \\ R[y_1, x_2] &= 3 \\ R[y_1, x_3] &= 1, \quad \text{and} \\ R[y_1, x_4] &= 4. \end{aligned}$$

For operation 4, we maintain an array M that is indexed by the elements in $X \cup Y$ to keep track of all the matched pairs. If $\{x_i, y_j\}$ is a matched pair at any point during the execution of the algorithm, then

$$M[x_i] = y_j \quad \text{and} \quad M[y_j] = x_i.$$

If x_i is unmatched at any given time, then $M[x_i] = 0$. Similarly, if y_j is unmatched, then $M[y_j] = 0$. The array M can easily be updated when required.

Using the above-described data structures, it can easily be seen that the GALE-SHAPLEY algorithm can be implemented to run in $O(n^2)$ time. This includes the time to initially construct the required arrays, lists, etc.

5.10.3 Uniqueness

One very important and perhaps unexpected property of the GALE-SHAPLEY algorithm is that all possible executions of the algorithm result in the same matching. That is, it does not matter which x_i is chosen in each iteration of the **while** loop in Algorithm 5.7. So, for example, the queue, which was suggested in operation 1 above, could be replaced by a stack, and this would not affect the outcome of the algorithm.

The matching produced by the GALE-SHAPLEY algorithm can be characterized in an interesting way. For every x_i, define $\mathsf{best}(x_i) = y_j$ if there exists at least one stable matching in which x_i is paired with y_j, and there is no stable matching in which x_i is paired with a y_k that x_i prefers to y_j.

Here is an equivalent way to define $\mathsf{best}(x_i)$. Suppose that there is at least one stable matching that contains a pair $\{x_i, y_j\}$. Then we say that y_j is a ***stable partner*** of x_i. Then we define $\mathsf{best}(x_i) = y_j$ if and only if, among all the stable partners of x_i, y_j is the one that is preferred by x_i.

Thus, $\mathsf{best}(x_i)$ represents the "optimal match" for x_i, given that the resulting matching is required to be stable. It turns out that the matching resulting from any execution of Algorithm 5.7 is in fact

$$\{\{x_i, \mathsf{best}(x_i)\} : 1 \leq i \leq n\}.$$

We will prove that the GALE-SHAPLEY algorithm produces a matching in which every x_i receives their best possible match, where "possible" means "consistent with some stable matching."

First we state and prove a simple but useful lemma.

LEMMA 5.9 *Suppose $\mathcal{M}$ is a matching constructed from the* GALE-SHAPLEY *algorithm. Suppose $\{x_i, y_j\} \in \mathcal{M}$ and $y_j \neq \mathsf{best}(x_i)$. Then, during the execution of the* GALE-SHAPLEY *algorithm, x_i is rejected by every stable partner that they prefer to y_j. Conversely, if x_i is rejected by any stable partner during the execution of the* GALE-SHAPLEY *algorithm, then x_i ends up matched with someone other than* $\mathsf{best}(x_i)$.

PROOF Any x_i proposes to elements of Y in decreasing order of preference. If x_i ends up being matched with y_j, then x_i must have (earlier) been rejected by all

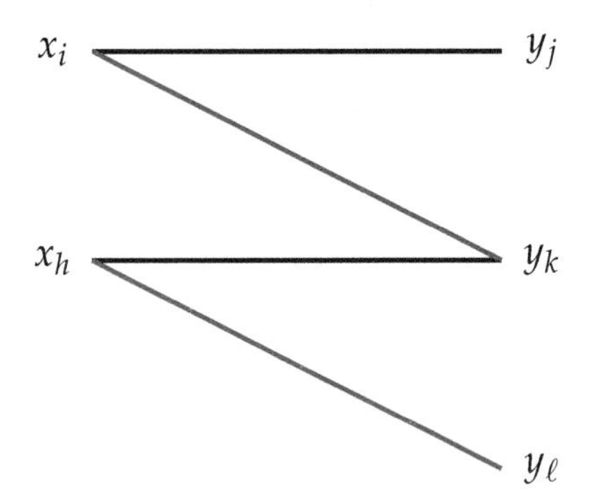

FIGURE 5.13: The GALE-SHAPLEY matching.

elements in Y that are higher on x_i's preference list, including all stable partners that x_i prefers to y_j.

Conversely, if x_i is rejected by any stable partner, then, in particular, x_i is rejected by $\mathsf{best}(x_i)$. This means that x_i is eventually matched with some $y_j \neq \mathsf{best}(x_i)$. ∎

We prove our main theorem now.

THEOREM 5.10 *The matching resulting from any execution of the* GALE-SHAPLEY *algorithm is*

$$\{\{x_i, \mathsf{best}(x_i)\} : 1 \leq i \leq n\}.$$

PROOF Let $\mathcal{M}$ be a matching constructed from the GALE-SHAPLEY algorithm. Our goal is to prove that there does not exist a pair $\{x_i, y_j\} \in \mathcal{M}$, where $y_j \neq \mathsf{best}(x_i)$.

Suppose such a pair occurs in $\mathcal{M}$. Let $y_k = \mathsf{best}(x_i)$ (thus it follows that x_i prefers y_k to y_j). Let $\mathcal{M}'$ be a stable matching where $\{x_i, y_k\} \in \mathcal{M}'$. We will obtain a contradiction, which proves that $y_j = \mathsf{best}(x_i)$.

It follows from Lemma 5.9 that, in $\mathcal{M}$, y_k must have rejected x_i in favor of some x_h who y_j prefers to x_i. We will assume that this rejection of x_i by y_k is the *first* rejection received by any element of X from a stable partner in the given execution of the GALE-SHAPLEY algorithm. (If not, then a prior rejection of this type would correspond to a different "non-optimal pair" $\{x_h, y_\ell\} \in \mathcal{M}$, and we could use this pair to begin the argument in place of $\{x_i, y_k\}$.)

Let $\{x_h, y_\ell\} \in \mathcal{M}'$. The pairs we will consider are depicted in Figure 5.13, where the two red edges are pairs in $\mathcal{M}'$. Suppose for the time being that x_h prefers y_ℓ to y_k. Then, in $\mathcal{M}$, x_h proposed to y_k after being rejected by y_ℓ. Also, y_k rejects x_i when the proposal from x_h is received. This implies that the rejection of x_h by y_ℓ occurs *before* the rejection of x_i by y_k. However, y_ℓ is a stable partner of x_h, so this contradicts the assumption that the rejection of x_i by y_k is the *first* rejection of an element of X by a stable partner.

Therefore, we conclude that x_h prefers y_k to y_ℓ. Then, in the matching $\mathcal{M}'$, the pair $\{x_h, y_k\}$ is an instability, because

1. $\{x_h, y_k\} \notin \mathcal{M}'$
2. $\{x_h, y_\ell\}, \{x_i, y_k\} \in \mathcal{M}'$
3. y_k prefers x_h to x_i, and
4. x_h prefers y_k to y_ℓ.

It follows that the matching $\mathcal{M}'$ is not stable, which is a contradiction. This completes the proof. ∎

The matching constructed by the GALE-SHAPLEY algorithm can be characterized in a different way, namely, as

$$\{\{y_j, \mathsf{worst}(y_j)\} : 1 \le j \le n\}.$$

That is, every y_j receives their "worst possible" match. Thus the GALE-SHAPLEY algorithm is completely biased in favor of the proposers!

5.11 Notes and References

The greedy algorithm to solve the **Rational Knapsack** problem is due to Dantzig [17]; it was published in 1957.

The bounds proven in Section 5.9.1, concerning the relative error of the greedy algorithm for **Multiprocessor Scheduling**, are from a 1969 paper due to Graham [28].

The GALE-SHAPLEY algorithm was first described in 1962 in [22]. Alvin Roth and Lloyd Shapley were jointly awarded the 2012 Nobel Prize in Economics "for the theory of stable allocations and the practice of market design." The press release[1] said the following of Shapley's work:

> A key issue is to ensure that a matching is stable in the sense that two agents cannot be found who would prefer each other over their current counterparts. Shapley and his colleagues derived specific methods—in particular, the so-called Gale-Shapley algorithm—that always ensure a stable matching.

Actually, entire books have been written on the **Stable Matching** problem, for example, Gusfield and Irving [31].

---

[1] `https://www.nobelprize.org/prizes/economic-sciences/2012/press-release/`

Exercises

5.1 The purpose of this exercise is to provide an alternate proof that Algorithm 5.3 always finds the optimal solution to Problem 5.3. Let $\mathcal{G} = [y_1, \dots, y_m]$ be the greedy solution. Provide details to prove the following.

(a) Define
$$\mathcal{H} = \{y_i - D : 1 \leq i \leq m\}.$$
Then $\mathcal{H} \subseteq \{x_1, \dots, x_n\}$.

(b) The distance between any two elements of $\mathcal{H}$ is greater than D.

(c) Therefore any feasible solution requires at least m wireless access points, and the greedy solution is optimal.

5.2 Suppose Alice wants to see n different movies at a certain movie theater complex. Each movie is showing on specified dates. For $1 \leq j \leq n$, suppose that movie M_j is showing from day a_j until day b_j inclusive (you can assume that a_j and b_j are positive integers such that $a_j \leq b_j$). The objective is to determine the minimum number of trips to the theater that will be required in order to view all n movies. For simplicity, assume that it is possible to view any number of movies in a given day.

(a) Design a greedy algorithm to solve this problem, and prove your algorithm is correct. Make sure to record the list of days during which the theatre will be visited. Determine the complexity of your algorithm and prove that it is correct.

(b) Illustrate the execution of your algorithm step-by-step on the problem instance consisting of the following movies $M_j = [a_j, b_j]$:

$$\begin{array}{cccccc} [7,9], & [14,23], & [18,32], & [32,36], & [2,6], & [10,19] \\ [15,34], & [8,15], & [22,31], & [33,37], & [4,7], & [7,24] \\ [24,38], & [3,10], & [8,18], & [30,40]. & & \end{array}$$

5.3 A lab performs various types of blood tests. For a given sample, a *chemical analysis* is done using a piece of expensive equipment; this is *Stage 1* of the testing. Then a *mathematical analysis* is done by a lab technician on their personal laptop; this is *Stage 2*. There are n tests to be done. The stage 1 analyses must be done *sequentially*, but any number of stage two analyses can be done in *parallel* (e.g., using different laptop computers). For each test T_i ($1 \leq i \leq n$), the time required for the stage one and stage two analyses is specified; these are denoted by c_i and m_i, respectively. These values are assumed to be positive integers.

A *schedule* is specified by an *ordering* of the n tests. The objective is to *minimize* the *maximum completion time*.

For example, suppose there are two tests to be performed, where we have $T_1 = (10, 4)$ and $T_2 = (2, 1)$. If the schedule is (T_1, T_2), then the completion time of T_1 is $10 + 4 = 14$ and the completion time of T_2 is $10 + 2 + 1 = 13$. The maximum completion time is 14. If the schedule is (T_2, T_1), then the completion time of T_1 is $2 + 10 + 4 = 16$ and the completion time of T_2 is $2 + 1 = 3$. The maximum completion time is 16. Therefore the schedule (T_1, T_2) is the optimal schedule.

(a) Suppose we have the following instance I comprising three tests: $T_1 = (10, 6)$, $T_2 = (7, 8)$ and $T_3 = (4, 2)$. Here are six "plausible" greedy strategies: order the jobs in *increasing* or *decreasing* order by the values c_i, m_i or $c_i + m_i$. For the given instance I, determine the schedule arising by applying each of these six strategies, and thereby identify which strategies are non-optimal for this instance. Show all your work.

HINT You should be able to establish that five of the six strategies are non-optimal.

(b) It turns out that one of the above strategies is in fact an optimal strategy. Prove this!

HINT Use an exchange argument.

5.4 Suppose we are given two lists of n integers, $A = [a_1, \ldots, a_n]$ and $B = [b_1, \ldots, b_n]$. We want to find a permutation π of the indices $\{1, \ldots, n\}$ such that the quantity

$$\sum_{i=1}^{n} (a_i - b_{\pi(i)})^2$$

is minimized. Devise a greedy algorithm to solve this problem and prove that it is correct.

5.5 Prove that Algorithm 5.5 always finds the optimal solution for any instance of the **Coin Changing** problem where $d_j \mid d_{j-1}$ for $2 \leq j \leq n$. Suppose the greedy solution is $X = [x_1, \ldots, x_n]$ and $X^* = [x_1^*, \ldots, x_n^*]$ is any optimal solution.

(a) For $2 \leq i \leq n$, prove that

$$0 \leq x_i \leq \frac{d_{i-1}}{d_i} - 1.$$

(b) For $2 \leq i \leq n$, prove that

$$0 \leq x_i^* \leq \frac{d_{i-1}}{d_i} - 1.$$

(c) Suppose that $X \neq X^*$. Let i be the last (i.e., highest) index such that $x_i^* \neq x_i$. Using the results proven in (a) and (b), derive a contradiction, thus proving that $X = X^*$.

5.6 We consider a variation of the **House Construction** problem (Problem 5.4). Suppose there is an integer constant $c > 0$ that $\ell_i = c$ for all i (i.e., all construction times are the same, namely, c days). Design a greedy algorithm to find the *maximum size* feasible set of houses. The algorithm will involve a suitable preprocessing step. Prove that your greedy algorithm is correct and determine its complexity.

5.7 A server has n customers, waiting to be served. It takes the server t_i minutes to serve customer i, and the server can only serve one customer at a time. The server wants to find an order to serve the customers so as to minimize the total waiting time of the n customers.

For example, given five customers with t_i values $6, 2, 1, 8, 9$, serving them in the given order requires total waiting time

$$T = 6 + (6 + 2) + (6 + 2 + 1) + (6 + 2 + 1 + 8) = 40,$$

but this is not optimal.

We want to reorder (permute) the numbers $t_1, t_2, \ldots, t_n$ so as to minimize the value

$$T = t_1 + (t_1 + t_2) + \cdots + (t_1 + t_2 + \cdots + t_{n-1}).$$

You may assume that the t_i values are all distinct. Design a greedy algorithm to solve this problem. Prove that your greedy algorithm is correct and determine its complexity.

5.8 We are given an instance of the **Stable Matching** problem where we are matching three interns, denoted d_1, d_2, d_3 to three hospitals, denoted h_1, h_2, h_3. The instance I is specified by the following preference lists, where "$>$" means "prefers":

$$\begin{array}{ll}
\text{preference list for } d_1: & h_1 > h_2 > h_3 \\
\text{preference list for } d_2: & h_1 > h_3 > h_2 \\
\text{preference list for } d_3: & h_3 > h_2 > h_1
\end{array}$$

$$\begin{array}{ll}
\text{preference list for } h_1: & d_2 > d_1 > d_3 \\
\text{preference list for } h_2: & d_2 > d_3 > d_1 \\
\text{preference list for } h_3: & d_1 > d_2 > d_3
\end{array}$$

(a) There are six possible matchings of the three interns with the three hospitals. For each matching, determine all the instabilities that exist. How many of the six matchings are stable?

(b) Show how the GALE-SHAPLEY algorithm would execute on the instance I. Assume that interns "propose" to hospitals. Show all the steps in the algorithm.

(c) Show how the GALE-SHAPLEY algorithm would execute on the instance I, if hospitals "proposed" to interns. Show all the steps in the algorithm.

Chapter 6

Dynamic Programming Algorithms

The dynamic programming technique was invented by Richard Bellman in 1950. It involves filling in a table, in a carefully specified order, to solve a problem. We give several examples of dynamic programming algorithms in this chapter. A recursive variation of this method, called memoization, is also presented.

6.1 Fibonacci Numbers

A ***dynamic programming algorithm*** is a method of filling in a table in an efficient way to solve a problem. Initially, we start with a recurrence relation, but the algorithm is not implemented recursively. Dynamic programming is most often applied to optimization problems. However, we introduce the basic idea by considering how to compute Fibonacci numbers based on their defining recurrence relation.

The ***Fibonacci numbers*** satisfy the following recurrence relation:

$$\begin{aligned} f_0 &= 0 \\ f_1 &= 1 \\ f_n &= f_{n-1} + f_{n-2} \qquad \text{if } n > 1. \end{aligned} \tag{6.1}$$

Thus, the first few Fibonacci numbers are

$$0, 1, 1, 2, 3, 5, 8, 13, 21, 34, 55 \ldots.$$

Suppose we write a recursive algorithm to compute f_n using the recurrence relation (6.1). The result is Algorithm 6.1. This algorithm is correct but it is very inefficient. To see why the algorithm is so slow, we look at the recursion tree to evaluate f_5, which is presented in Figure 6.1.

In general, the recursion tree has f_n leaf nodes with the value 1 and f_{n-1} leaf nodes with the value 0. So there are a total of

$$f_n + f_{n-1} = f_{n+1}$$

leaf nodes. Also, the number of interior nodes is $f_{n+1} - 1$.

Thus, in the unit cost model where we assume that an addition takes $\Theta(1)$ time, the complexity of computing f_n is $\Theta(f_{n+1})$. So the complexity of Algorithm

DOI: 10.1201/9780429277412-6

Algorithm 6.1: BADFIB(n)

comment: $n \geq 0$ is an integer
if $n = 0$
  then $f \leftarrow 0$
 else if $n = 1$
  then $f \leftarrow 1$
  else $\begin{cases} f_1 \leftarrow \text{BADFIB}(n-1) \\ f_2 \leftarrow \text{BADFIB}(n-2) \\ f \leftarrow f_1 + f_2 \end{cases}$
return (f)

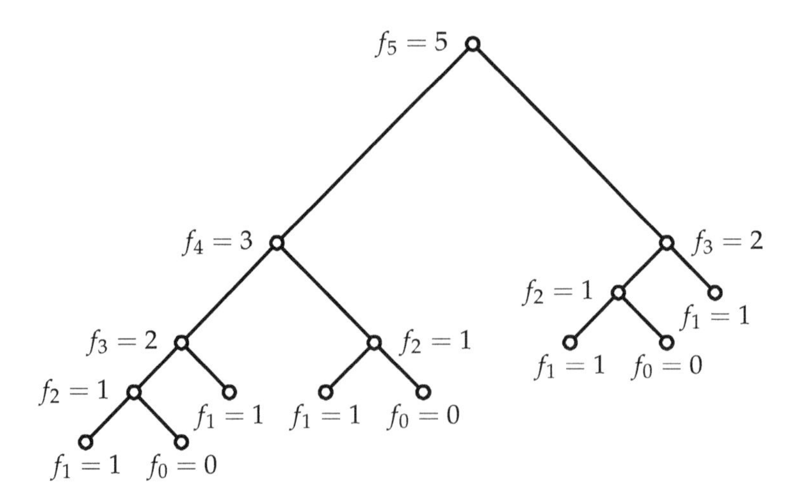

FIGURE 6.1: The recursion tree to evaluate f_5.

6.1 (to compute f_n) is proportional to the value of the $(n+1)$st Fibonacci number. Thus it is of interest to ask how quickly f_n grows as a function of n. This can be determined by using a non-recursive formula for f_n.

Let $\phi = (1+\sqrt{5})/2$; the value $\phi \approx 1.6$ is known as the ***golden ratio***. The following formula for f_n is expressed in terms of ϕ.

$$f_n = \frac{\phi^n - (-\phi)^{-n}}{\sqrt{5}} = \left\lfloor \frac{\phi^n}{\sqrt{5}} + \frac{1}{2} \right\rfloor . \tag{6.2}$$

From (6.2), we see that $f_n \in \Theta(\phi^n)$ and hence we also have $f_{n+1} \in \Theta(\phi^n)$. Therefore, the time to compute f_n is *exponential* as a function of n in the unit cost model.

Actually, we can easily prove that the Fibonacci numbers grow exponentially quickly without appealing to the formula (6.2). It is not hard to prove by induction, directly from the recurrence relation (6.1), that $f_n \geq 2^{n/2}$ for all $n \geq 6$ (we leave this as an exercise for the reader). Therefore, $f_n \in \Omega((\sqrt{2})^n)$.

Algorithm 6.1 is an inefficient use of recursion because we have to solve two subproblems f_{n-1} and f_{n-2}, both of which are almost as large as the original problem instance f_n. The recursion tree ends up being of *linear* depth and *exponential* size (as a function of n). In a divide-and-conquer algorithm, we typically solve "small" subproblems, e.g., of size $n/2$, to solve the given instance of size n. In these situations, the recursion tree is of *logarithmic* depth and *polynomial* size.

We can also see from the recursion tree in Figure 6.1 that many subproblems are solved multiple times. For instance, f_2 is computed three times and f_3 is computed twice.

An alternative approach is to compute the Fibonacci numbers non-recursively (i.e., ***bottom-up***), by computing $f_0, f_1, \ldots, f_n$ in order. See Algorithm 6.2. Because each Fibonacci number is just the sum of the two previous numbers in the sequence, we compute each f_i only once by using this method.

Algorithm 6.2: BETTERFIB(n)

```
f[0] ← 0
f[1] ← 1
for i ← 2 to n
  do f[i] ← f[i − 1] + f[i − 2]
return (f[n])
```

It is easy to modify Algorithm 6.2 so it does not store the entire sequence of Fibonacci numbers that are computed, since we only need to store the last two numbers in the sequence in order to compute the next one.

In Algorithm 6.2, we perform $n-1$ additions, so the complexity is $\Theta(n)$ in the unit cost model where we assume $\Theta(1)$ cost for addition.

However, we should point out that the unit cost model is not realistic for the problem of computing Fibonacci numbers. The addition $f_{n-1} + f_{n-2}$ to compute f_n involves two $\Theta(n)$-bit integers, so this computation actually takes $\Theta(n)$ time. In Algorithm 6.2, we successively compute $f_2, f_3, \ldots, f_n$, so the bit complexity of this algorithm is

$$\begin{aligned}\sum_{i=2}^{n} \Theta(i) &= \Theta\left(\sum_{i=2}^{n} i\right)\\ &= \Theta(n^2).\end{aligned}$$

Perhaps surprisingly, it can be shown that the bit complexity of Algorithm 6.1 is $\Theta(\phi^n)$. Thus, it is the same as it was in the unit cost model. This is because the cost of the arithmetic operations is dominated by the number of recursive calls.

There is one other subtlety that we should address. An instance I of the **Fibonacci number** problem consists of just the integer n. This integer, represented in binary, requires $\Theta(\log n)$ bits. Thus $\mathsf{size}(I) \in \Theta(\log n)$. Therefore, the complexity $\Theta(n^2)$ is an exponential function of $\mathsf{size}(I)$. However, it is unavoidable that computing f_n will take exponential time, because the result of the computation is an integer consisting of $\Theta(n)$ bits. So an algorithm that requires only polynomial time (as a function of $\mathsf{size}(I)$) is clearly impossible.

6.2 Design Strategy

We now discuss the general design strategy for dynamic programming algorithms for optimization problems. To start with, we need to identify an appropriate optimal structure in the problem we are studying. Here are the main steps in developing a dynamic programming algorithm.

optimal structure

Examine the structure of an optimal solution to a problem instance I, and determine if an optimal solution for I can be expressed in terms of optimal solutions to one or more ***subproblems*** of I.

define subproblems

Define a set of subproblems $\mathcal{S}(I)$ of the instance I, the solution of which enables the optimal solution of I to be computed. I will be the last or largest instance in the set $\mathcal{S}(I)$.

recurrence relation

Derive a ***recurrence relation*** on the optimal solutions to the instances in $\mathcal{S}(I)$. This recurrence relation should be completely specified in terms of optimal solutions to (smaller) instances in $\mathcal{S}(I)$ and/or base cases.

compute optimal solutions

Compute the optimal solutions to all the instances in $\mathcal{S}(I)$. Compute these solutions using the recurrence relation in a ***bottom-up*** fashion, filling in a table of values containing these optimal solutions. Whenever a particular table entry is filled in using the recurrence relation, the optimal solutions of relevant subproblems can be looked up in the table (because they have been computed already). The final table entry is the solution to I.

In the Fibonacci number problem, we just want to compute a particular Fibonacci number f_n. We are given the recurrence relation (6.1) to start with, and the "table" we compute is just the list of numbers $f_0, f_1, \dots, f_n$.

We will see many examples of more complicated dynamic programming algorithms in the rest of this chapter. Most of these algorithms will be developed to solve optimization problems.

Before proceeding further, we have a few general comments.

- The term ***dynamic programming*** refers to computing the solutions to all subproblems in a bottom-up fashion.
- Sometimes careful thought needs to be given to the appropriate order to compute the table entries. It is important that each table entry depends only on entries that have already been computed.
- ***Memoization*** is a different method from dynamic programming. It is top-down (i.e., recursive) but it *remembers* the solutions to subproblems in order to avoid re-computation of the same things over and over again. We will discuss memoization in Section 6.9.
- Memoization is not dynamic programming! It is an *alternative* to dynamic programming.

6.3 Treasure Hunt

We begin with a dynamic programming algorithm for a simple optimization problem that we call **Treasure Hunt**; see Problem 6.1.

Problem 6.1: Treasure Hunt

Instance: An instance consists of n treasures, whose values are recorded in the list

$$V = [v_1, \dots, v_n].$$

These values are all positive integers.

Feasible solution: A subset of treasures, such that no two "consecutive" treasures are chosen. We can phrase this using mathematical notation by defining a feasible solution to be an n-tuple

$$X = [x_1, \dots, x_n] \in \{0,1\}^n$$

such that no two consecutive x_i's both equal 1.

Find: A feasible solution that achieves the maximum possible total value of the selected treasures. That is, the optimal solution is the feasible solution X that maximizes

$$\mathsf{profit}(X) = \sum_{i=1}^{n} v_i x_i.$$

We design a dynamic programming algorithm for **Treasure Hunt** by following the four steps enumerated in Section 6.2.

optimal structure

Suppose $X = [x_1, \ldots, x_n]$ is an optimal solution to an instance I.

If $x_n = 0$, then we ignore the last treasure and consider the first $n-1$ treasures. Thus $X' = [x_1, \ldots, x_{n-1}]$ is the optimal solution to the instance (subproblem) with $n-1$ treasures, having values $v_1, \ldots, v_{n-1}$. In this case, we have

$$\mathsf{profit}(X) = \mathsf{profit}(X').$$

If $x_n = 1$, then we are including the nth treasure, which has the value v_n. Since $x_n = 1$, we must define $x_{n-1} = 0$.

Then $X' = [x_1, \ldots, x_{n-2}]$ is the optimal solution to the instance (subproblem) with $n-2$ treasures, having values $v_1, \ldots, v_{n-2}$. In this case,

$$\mathsf{profit}(X) = \mathsf{profit}(X') + v_n.$$

subproblems

We have seen above that the optimal solution to the original problem instance consisting of n treasures can be expressed in terms of optimal solutions to two subproblems, one of which is defined on the first $n-1$ treasures and the other is defined on the first $n-2$ treasures.

If we apply the same analysis recursively, it is natural to consider subproblems consisting of the first i treasures (having values $v_1, \ldots, v_i$), for $i = 1, 2, \ldots, n$.

Let $P[i]$ denote the optimal profit for the subproblem defined above. Then $P[n]$ is the final answer we are seeking.

recurrence relation

We can now derive a recurrence relation for the values $P[i]$. We need to specify base cases as well as the "general" form of the recurrence relation.

The base cases will correspond to $i = 0, 1$. Clearly

$$P[0] = 0 \qquad \text{and} \qquad P[1] = v_1.$$

In the general case, $P[i]$ depends on two "previous" values, namely, $P[i-1]$ and $P[i-2]$. In the first case, we are not selecting the ith treasure. In the second case, we are including it (so its value, v_i, is added to the optimal solution for the relevant subproblem).

We obtain the following recurrence relation:

$$\begin{aligned} P[i] &= \max\{P[i-1], v_i + P[i-2]\} \qquad \text{if } i \geq 3 \\ P[0] &= 0 \\ P[1] &= v_1. \end{aligned} \tag{6.3}$$

compute optimal solutions

In this dynamic programming algorithm, we are filling in a list of values $P[i]$, for $i = 1, 2, \ldots, n$. The base cases are when $i = 1, 2$.

In general, an entry $P[i]$ depends on the two previous entries, $P[i-1]$ and $P[i-2]$, and the final answer is the value $P[n]$.

The resulting algorithm, which is based on (6.3), is presented as Algorithm 6.3.

Algorithm 6.3: TREASURE(V)

```
comment: V = [v_1, ..., v_n] is a list of n values
P[1] ← v_1
P[0] ← 0
for i ← 2 to n
       ⎧ z_0 ← P[i − 1]
   do  ⎨ z_1 ← P[i − 2] + v_i
       ⎩ P[i] ← max{z_0, z_1}
return (P)
```

We are just filling a list of n values $P[0], P[1], \ldots, P[n]$ in sequential order, and each $P[i]$ can be computed in $\Theta(1)$ time. So the complexity of Algorithm 6.3 is $\Theta(n)$.

It is straightforward to actually construct the optimal solution X after we have computed all the values $P[0], P[1], \ldots, P[n]$. See Algorithm 6.4. The idea is to proceed "backwards" from the end to the beginning, so we start by determining x_n. If $p = P[n-1]$, then the optimal profit does not change when we consider the nth treasure, so $x_n = 0$. Otherwise $x_n = 1$. Once we have determined the value of x_n, we adjust p accordingly and consider x_{n-1}. This pattern continues for $i = n, n-1, \ldots, 1$.

Algorithm 6.4: OPTIMALTREASURE(V, P)

```
comment: the array P is computed by TREASURE
p ← P[n]
for i ← n downto 1
       ⎧ if p = P[i − 1]
       ⎪   then x_i ← 0
   do  ⎨
       ⎪   else ⎧ x_i ← 1
       ⎩        ⎩ p ← p − v_i
return (X = [x_1, ..., x_n])
```

It is interesting to compare Algorithm 6.3 to a recursive algorithm that implements the recurrence relation (6.3). A recursive algorithm will basically enumerate all the feasible solutions, so the relevant question is how many feasible solutions there are. Let us denote the number of feasible solutions by s_n. The analysis that led to the recurrence relation (6.3) also gives rise to a recurrence relation for the values s_n:

$$\begin{aligned} s_n &= s_{n-1} + s_{n-2} \qquad \text{if } n \geq 3 \\ s_2 &= 2 \\ s_1 &= 1. \end{aligned} \tag{6.4}$$

Does this recurrence relation look familiar? This is basically the Fibonacci recurrence! In fact, $s_n = f_{n+1}$ for all $n \geq 1$. We have already observed in Section 6.1 that the Fibonacci numbers grow exponentially quickly. For example, we proved that $f_n \geq 2^{n/2}$, which immediately implies that $s_n \in \Omega(2^{n/2})$. Hence, any algorithm that tests all feasible solutions, including the basic recursive algorithm, will have complexity $\Omega(2^{n/2})$. So the dynamic programming approach has made a dramatic improvement in the complexity.

6.4 0-1 Knapsack

In this section, we derive a dynamic programming algorithm for the **0-1 Knapsack** problem, which we introduced in Section 5.7. We recall the statement of this problem in Problem 6.2.

Problem 6.2: 0-1 Knapsack

Instance: An instance consists of n objects, having

profits	$\mathcal{P} = [p_1, \dots, p_n]$	and
weights	$\mathcal{W} = [w_1, \dots, w_n]$,	as well as a
knapsack capacity	M.	

These values are all positive integers.

Feasible solution: An n-tuple $X = [x_1, \dots, x_n]$ where $0 \leq x_i \leq 1$ for $1 \leq i \leq n$ and

$$\sum_{i=1}^{n} w_i x_i \leq M.$$

Find: A feasible solution X that maximizes

$$\mathsf{profit}(X) = \sum_{i=1}^{n} p_i x_i.$$

We again design our dynamic programming algorithm by following the four steps enumerated in Section 6.2.

optimal structure

Suppose $X = [x_1, \ldots, x_n]$ is an optimal solution to an instance I.

If $x_n = 0$, then we just "throw away" the last object and consider the first $n-1$ items. Thus $X' = [x_1, \ldots, x_{n-1}]$ is the optimal solution to the instance (subproblem) with profits $p_1, \ldots, p_{n-1}$, weights $w_1, \ldots, w_{n-1}$ and capacity M. In this case, we have

$$\mathsf{profit}(X) = \mathsf{profit}(X').$$

If $x_n = 1$, then we are including the nth object, which contributes a profit of p_n and uses up w_n of the capacity. Then $X' = [x_1, \ldots, x_{n-1}]$ is the optimal solution to the instance (subproblem) with profits $p_1, \ldots, p_{n-1}$, weights $w_1, \ldots, w_{n-1}$ and capacity $M - w_n$. Here,

$$\mathsf{profit}(X) = \mathsf{profit}(X') + p_n.$$

subproblems

We have seen above that the optimal solution to the original problem instance consisting of n objects can be expressed in terms of optimal solutions to two subproblems defined on the first $n-1$ objects. These subproblems do not alter profits or weights of items, but they can modify the capacity of the knapsack.

If we apply the same analysis recursively, it is natural to consider subproblems consisting of the first i objects (having profits $p_1, \ldots, p_i$ and weights $w_1, \ldots, w_i$), for $i = 1, 2, \ldots, n$, as well as a variable capacity m, for all $0 \leq m \leq M$.

Let $P[i, m]$ denote the optimal profit for the subproblem defined above. Then $P[n, M]$ is the final answer we are looking for.

recurrence relation

We can now derive a recurrence relation for the values $P[i, m]$. We need to specify base cases as well as the "general" form of the recurrence relation.

The base cases will correspond to $i = 1$. In this case, we only have the first object, and we will include it in the knapsack if its weight $w_1 \leq m$.

In general, $P[i, m]$ depends on two "smaller" values, namely, $P[i-1, m]$ and $P[i-1, m-w_i]$. In the first case, we are not including the ith object in the knapsack, and in the second case, we are including it. However, if $w_i > m$, then we cannot include the ith object in the knapsack because there is not enough capacity to do so.

Taking into account all the possible cases, we obtain the following recurrence relation:

$$\begin{aligned} P[i,m] &= \max\{P[i-1,m], p_i + P[i-1,m-w_i]\} \\ &\qquad \text{if } i \geq 2,\ m \geq w_i \\ P[i,m] &= P[i-1,m] \qquad \text{if } i \geq 2,\ m < w_i \\ P[1,m] &= p_1 \qquad \text{if } m \geq w_1 \\ P[1,m] &= 0 \qquad \text{if } m < w_1. \end{aligned} \tag{6.5}$$

compute optimal solutions

In this dynamic programming algorithm, we are filling in a two-dimensional table of values $P[i,m]$, as opposed to the one-dimensional array that sufficed for computing the Fibonacci numbers or solving the **Treasure Hunt** problem.

The base cases comprise the first row of the table, so we will fill them in first.

In general, an entry $P[i,m]$ depends on two entries in row $i-1$. This suggests that we fill in the rows of the table in order, i.e., row 1, then row 2, etc. For convenience, we will fill in each row of the table from left to right (though other possibilities would also work).

For the last row, we only need to compute the single value $P[n,M]$ in the bottom right corner of the table.

It is now straightforward to write out the dynamic programming algorithm to solve the **0-1 Knapsack** problem; see Algorithm 6.5. We also present a small example to illustrate the working of the algorithm.

```
Algorithm 6.5: 0-1KNAPSACK(P, W, M)

comment: P and W are lists of n profits and weights, resp.
for m ← 0 to M
  do { if m ≥ w_1
         then P[1, m] ← p_1
         else P[1, m] ← 0
for i ← 2 to n
  do { for m ← 0 to M
         do { if m < w_i
                then P[i, m] ← P[i − 1, m]
                else { z_0 ← P[i − 1, m]
                       z_1 ← P[i − 1, m − w_i] + p_i
                       P[i, m] ← max{z_0, z_1}
return (P[n, M])
```

Example 6.1 Suppose we have six objects, with profits $\mathcal{P} = [1,2,3,5,7,10]$ and weights $\mathcal{W} = [2,3,5,8,13,16]$, and the capacity of the knapsack is $M = 22$. The following table is computed:

	0	1	2	3	4	5	6	7	8	9	10	11	12	13	14	15	16	17	18	19	20	21	22
1	0	0	1	1	1	1	1	1	1	1	1	1	1	1	1	1	1	1	1	1	1	1	1
2	0	0	1	2	2	3	3	3	3	3	3	3	3	3	3	3	3	3	3	3	3	3	3
3	0	0	1	2	2	3	3	4	5	5	6	6	6	6	6	6	6	6	6	6	6	6	6
4	0	0	1	2	2	3	3	4	5	5	6	7	7	8	8	9	10	10	11	11	11	11	11
5	0	0	1	2	2	3	3	4	5	5	6	7	7	8	8	9	10	10	11	11	11	12	12
6	-	-	-	-	-	-	-	-	-	-	-	-	-	-	-	-	-	-	-	-	-	-	13

The last entry, $P[6,22]$, is computed as follows:

$$\begin{aligned} P[6,22] &= \max\{P[5,22], P[5,6]+10\} \\ &= \max\{12, 3+10\} \\ &= \max\{12, 13\} \\ &= 13. \end{aligned}$$

We have highlighted the three relevant table entries in red that are used in this calculation. Since this is the "final" entry that is computed, we have established that the optimal profit for the given instance is 13. □

We have developed an algorithm to compute the maximum profit of a given problem instance. However, in practice, we might also want to determine the n-tuple $X = [x_1, \ldots, x_n] \in \{0,1\}^n$ that attains this maximum profit. It turns out that, after we compute the table of all the values $P[i,m]$, the optimal solution X can be obtained by what we call a ***traceback*** process. This basically consists of "tracing back" through the table. We illustrate the technique by returning to the instance considered in Example 6.1.

Example 6.2 We computed the following table:

	0	1	2	3	4	5	6	7	8	9	10	11	12	13	14	15	16	17	18	19	20	21	22
1	0	0	1	1	1	1	1	1	1	1	1	1	1	1	1	1	1	1	1	1	1	1	1
2	0	0	1	2	2	3	3	3	3	3	3	3	3	3	3	3	3	3	3	3	3	3	3
3	0	0	1	2	2	3	3	4	5	5	6	6	6	6	6	6	6	6	6	6	6	6	6
4	0	0	1	2	2	3	3	4	5	5	6	7	7	8	8	9	10	10	11	11	11	11	11
5	0	0	1	2	2	3	3	4	5	5	6	7	7	8	8	9	10	10	11	11	11	12	12
6	-	-	-	-	-	-	-	-	-	-	-	-	-	-	-	-	-	-	-	-	-	-	13

We start with the entry $P[6,22] = 13$. Recall that

$$P[6,22] = \max\{P[5,22], P[5,6]+10\},$$

where the first term is 12 and the second term is 13. This means that we include the last object in the optimal knapsack, i.e., $x_6 = 1$.

More succinctly, we observe that we can set

$$\begin{aligned} x_6 &= 1 \qquad \text{if } P[6,22] > P[5,22] \\ x_6 &= 0 \qquad \text{if } P[6,22] = P[5,22]. \end{aligned}$$

Having set $x_6 = 1$, the remaining capacity is $22 - 16 = 6$ and the optimal solution for the first five objects is given by the entry $P[5,6] = 3$. We have

$$\begin{aligned} P[5,6] &= P[4,6], && \text{so } x_5 = 0, \\ P[4,6] &= P[3,6], && \text{so } x_4 = 0, \\ P[3,6] &= P[2,6], && \text{so } x_3 = 0, \text{ and} \\ P[2,6] &> P[1,6], && \text{so } x_2 = 1. \end{aligned}$$

Now, the remaining capacity is $6 - 3 = 3$ and the optimal solution for the first object is given by the entry $P[1,3] = 1$. We have

$$P[1,3] > 0, \qquad \text{so } x_1 = 1.$$

The optimal solution is $X = [1,1,0,0,0,1]$. □

The process described in Example 6.2 works in general, and it leads to Algorithm 6.6. This algorithm takes the table P as input and computes the optimal solution $X = [x_1, \dots, x_n]$.

Algorithm 6.6: OPTIMALKNAPSACK($\mathcal{P}, \mathcal{W}, M, P$)

```
comment: P is computed by 0-1KNAPSACK
m ← M
p ← P[n, M]
for i ← n downto 2
  do { if p = P[i − 1, m]
         then x_i ← 0
         else { x_i ← 1
                p ← p − p_i
                m ← m − w_i
if p = 0
  then x_1 ← 0
  else x_1 ← 1
return (X = [x_1, ..., x_n])
```

Clearly we need the entire table of values $P[i,m]$ in order to compute the optimal solution X. If we just want to determine the value of the optimal profit, however, then we do not need to store the entire table. In Algorithm 6.5, it would suffice to keep track of the "current" row and the "previous" row of the table at any given time.

We now analyze the complexity of the dynamic programming algorithm that we have developed. Suppose, for convenience, we first carry out an analysis in the unit cost model. Thus we assume that additions and subtractions take time $O(1)$.

It is obvious that the complexity to initialize and construct the table of values $P[i, m]$ in Algorithm 6.5 is $\Theta(nM)$. The "traceback" in Algorithm 6.6 (given the table of values $P[i, m]$) only takes time $\Theta(n)$.

It is instructive to consider if Algorithm 6.5 is a polynomial-time algorithm. Certainly the complexity $\Theta(nM)$ looks like it is a polynomial. However, we need to remember that an algorithm is polynomial-time only if the complexity is a polynomial function of the size of the problem instance. For the **0-1 Knapsack** problem, an instance is specified by a collection of integers. Assuming that the integers are represented in binary, we have

$$\mathsf{size}(I) = \log_2 M + \sum_{i=1}^{n} \log_2 w_i + \sum_{i=1}^{n} \log_2 p_i.$$

The complexity of Algorithm 6.5 is $\Theta(nM)$. Note, in particular, that M is exponentially large compared to the term $\log_2 M$ in the expression for $\mathsf{size}(I)$. So constructing the table is not a polynomial-time algorithm, even in the unit cost model.

The complexity $\Theta(nM)$ for the dynamic programming algorithm is termed ***pseudo-polynomial***. This means that the complexity is a polynomial function of the *value* of the largest integer in the problem instance. Roughly speaking, an algorithm with pseudo-polynomial complexity will be efficient if all the integers in the problem instance are "small." On the other hand, in order to achieve polynomial complexity, we would require that the complexity of the algorithm is a polynomial function of the *bit-length* of the largest integer. Thus, as we see in the current situation, there is an exponential gap between polynomial and pseudo-polynomial complexity.

Another interesting comparison would be to a recursive algorithm that implements the recurrence relation (6.5) directly. Each $P[i, m]$ depends on two values $P[i - 1, -]$, so we would have two recursive calls each time the recursive algorithm is invoked. Since the recursion has depth n, a recursive algorithm would have complexity $\Theta(2^n)$ in the unit cost model.

The interesting point of comparison is that a recursive algorithm is practical if n is sufficiently small (e.g., say $n = 20$) even if the profits and weights are very large. In contrast, dynamic programming can easily handle much larger values of n (e.g., say $n = 1000$), as long as the profits and weights are quite small. Thus, there are instances that can be solved in a reasonable amount of time by dynamic programming, but not by recursion; and there are also instances that can be solved quickly by recursion, but not by dynamic programming.

6.5 Rod Cutting

We now study a problem that involves cutting a rod into pieces so as to maximize the total profit realized by the pieces.

Problem 6.3: Rod Cutting

Instance: An instance consists of a rod of length n, and a list of n values, $V = [v_1, \ldots, v_n]$. These are all positive integers.

Feasible solution: A subdivision (or partition) of the rod into one or more pieces, each of which has a length that is a positive integer.

Find: A feasible solution that achieves the maximum possible total value (or profit) realized by the pieces.

Example 6.3 Suppose that $n = 5$ and the values are

$$V = [3, 7, 12, 14, 16].$$

We list the seven feasible solutions and the profit achieved by each solution.

partition	profit
5	16
$4+1$	$14+3=17$
$3+2$	$12+7=19$
$3+1+1$	$12+3+3=18$
$2+2+1$	$7+7+3=17$
$2+1+1+1$	$7+3+3+3=16$
$1+1+1+1+1$	$3+3+3+3+3=15$

Hence, the optimal solution is given by the partition $3+2$, yielding profit 19. □

The number of feasible solutions is equal to the number of partitions of the integer n into positive integers. This ***partition number*** is often denoted by p_n. As can be seen from Example 6.3, $p_5 = 7$. The first ten partition numbers (starting from $p_1 = 1$) are as follows:

$$1, 2, 3, 5, 7, 11, 15, 22, 30, 42.$$

It may not be obvious from looking at small examples, but the partition numbers grow exponentially quickly as a function of n. In fact, it was shown by Hardy and Littlewood in 1918 that

$$p_n \sim \frac{e^{\pi\sqrt{2n/3}}}{4n\sqrt{3}}.$$

Clearly we should not attempt to enumerate all the possible solutions! However, dynamic programming can be applied to solve this problem quite successfully.

Suppose we focus on the rightmost cut of the rod. Say that this cut yields a piece having length k, where $1 \leq k \leq n$. The remaining (initial) portion of the rod thus has length $n-k$, and this could recursively be partitioned into smaller pieces. (Note that we allow $k = n$. Then the remaining portion has length 0, and we would be finished.) So we have one piece of length k and one subproblem of size $n-k$. The solution to the subproblem should be optimal if we want an optimal solution to the original problem instance.

Of course, we do not know which value of k will ultimately lead to the optimal solution, so we need to consider all possible values of k (i.e., $k = 1, \dots, n$).

Continuing this process recursively, the subproblems will involve rods of lengths $0, 1, \dots, n$, where the list of profits is fixed. Let $P[j]$ denote the maximum profit achievable from a rod of length j, where $0 \leq j \leq n$. Of course we eventually want to compute $P[n]$.

The recurrence relation can be written as follows.

$$\begin{aligned} P[j] &= \max\{v_k + P[j-k] : 1 \leq k \leq j\} \qquad \text{if } 1 \leq j \leq n \\ P[0] &= 0. \end{aligned} \tag{6.6}$$

The dynamic programming algorithm based on (6.6) would just compute $P[0], P[1], \dots, P[n]$, in that order. We illustrate by returning to Example 6.3.

Example 6.4 We again consider the list of values

$$V = [3, 7, 12, 14, 16].$$

We show the computation of each $P[j]$ in turn, using the recurrence relation (6.6):

j	k	profit	$P[j]$
0			$P[0] = 0$
1	1	$v_1 + P[0] = 3 + 0 = 3$	$P[1] = 3$
2	1	$v_1 + P[1] = 3 + 3 = 6$	
2	2	$v_2 + P[0] = 7 + 0 = 7$	$P[2] = \max\{6, 7\} = 7$
3	1	$v_1 + P[2] = 3 + 7 = 10$	
3	2	$v_2 + P[1] = 7 + 3 = 10$	
3	3	$v_3 + P[0] = 12 + 0 = 12$	$P[3] = \max\{10, 10, 12\} = 12$
4	1	$v_1 + P[3] = 3 + 12 = 15$	
4	2	$v_2 + P[2] = 7 + 7 = 14$	
4	3	$v_3 + P[1] = 12 + 3 = 15$	
4	4	$v_4 + P[0] = 14 + 0 = 14$	$P[4] = \max\{15, 14, 15, 14\} = 15$
5	1	$v_1 + P[4] = 3 + 15 = 18$	
5	2	$v_2 + P[3] = 7 + 12 = 19$	
5	3	$v_3 + P[2] = 12 + 7 = 19$	
5	4	$v_4 + P[1] = 14 + 3 = 17$	
5	5	$v_5 + P[0] = 16 + 0 = 16$	$P[5] = \max\{18, 19, 19, 17, 16\} = 19$

□

It is straightforward to present a pseudocode description of the algorithm to compute the profits $P[0], \ldots, P[n]$. However, we might also want to determine the partition that yields the optimal profit. It is easy to do this using a "traceback" algorithm if we keep track of the optimal k-values, in a separate list L, as we proceed. See Algorithm 6.7.

```
Algorithm 6.7: RODCUTTING(V)

comment: V = [v_1, ..., v_n] is a list of n positive integers
P[0] ← 0
for j ← 1 to n
  do { max ← v_1 + P[j − 1]
       maxlen ← 1
       for k ← 2 to j
         do { if v_k + P[j − k] > max
                then { max ← v_k + P[j − k]
                       maxlen ← k
       P[j] ← max
       L[j] ← maxlen
return (V, L)
```

Now, given the list L, we can find the optimal partition of the rod. This is done in Algorithm 6.8. Here is an example to illustrate.

```
Algorithm 6.8: OPTIMALROD(L)

comment: L is computed by RODCUTTING
partition ← ()
len ← n
while len > 0
  do { partition ← L[len] || partition
       len ← len − L[len]
return (partition)
```

Example 6.5 We consider the list of values

$$V = [3, 7, 12, 14, 16]$$

yet one more time. Looking at the calculations done in Example 6.4, we would have

$$\begin{aligned} L[1] &= 1 \\ L[2] &= 2 \\ L[3] &= 3 \\ L[4] &= 1 \\ L[5] &= 2. \end{aligned}$$

Then the optimal partition would be computed, using Algorithm 6.8, as follows:

$$\begin{aligned} partition &\leftarrow () \\ len &\leftarrow 5 \\ partition &\leftarrow (L[5]) = (2) \\ len &\leftarrow 5 - 2 = 3 \\ partition &\leftarrow (L[3], 2) = (3, 2) \\ len &\leftarrow 3 - 3 = 0. \end{aligned}$$

□

Finally, it is not hard to see that Algorithm 6.7 has complexity $\Theta(n^2)$ and Algorithm 6.8 has complexity $\Theta(n)$.

6.6 Coin Changing

We recall the **Coin Changing** problem that we studied in Section 5.8. See Problem 6.4.

Note that we are assuming that smallest denomination coin is $d_1 = 1$. This will be convenient for the "base cases" we consider. We would probably also consider the coins in increasing order of denomination, i.e., we would presort the coins so that

$$d_1 < \cdots < d_n,$$

although the algorithm will work correctly even if this is not done.

The **Coin Changing** problem has some similarities to the **0-1 Knapsack** problem and a dynamic programming algorithm to solve it has a similar flavor. In developing the algorithm, we will proceed by following the usual four steps: identify optimal structure, define subproblems, derive a recurrence relation, and fill in a table.

Suppose we begin by considering the number of coins of denomination d_n that we might use. Evidently, we could have several choices. We can use j coins of this denomination, for any integer j such that

$$0 \leq j \leq \left\lfloor \frac{T}{d_n} \right\rfloor.$$

Problem 6.4: Coin Changing

Instance: A list of ***coin denominations,***

$$D = [d_1, d_2, \ldots, d_n],$$

and a positive integer T, which is called the ***target sum***. We assume that $d_1 = 1$.

Find: An n-tuple of non-negative integers, say

$$A = [a_1, \ldots, a_n],$$

such that

$$T = \sum_{i=1}^{n} a_i d_i$$

and such that

$$N = \sum_{i=1}^{n} a_i$$

is minimized.

Given that we use exactly j coins of denomination d_n, the target sum is reduced from T to $T - jd_n$. This new target sum should be achieved using coins of the first $n - 1$ denominations, and clearly we would want to use an optimal number of coins in order to do so. Therefore, the optimal solution for the given problem instance (consisting of coins of n denominations) depends on optimal solutions to instances with coins of the first $n - 1$ denominations and varying target sums.

If we continue this process recursively, we are led to consider subproblems consisting of the first i coin denominations, for target sums ranging from 0 to T. Therefore, for $1 \leq i \leq n$ and $0 \leq t \leq T$, we define $N[i, t]$ to denote the optimal number of coins used to solve to the subproblem consisting of the first i coin denominations (namely, $d_1, \ldots, d_i$) and target sum t. These are all the subproblems that we will consider in the solution of the given problem instance.

Now we can develop the recurrence relation for the quantities $N[i, t]$. In the base case, we assume that $d_1 = 1$, and we immediately have $N[1, t] = t$ for all t.

For $i \geq 2$, the number of coins of denomination d_i will be an integer j where $0 \leq j \leq \lfloor t/d_i \rfloor$. If we use j coins of denomination d_i, then the target sum is reduced to $t - jd_i$, which we must achieve using the first $i - 1$ coin denominations. We must also take into account the j coins of denomination d_i.

Thus we have the following recurrence relation:

$$\begin{aligned} N[i,t] &= \min\left\{ j + N[i-1, t - jd_i] : 0 \le j \le \left\lfloor \frac{t}{d_i} \right\rfloor \right\} \\ &\qquad \text{if } i \ge 2, 0 \le t \le T \\ N[1,t] &= t \qquad \text{for } 0 \le t \le T. \end{aligned} \tag{6.7}$$

It will be useful to keep track of the number of coins of denomination d_i used in the optimal solution to each subproblem $N[i,t]$. We denote this number of coins by $A[i,t]$. That is, when we compute

$$\min\left\{ j + N[i-1, t - jd_i] : 0 \le j \le \left\lfloor \frac{t}{d_i} \right\rfloor \right\}$$

in (6.7), $A[i,t]$ is defined to be the integer j that yields the minimum value.

We need to fill the table of values $A[i,t]$ (and $N[i,t]$). Similar to the **0-1 Knapsack** problem, we will fill in the table from top to bottom, filling in each row from left to right. The entry in the bottom right corner, $N[n,T]$ is the solution to the given problem instance.

We now present our dynamic programming algorithm for the **Coin Changing** problem as Algorithm 6.9.

```
Algorithm 6.9: DPCoinChanging(D, T)

comment: D = [d_1, ..., d_n] is a list of n coin values and d_1 = 1
for t ← 0 to T
  do { N[1, t] ← t
       A[1, t] ← t
for i ← 2 to n
  do { for t ← 0 to T
         do { N[i, t] ← N[i − 1, t]
              A[i, t] ← 0
              for j ← 1 to ⌊t/d_i⌋
                do { if j + N[i − 1, t − jd_i] < N[i, t]
                       then { N[i, t] ← j + N[i − 1, t − jd_i]
                              A[i, t] ← j
return (N, A)
```

There are $\Theta(nT)$ table entries to compute, so the complexity of the algorithm is certainly $\Omega(nT)$ (in the unit cost model). The size of the problem instance is

$$\mathsf{size}(I) = \log_2 T + \sum_i \log_2 d_i.$$

This is therefore not a polynomial-time algorithm, because the term T in $\mathsf{size}(I)$

requires only $\log_2 T$ bits to specify and T is exponential in $\log_2 T$. It is a pseudo-polynomial-time algorithm, however.

Let's examine the complexity in a bit more detail. We will still assume the unit cost model, for simplicity. The exact complexity is not $\Theta(nT)$, because we cannot compute a table entry in constant time (we need to consider $\lfloor (t/d_i) \rfloor$ possible values of j in the innermost **for** loop).

We focus on the three nested loops, starting with the innermost loop and proceeding outwards, as usual.

- The innermost **for** loop (on j) has complexity $\Theta(t/d_i)$.
- The **for** loop on t has complexity

$$\begin{aligned}\sum_{t=0}^{T} \Theta\left(\frac{t}{d_i}\right) &= \Theta\left(\sum_{t=0}^{T} \frac{t}{d_i}\right) \\ &= \Theta\left(\frac{T^2}{d_i}\right).\end{aligned}$$

- The outer **for** loop (on i) has complexity

$$\begin{aligned}\sum_{i=1}^{n} \Theta\left(\frac{T^2}{d_i}\right) &= \Theta\left(T^2 \sum_{i=1}^{n} \frac{1}{d_i}\right) \\ &= \Theta(DT^2),\end{aligned}$$

  where

$$D = \sum_{i=1}^{n} \frac{1}{d_i}.$$

- If the set of coins is fixed, then the complexity is $O(T^2)$.

In order to compute the optimal set of coins, we trace back, making use of the information stored in the auxiliary table A. The value $A[i,t]$ is just the number of coins of denomination d_i in the optimal solution to the corresponding subproblem. We begin by selecting $A[n,T]$ coins of denomination d_n. Then we adjust T appropriately and look at the relevant entry in the previous row of the table. This process is continued until we have the entire list of coins in the optimal solution. See Algorithm 6.10.

Algorithm 6.10: COINCHANGINGTRACEBACK(D, T, A)

comment: A is computed by DPCOINCHANGING
$t \leftarrow T$
for $i \leftarrow n$ **downto** 1
  do $\begin{cases} a_i \leftarrow A[i,t] \\ t \leftarrow t - a_i d_i \end{cases}$
return $(a_1, \ldots, a_n)$

It is interesting that we only need the $A[i,t]$ values to do the traceback.

6.7 Longest Common Subsequence

We now study a problem in string matching which is presented as Problem 6.5.

Problem 6.5: Longest Common Subsequence

Instance: Two sequences $X = (x_1, \ldots, x_m)$ and $Y = (y_1, \ldots, y_n)$ over some finite alphabet Γ.

Find: A maximum length sequence Z that is a subsequence of both X and Y. This sequence is called the ***longest common subsequence*** (or **LCS**) of X and Y.

Problem 6.5 involves subsequences of a sequence. Here is the definition of this concept. We say that $Z = (z_1, \ldots, z_\ell)$ is a ***subsequence*** of X if there exist indices

$$1 \leq i_1 < \cdots < i_\ell \leq m$$

such that

$$z_j = x_{i_j},$$

for $1 \leq j \leq \ell$. Similarly, Z is a subsequence of Y if there exist (possibly different) indices

$$1 \leq h_1 < \cdots < h_\ell \leq n$$

such that

$$z_j = y_{h_j},$$

for $1 \leq j \leq \ell$.

As an example, suppose

$$X = \texttt{gdvegta} \quad \text{and} \quad Y = \texttt{gvcekst}.$$

The LCS is of X and Y is `gvet`, having length four.

We begin by identifying optimal structure. Let $\mathsf{LCS}(X, Y)$ denote the length of the LCS of the two strings X and Y. We have two observations that we state as lemmas.

LEMMA 6.1 *Let $X = (x_1, \ldots, x_m)$ and $Y = (y_1, \ldots, y_n)$. If $x_m = y_n$, then*

$$\mathsf{LCS}(X, Y) = 1 + \mathsf{LCS}(X', Y').$$

Here the LCS of X and Y ends with $x_m = y_n$.

PROOF Let $Z = (z_1, \ldots, z_\ell)$ be the LCS of X and Y. If $z_\ell \neq x_m$, then Z is a substring of X' and Y'. But then Z cannot be the LCS, because $Z \parallel x_m$ would be a longer substring of X and Y. Therefore,

$$z_\ell = x_m = y_n,$$

and it is clear that Z is formed by taking the LCS of X' and Y' and then concatenating x_m. ∎

LEMMA 6.2 *Let $X = (x_1, \ldots, x_m)$ and $Y = (y_1, \ldots, y_n)$. If $x_m \neq y_n$, then*

$$\mathsf{LCS}(X,Y) = \max\{\mathsf{LCS}(X,Y'), \mathsf{LCS}(X',Y)\},$$

where $X' = (x_1, \ldots, x_{m-1})$ and $Y' = (y_1, \ldots, y_{n-1})$.

PROOF Let $Z = (z_1, \ldots, z_\ell)$ be the LCS of X and Y. We identify three cases.

1. If $z_\ell = y_n$, then Z is the LCS of X' and Y, because $z_\ell \neq x_m$ in this case.
2. Similarly, if $z_\ell = x_m$, then Z is the LCS of X and Y', because here $z_\ell \neq y_n$.
3. If $z_\ell \notin \{x_m, y_n\}$, then Z is the LCS of X' and Y'.

Thus, we have

$$\mathsf{LCS}(X,Y) = \max\{\mathsf{LCS}(X,Y'), \mathsf{LCS}(X',Y), \mathsf{LCS}(X',Y')\}.$$

It is obvious that $\mathsf{LCS}(X',Y) \geq \mathsf{LCS}(X',Y')$, so

$$\mathsf{LCS}(X,Y) = \max\{\mathsf{LCS}(X,Y'), \mathsf{LCS}(X',Y)\}.$$

∎

If we apply Lemmas 6.1 and 6.2 recursively, we are led to consider subproblems consisting of all possible ***prefixes*** of X and Y. Therefore, we define $c[i,j]$ to be the length of the LCS of $(x_1, \ldots, x_i)$ and $(y_1, \ldots, y_j)$. If $i = 0$ or $j = 0$, then we are considering the "empty prefix" of X or Y (respectively). The optimal solution to the original problem instance will be $c[m,n]$.

Using the same analysis that was done in Lemmas 6.1 and 6.2, we obtain the following recurrence relation:

$$c[i,j] = \begin{cases} c[i-1,j-1]+1 & \text{if } i,j \geq 1 \text{ and } x_i = y_j \\ \max\{c[i-1,j], c[i,j-1]\} & \text{if } i,j \geq 1 \text{ and } x_i \neq y_j \\ 0 & \text{if } i = 0 \text{ or } j = 0. \end{cases} \tag{6.8}$$

Let's consider how to fill in the table of values $c[i,j]$. The base cases consist of the top (zeroth) row and the leftmost (zeroth) column. In general, when we fill in a value $c[i,j]$ with $i,j \geq 1$, we need to look up the values $c[i-1,j]$ and $c[i,j-1]$ (if $x_i \neq y_j$) or the value $c[i-1,j-1]$ (if $x_i = y_j$). Therefore, if we fill in the table row by row, from top to bottom, and we fill in each row from left to right, all values that we might need to look up will have already been computed. It is now straightforward to write out a dynamic programming algorithm to compute $\mathsf{LCS}(X,Y)$; see Algorithm 6.11, which has complexity $\Theta(mn)$.

```
Algorithm 6.11: LCS1(X, Y)

comment: X and Y are strings of lengths m and n, respectively
for i ← 0 to m
  do c[i, 0] ← 0
for j ← 0 to n
  do c[0, j] ← 0
for i ← 1 to m
  do { for j ← 1 to n
         do { if x_i = y_j
                then c[i, j] ← c[i − 1, j − 1] + 1
                else c[i, j] ← max{c[i, j − 1], c[i − 1, j]}
return (c);
```

Of course we might want to find the actual LCS of X and Y, not just its length. To facilitate this, we keep track of three possible cases that can arise in the recurrence relation (6.8) as the algorithm proceeds:

- if $x_i = y_j$, we include this symbol in the LCS and denote this case by **UL**,
- if $x_i \neq y_j$ and $c[i, j-1] > c[i-1, j]$, denote this case by **L**,
- if $x_i \neq y_j$ and $c[i, j-1] \leq c[i-1, j]$, denote this case by **U**.

As a mnemonic aid, **U** means "up," **L** means "left" and **UL** is an abbreviation for "up and to the left."

The modified algorithm, Algorithm 6.12, also constructs a table of values $\pi[i, j]$ which stores the relevant "directions" **U** , **L** and **UL**.

```
Algorithm 6.12: LCS2(X, Y)

comment: X and Y are strings of lengths m and n, respectively
for i ← 0 to m
  do c[i, 0] ← 0
for j ← 0 to n
  do c[0, j] ← 0
for i ← 1 to m
  do { for j ← 1 to n
         do { if x_i = y_j
                then { c[i, j] ← c[i − 1, j − 1] + 1
                       π[i, j] ← UL
              else if c[i, j − 1] > c[i − 1, j]
                then { c[i, j] ← c[i, j − 1]
                       π[i, j] ← L
                else { c[i, j] ← c[i − 1, j]
                       π[i, j] ← U
return (c, π);
```

Now we can use the table of values $\pi[i,j]$ in Algorithm 6.13, which is a simple traceback algorithm to construct the LCS of X and Y. The three cases are handled as follows:

- if $\pi[i,j] = \mathbf{UL}$, we include the current symbol in the LCS (prepending it onto the LCS we are constructing), and decrease the values of i and j by 1,
- if $\pi[i,j] = \mathbf{L}$, we decrease the value of j by 1, and
- if $\pi[i,j] = \mathbf{U}$, we decrease the value of i by 1.

Algorithm 6.13: FINDLCS(π)

```
comment: π is computed by LCS2
seq ← ()
i ← m
j ← n
while min{i, j} > 0
      ⎧ if π[i, j] = UL
      ⎪          ⎧ seq ← x_i || seq
      ⎪   then   ⎨ i ← i − 1
  do  ⎨          ⎩ j ← j − 1
      ⎪ else if π[i, j] = L
      ⎪   then j ← j − 1
      ⎩   else i ← i − 1
return (seq)
```

Observe that Algorithm 6.13 only requires the table of values $\pi[i,j]$ to perform the traceback. The complexity of this algorithm is $\Theta(n+m)$, where $|X| = m$ and $|Y| = n$.

We work out a small example to illustrate the traceback process.

Example 6.6 Suppose $X =$ `gdvegta` and $Y =$ `gvcekst`. Table 6.1 superimposes the values $c[i,j]$ and $\pi[i,j]$. Also, to make the trackback more appealing visually, we use arrows as follows: **UL** is replaced by $\nwarrow$, **L** is replaced by $\leftarrow$, and **U** is replaced by $\uparrow$.

The traceback now just consists of following the path traced out by the table entries that are highlighted in red, starting from the lower right corner. Every time we encounter a **UL** arrow, we include the corresponding symbol in the LCS. Note that the LCS is constructed in reverse order, from the end to the beginning. In this example, the LCS is `gvet`. □

TABLE 6.1: Finding a longest common subsequence.

	X		g	d	v	e	g	t	a
Y		$i=0$	1	2	3	4	5	6	7
	$j=0$	0	0	0	0	0	0	0	0
g	1	0	$\nwarrow 1$	$\leftarrow 1$	$\leftarrow 1$	$\leftarrow 1$	$\nwarrow 1$	$\leftarrow 1$	$\leftarrow 1$
v	2	0	$\uparrow 1$	$\uparrow 1$	$\nwarrow 2$	$\leftarrow 2$	$\leftarrow 2$	$\leftarrow 2$	$\leftarrow 2$
c	3	0	$\uparrow 1$	$\uparrow 1$	$\uparrow 2$	$\uparrow 2$	$\uparrow 2$	$\uparrow 2$	$\uparrow 2$
e	4	0	$\uparrow 1$	$\uparrow 1$	$\uparrow 2$	$\nwarrow 3$	$\leftarrow 3$	$\leftarrow 3$	$\leftarrow 3$
k	5	0	$\uparrow 1$	$\uparrow 1$	$\uparrow 2$	$\uparrow 3$	$\uparrow 3$	$\uparrow 3$	$\uparrow 3$
s	6	0	$\uparrow 1$	$\uparrow 1$	$\uparrow 2$	$\uparrow 3$	$\uparrow 3$	$\uparrow 3$	$\uparrow 3$
t	7	0	$\uparrow 1$	$\uparrow 1$	$\uparrow 2$	$\uparrow 3$	$\uparrow 3$	$\nwarrow 4$	$\leftarrow 4$

6.8 Minimum Length Triangulation

In this section, we study a geometric problem concerning dividing a polygon into triangles. We begin by defining two closely related versions of the problem as Problems 6.6 and 6.7.

Problem 6.6: Minimum Length Triangulation v1

Instance: n points $q_1, \dots, q_n$ in the Euclidean plane that form a convex n-gon P, where $n \geq 3$.

Find: A triangulation of P such that the sum S_c of the lengths of the $n-3$ chords is minimized.

Problem 6.7: Minimum Length Triangulation v2

Instance: n points $q_1, \dots, q_n$ in the Euclidean plane that form a convex n-gon P, where $n \geq 3$.

Find: A triangulation of P such that the sum S_p of the perimeters of the $n-2$ triangles is minimized.

A couple of comments are in order. First, the triangulation is formed from chords, each of which joins two vertices of the polygon. Also, the chords are not allowed to "cross." It follows that if we have $n-3$ chords of an n-gon, then the polygon is divided into $n-2$ disjoint triangles.

Let L denote the length of the perimeter of the polygon P. Then we have that

$$S_p = L + 2S_c.$$

This is because each chord occurs in two adjacent triangles, and each edge of the polygon occurs in one triangle of the triangulation. Hence the two versions of the **Minimum Length Triangulation** problem are equivalent in the sense that they have the same optimal solutions.

The number of triangulations of a convex n-gon is equal to the $(n-2)$nd ***Catalan number***, where the nth Catalan number is given by the formula

$$C_n = \frac{1}{n+1}\binom{2n}{n}.$$

The sequence of Catalan numbers begins

$$\begin{array}{lll} C_1 = 1 & C_2 = 2 & C_3 = 5 \\ C_4 = 14 & C_5 = 42 & C_6 = 132 \\ C_7 = 429 & C_8 = 1430 & C_9 = 4862. \end{array}$$

Thus there is one triangulation of a triangle, two triangulations of a quadrilateral, five triangulations of a pentagon, etc.

Using Stirling's formula (1.17), it can be shown that $C_n \in \Theta(4^n/n^{3/2})$. Thus the number of possible triangulations of an n-gon grows exponentially, and an algorithm that enumerates all the possible triangulations would not be feasible for large n. However, it turns out that we can develop an efficient dynamic programming algorithm for the **Minimum Length Triangulation** problem.

We will focus on version 2 of the problem. First we identify the optimal structure. The edge q_nq_1 is in a triangle with a third vertex q_k, where $k \in \{2, \dots, n-1\}$. For a given vertex k, we can consider the original polygon to be decomposed into three pieces, as shown in Figure 6.2:

(1) the triangle $q_1q_kq_n$,

(2) the polygon with vertices $q_1, \dots, q_k$,

(3) the polygon with vertices $q_k, \dots, q_n$.

The polygon in (2) is trivial if $k = 2$, and the polygon in (3) is trivial if $k = n-1$. These "polygons" do not lead to triangles in the resulting triangulation, and thus they can be considered to have perimeters of length equal to 0.

The optimal solution to the given problem instance will consist of optimal solutions to the two subproblems defined in (2) and (3), along with the triangle in (1) (which will not be subdivided further). For the subproblem (2), we would focus on the "last" edge q_nq_k and pick a third point to form a triangle that includes that edge. Similarly, for the subproblem (3), we would choose a point to form a

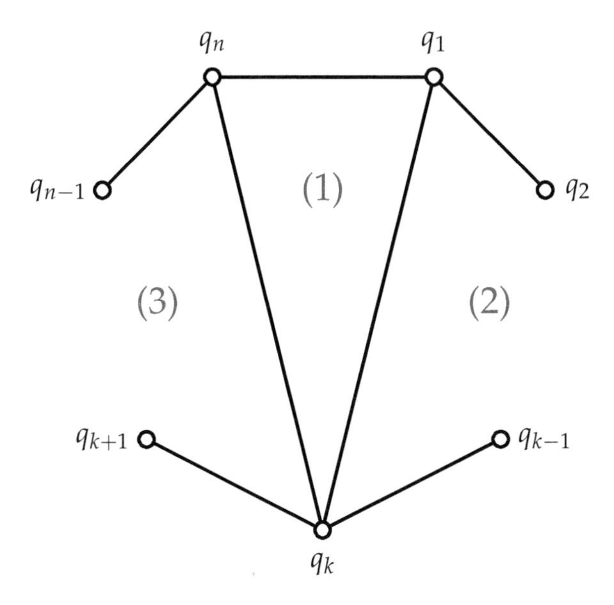

FIGURE 6.2: Optimal structure in a triangulation.

triangle with the edge $q_k q_1$. Recursively, we are led to consider all the possible subproblems that are defined by polygons having *consecutive vertices*, $q_i, \ldots, q_j$, where $1 \leq i < j \leq n$.

Therefore, for $1 \leq i < j \leq n$, we let $S[i, j]$ denote the optimal solution to the subproblem consisting of the polygon having vertices $q_i, \ldots, q_j$. How many subproblems are there? The pair of indices $\{i, j\}$ are chosen from $\{1, \ldots, n\}$, so the number of subproblems to be examined is $\binom{n}{2} \in \Theta(n^2)$. This is already quite promising, because an exponential number of triangulations can be tested by solving a quadratic number of subproblems.

It is also convenient to define $\Delta(q_i, q_k, q_j)$ to be the length of the perimeter of the triangle having vertices q_i, q_k and q_j. Then we have the recurrence relation

$$S[i, j] = \min\{\Delta(q_i, q_k, q_j) + S[i, k] + S[k, j] : i < k < j\}. \tag{6.9}$$

The base cases are given by

$$S[i, i+1] = 0$$

for all i.

So we need to fill in a table of values $S[i, j]$ for all $1 \leq i < j \leq n$. The question is, in what order should we fill these table entries? We would probably start with the base cases, where $j = i + 1$.

Next, we could consider cases with $j = i + 2$. For such cases, we can only take $k = j + 1$ in (6.9). The evaluation of $S[i, i+2]$ can be carried out by referring to the two base cases $S[i, i+1]$ and $S[i+1, i+2]$.

Continuing this pattern, we could next compute all $S[i, j]$ with $j = i + 3$, then all $S[i,j]$ with $j = i+4$, etc. That is, we fill in the table of values $S[i,j]$ one diagonal at a time. Then, whenever we compute an entry $S[i,j]$ using (6.9), we make use of entries in row i, to the left of $S[i,j]$, and in column j, below $S[i,j]$. Both of these entries are in "previous" diagonals and hence they have already been computed.

The cth diagonal, for $c = 1,2,3,\ldots,n-1$, consists of all values $S[i,j]$ with $j = i+c$. The solution to the original problem instance is $S[1,n]$, which is the unique entry in the $(n-1)$st diagonal.

The following diagram illustrates the idea for $n = 6$. Each cell (i,j) above the main diagonal contains the entry c, where $j = i+c$. We can see that there are five cells with $c = 1$, four cells with $c = 2$, etc. Finally, the cell $(1,6)$ (which has the entry $c = 5$) would contain the solution to the given problem instance.

	1	2	3	4	5
		1	2	3	4
			1	2	3
				1	2
					1

As we have already mentioned, we are filling in a table of $\Theta(n^2)$ entries. In order to determine the complexity of the algorithm, we need to take into account the time required to compute each entry $S[i,j]$.

Each of the $n-1$ base cases requires only $\Theta(1)$ time. This yields a total time $\Theta(n-1)$.

Now, suppose $j = i+c$ where $c > 1$. If we look at the recurrence (6.9), we can observe that computing the entry $S[i,i+c]$ involves finding the minimum of $c-1$ values, each of which can be computed in $\Theta(1)$ time. Also, there are $n-c$ entries of the form $S[i,i+c]$ that need to be computed. So the time to compute all the entries $S[i,i+c]$ is $\Theta((n-c)(c-1))$.

The complexity of the algorithm (including the base cases) is

$$\Theta(n-1) + \sum_{c=2}^{n-1} \Theta((n-c)(c-1)) = \Theta\left(\sum_{c=2}^{n-1} (n-c)(c-1)\right).$$

It is clear that the sum

$$\sum_{c=2}^{n-1} (n-c)(c-1) \in O(n^3),$$

since each of the $n-2$ terms in the sum is upper-bounded by n^2.

With a bit more work, using the formulas (1.1) and (1.21) from Chapter 1, it can be shown that the sum is actually in $\Theta(n^3)$. Thus the algorithm has cubic complexity.

Finally, we can consider how to find the optimal triangulation, given the table of values $S[i,j]$. Evidently, it would be useful for the purposes of a traceback algorithm to keep track of the "optimal" k-value as each entry $S[i,j]$ is computed. Each of these k-values indicates that a particular triangle ijk is part of the optimal triangulation. We leave the details for the reader to work out.

6.9 Memoization

The goal of dynamic programming is to eliminate solving subproblems more than once. ***Memoization*** is another way to accomplish the same objective. The term "memoization" was coined by Donald Michie in 1968 and is derived from the Latin word "memorandum."

When using memoization, the first three steps of developing the algorithm (namely, identify optimal structure, define subproblems, and derive a recurrence relation) are the same as in a dynamic programming algorithm. However, the fourth step is different. We do not fill in a table "bottom-up." Rather, memoization is a (top-down) recursive algorithm that is nevertheless based on the same recurrence relation that a dynamic programming algorithm would use.

The main idea is to remember which subproblems have been solved; if the same subproblem is encountered more than once during the recursion, the solution will be looked up in a table rather than being re-calculated. This is easy to do if we initialize a table of all possible subproblems having the value *undefined* in every entry. Whenever a subproblem is solved, the table entry is updated.

We will illustrate the basic technique by looking again at a couple of the problems we studied in earlier sections of this chapter.

6.9.1 Fibonacci Numbers

First, we revisit the problem of computing the Fibonacci numbers. As before, suppose we want to find the Fibonacci number f_n, based on the recurrence relation (6.1). This will ultimately require computing all the Fibonacci numbers f_i, for $0 \leq i \leq n$. We use Algorithm 6.1 as our starting point, but we make sure to never recompute a Fibonacci number that we have already computed. The following modifications will allow us to accomplish this goal.

1. We will create a list M of the Fibonacci numbers up to f_n.
2. Initially, all entries in M are "undefined." We will use the value -1 to indicate an undefined entry.
3. The list M does not contain the base cases f_0 or f_1.
4. If f_i has already been computed, we look it up in the list M.
5. If we compute a "new" f_i, we store it in the list M.

The resulting memoized algorithm is presented as Algorithm 6.14. After initializing the entries in M, we call the recursive function RECFIB. This function first checks to see if we are evaluating a base case. If not, it checks to see if the desired Fibonacci number has already been computed. If it has not previously been computed, then

- it invokes the recurrence relation (6.1) by making the two relevant recursive calls,
- it computes the desired Fibonacci number, and
- it adds the newly computed Fibonacci number to the list M.

Algorithm 6.14: MEMOFIB(n)

```
procedure RECFIB(n)
 if n = 0
   then f ← 0
 else if n = 1
   then f ← 1
 else if M[n] ≠ −1
   then f ← M[n]
        ⎧ f1 ← RECFIB(n − 1)
   else ⎨ f2 ← RECFIB(n − 2)
        ⎪ f ← f1 + f2
        ⎩ M[n] ← f
 return (f)

main
 for i ← 2 to n
   do M[i] ← −1
 return (RECFIB(n))
```

Memoization is very effective—it reduces the size of the recursion tree to $\Theta(n)$. We illustrate the application of Algorithm 6.14 in the case $n = 5$ by constructing the recursion tree in Figure 6.3. It is interesting and instructive to compare this tree to the one in Figure 6.1.

The red values in Figure 6.3 are table look-ups. Each of these look-ups replaces a subtree from Figure 6.1. It is not difficult to show that the total number of nodes in the memoized recursion tree to compute f_n is $2n - 2$. Thus the memoized recursion tree is exponentially smaller than the original tree.

6.9.2 0-1 Knapsack

In this section, we present a memoized algorithm to solve the **0-1 Knapsack** problem. This algorithm will be based on the recurrence relation (6.5), which we

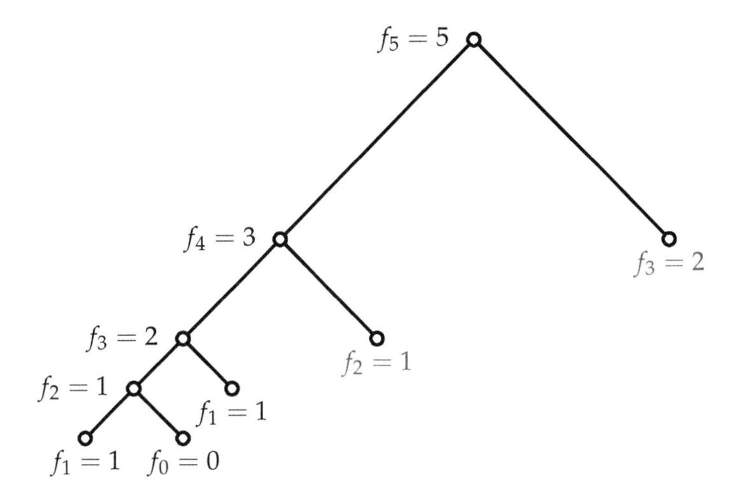

FIGURE 6.3: The memoized recursion tree to evaluate f_5.

restate now:

$$\begin{aligned} P[i,m] &= \max\{P[i-1,m], p_i + P[i-1,m-w_i]\} \\ &\qquad \text{if } i \geq 2,\ m \geq w_i \\ P[i,m] &= P[i-1,m] \qquad \text{if } i \geq 2,\ m < w_i \\ P[1,m] &= p_1 \qquad \text{if } m \geq w_1 \\ P[1,m] &= 0 \qquad \text{if } m < w_1. \end{aligned}$$

Algorithm 6.15: BADKNAP(i, m)

```
global 𝒫, 𝒲, M
comment: 𝒫 and 𝒲 are lists of n profits and weights, resp.
if i = 1
  then ⎧ if m < w_1
       ⎨   then P ← 0
       ⎩   else P ← p_1
 else if m < w_i
  then P ← BADKNAP(i − 1, m)
       ⎧ P_1 ← BADKNAP(i − 1, m)
  else ⎨ P_2 ← BADKNAP(i − 1, m − w_i) + p_i
       ⎩ P ← max{P_1, P_2}
return (P)
```

Algorithm 6.15 is an inefficient recursive algorithm based on this recurrence relation. Algorithm 6.16 is an improved, memoized version of the algorithm that makes the following changes:

- initialize the entries $P[i, m]$ of the table to be undefined, for $i \geq 2$ (i.e., ignore base cases);
- in the recursive algorithm, first check for base cases;
- if it is not a base case, then check to see if the relevant table entry has already been computed;
- if the table entry has not already been computed, proceed to the two recursive calls, and place the final result in the table.

Algorithm 6.16: MEMOKNAP($\mathcal{P}, \mathcal{W}, M$)

```
comment: P and W are lists of n profits and weights, resp.

procedure RECKNAP(i, m)
 global P, W, M
 if i = 1
   then { if m < w_1
            then P ← 0
            else P ← p_1
  else if P[i, m] ≠ −1
   then f ← P[i, m]
   else { P_1 ← RECKNAP(i − 1, m)
          P_2 ← RECKNAP(i − 1, m − w_i) + p_i
          P ← max{P_1, P_2}
          P[i, m] ← P
 return (P);

main
 for i ← 2 to n
   do { for m ← 0 to M
          do P[i, m] ← −1
 return (RECKNAP(n, M))
```

6.9.3 Comparison of Dynamic Programming and Memoization

It is not immediately clear which of dynamic programming or memoization is more efficient. They both have advantages and disadvantages. We list several points that can be considered:

- Memoization involves the use of recursion, which has a certain amount of overhead associated with it, in order to maintain the stack of recursive calls.
- For some problems, initializing the table when using memoization already has the same complexity that a dynamic programming algorithm would have. Of course, memoization still might turn out to be faster, depending on constant factors, if the running time of the two algorithms has the same complexity. In this case, it might be easiest to implement the two algorithms and perform some timings to see which one performs better in practice.
- Another approach to implement memoization is to store the solved subproblems in a *hash table*. This avoids having to initialize a table of all possible subproblems as "undefined." Of course, we would require a good hash function, along with a mechanism for collision resolution. Finally, the hash table itself would need to be initialized. Some techniques for hash tables were presented in Section 3.6.
- Dynamic programming solves every possible subproblem. In some situations, memoization might turn out to be more efficient if a relatively small number of the "possible" subproblems actually need to be solved.
- A traceback algorithm to find an optimal solution is similar in the case of memoization as it was for dynamic programming. The memoized algorithm constructs a table of solutions of all the relevant subproblems. If we record possibly additional relevant information as the table is constructed, then the traceback algorithm will be identical to the one used in the dynamic programming setting.

6.10 Notes and References

The motivation for the term "dynamic programming" is quite interesting. Richard Bellman, who invented dynamic programming in 1950, related the following, which is excerpted from his autobiography [19]:

> What title, what name, could I choose? In the first place I was interested in planning, in decision making, in thinking. But planning, is not a good word for various reasons. I decided therefore to use the word, 'programming.' I wanted to get across the idea that this was dynamic, this was multistage, this was time-varying—I thought, let's kill two birds with one stone. Let's take a word that has an absolutely precise meaning, namely dynamic, in the classical physical sense. It also has a very interesting property as an adjective, and that is it's impossible to use the word, dynamic, in a pejorative sense. Try thinking of some

> combination that will possibly give it a pejorative meaning. It's impossible. Thus, I thought dynamic programming was a good name. It was something not even a Congressman could object to. So I used it as an umbrella for my activities.

It is also interesting to read Bellman's definition of optimal structure [4]:

> PRINCIPLE OF OPTIMALITY. An optimal policy has the property that whatever the initial state and initial decisions are, the remaining decisions must constitute an optimal policy with regard to the state resulting from the first decisions.

Bergroth, Hakonen and Raita [7] is a survey on the **Longest Common Subsequence** problem and related problems.

The dynamic programming algorithm for **Minimum Length Triangulation** was discovered relatively recently, in 1980 (see Klincsek [40]), despite the fact that it is quite a simple algorithm.

Donald Michie introduced the idea of memoization in a 1968 article published in *Nature* [49]. The abstract of this article states:

> It would be useful if computers could learn from experience and thus automatically improve the efficiency of their programs during execution. A simple but effective rote-learning facility can be provided within the framework of a suitable programming language.

Exercises

6.1 Use induction to prove that $f_n \geq 2^{n/2}$ for all $n \geq 6$.

6.2 Use equation (1.16) to prove that $C_n \in \Theta(4^n / n^{3/2})$.

6.3 We define the **Weighted Interval Selection** problem:

> **Problem 6.8: Weighted Interval Selection**
>
> **Instance:** We are given n intervals, say $I_j = [s_j, f_j]$ for $1 \leq j \leq n$, and for each I_j we have a weight $w_j > 0$.
>
> **Find:** A subset of pairwise disjoint intervals, such that the sum of the weights of the chosen intervals is maximized.

Suppose that the intervals are pre-sorted by finishing time, i.e., $f_1 \leq \cdots \leq f_n$. For each j, $1 \leq j \leq n$, let

$$b(j) = \max\{i : f_i \leq s_j\}.$$

If $f_i > s_j$ for all i, define $b(j) = 0$. Also, let $W(j)$ denote the maximum achievable weight for the subproblem consisting of the first j intervals. Thus, $W(n)$ is the solution for the given problem instance.

(a) It is easy to compute all n values $b(1), \ldots, b(n)$ in time $O(n^2)$. Give a high-level description of a method to compute these n values in time $O(n \log n)$.

(b) Give a recurrence relation that can be used to compute the values $W(j)$, $j = 1, \ldots, n$. This part of the algorithm should take time $\Theta(n)$. In general, $W(j)$ will depend on two previous W-values. The $b(j)$ values will be used in the recurrence relation, and you need to handle the situation where $b(j) = 0$ correctly. You should give a brief explanation justifying the correctness of your recurrence relation.

(c) Solve the following problem instance by dynamic programming, using the recurrence relation that you presented in part (b). Give the values $b(1), \ldots, b(n)$ as well as $W(1), \ldots, W(n)$ and show all the steps carried out to perform these computations:

interval	$[s_j, f_j]$	w_j	interval	$[s_j, f_j]$	w_j
I_1	$[2,4]$	3	I_2	$[1,5]$	4
I_3	$[2,6]$	3	I_4	$[5,8]$	4
I_5	$[4,10]$	6	I_6	$[7,12]$	5
I_7	$[9,13]$	5	I_8	$[6,14]$	6
I_9	$[11,16]$	3	I_{10}	$[12,17]$	4
I_{11}	$[8,18]$	5	I_{12}	$[10,20]$	3

6.4 A ***palindrome*** is a string that is the same backwards as forwards, e.g., `kayak` or `abbaabba`. The **Longest Palindromic Subsequence** problem asks for the *longest palindromic subsequence* (or *LPS*) of a given string. For example, the longest LPS of the string `character` is `carac`.

The **Longest Palindromic Subsequence** problem can be solved using dynamic programming. Suppose that $(a_1, \ldots, a_n)$ is the string for which we want to find the LPS. For $1 \leq i \leq j \leq n$, define $L[i, j]$ to be the length of the LPS of $(a_i, \ldots, a_j)$.

(a) Write down a recurrence relation to compute the values $L[i, j]$. You can treat $j = i$ and $j = i + 1$ as base cases. The general case is when $j \geq i + 2$.

(b) Construct a dynamic programming algorithm using the recurrence relation from part (a). There will be n iterations. In iteration k (where $0 \leq k \leq n - 1$), compute all the values $L[i, i + k]$, where $1 \leq i \leq n - k$. The final answer is $L[1, n]$. The complexity of computing the table of values $L[i, j]$ should be $\Theta(n^2)$.

(c) Include a "trace-back" to actually compute the LPS (not just its length). When you perform the trace-back, you need to decide what to do in the situations where $L[i+1,j] = L[i,j-1]$. Suppose we use the convention of choosing $L[i,j-1]$ if $L[i+1,j] = L[i,j-1]$ and we are in this case of the recurrence. Similarly, we choose $L[i]$ when $j = i+1$ and $L[j] \neq L[i]$. The complexity of the trace-back should be $O(n)$.

6.5 We consider a more complicated version of the house-building problem, which was Problem 5.4.

> **Problem 6.9: House Construction Optimization**
>
> **Instance:** Three lists of n positive integers,
>
> $$D = [d_1, \ldots, d_n], \quad L = [\ell_1, \ldots, \ell_n] \quad \text{and} \quad P = [p_1, \ldots, p_n].$$
>
> The d_i's are *deadlines*, the ℓ_i's are *construction times* and the p_i's are *profits* for n houses.
>
> **Find:** A feasible schedule to build a subset of the houses (i.e., a schedule such that the houses in the given subset are completed on time), and such that the total profit of these houses is maximized.

For example, suppose that we have $n = 3$, $D = [3,4,3]$, $L = [2,3,1]$ and $P = [8,10,6]$. It is not hard to see that the optimal solution is to build house 3 and then house 2, obtaining profit $6 + 10 = 16$.

(a) Let $M[d,i]$ denote the maximum possible profit for a schedule only considering up to time d (i.e., all houses in the schedule must be completed by time d) and only considering houses 1 to i. Give a recurrence relation for computing the values $M[d,i]$.

HINT In deriving the recurrence relation, it may be useful to preprocess the input in a particular way.

(b) Obtain a dynamic programming algorithm based on the recurrence relation from part (a). What is the complexity of your algorithm?

(c) Create the table $M[d,i]$ which computes the profit of the optimum schedule for the following input:

$$D = [2,3,4,6,6]$$
$$L = [2,3,2,3,2]$$
$$P = [15,30,25,30,25].$$

6.6 We introduce a new problem, **Positive Split**. We are given a list of an even number of integers, such that their sum is positive. We seek a *positive split* (i.e., we want to split the list T into two equal-sized sublists such that both sublists have positive sum). See Problem 6.10 for a mathematical description of the problem.

Problem 6.10: Positive Split

Instance: A list of n integers,

$$T = [t_1, \dots, t_n],$$

such that n is even and such that

$$S = \sum_{i=1}^{n} t_i > 0.$$

Find: A subset $I \subseteq \{1, \dots, n\}$ such that $|I| = n/2$,

$$\sum_{i \in I} t_i > 0 \quad \text{and} \quad \sum_{i \in \{1,\dots,n\} \setminus I} t_i > 0.$$

For example, suppose we have $n = 4$ and $T = [10, -14, 20, -6]$. Then $S = 10$, and we can split T into two sublists of size two having positive sums, as required: $[10, -6]$ has sum 4 and $[-14, 20]$ has sum 6. On the other hand, if $T = [5, -14, 20, -6]$, then $S = 5$ but there is no positive split.

(a) Suppose that n is even and $S > 0$. Prove that T has a positive split if and only if there exists $I \subseteq \{1, \dots, n\}$ such that $|I| = n/2$ and

$$\sum_{i \in I} t_i \in \{1, \dots, s - 1\}.$$

(b) Suppose we define a three-dimensional Boolean array A, where $A[m, r, u] = \mathit{true}$ if and only if there exists $J \subseteq \{1, \dots, m\}$ such that $|J| = r$ and

$$\sum_{i \in J} t_i = u.$$

Here $1 \leq m \leq n, 0 \leq r \leq n/2$ and $S^- \leq u \leq S^+$, where

$$S^- = \sum_{i:t_i<0} t_i \quad \text{and} \quad S^- = \sum_{i:t_i<0} t_i.$$

Write down a recurrence relation, including base cases, for the values $A[m, r, u]$.

(c) Using the recurrence relation from part (b), obtain a dynamic programming algorithm to solve **Positive Split**. What is the complexity of your algorithm?

6.7 Suppose that two teams play a series of games. For $i = 1, 2, \dots,$ the probability that team 1 wins the ith game is p_i. The teams play until one team wins W games.

(a) Let $T[j, k]$ denote the probability that team 1 wins exactly j of the first k games, where

$$0 \leq j \leq k, \quad j \leq W \quad \text{and} \quad k - j \leq W.$$

Write down a recurrence relation for the values $T[j,k]$, including base cases.

(b) Using the recurrence relation from part (a), obtain a dynamic programming algorithm to determine the probability that team 1 wins the series.

6.8 In the **Coin Changing** problem (Problem 5.6), we defined $N[i,t]$ to denote the optimal number of coins used to solve to the subproblem consisting of the first i coin denominations (namely, $d_1, \ldots, d_i$) and target sum t. The recurrence relation (6.7) gives a formula for $N[i,t]$ that requires us to compute the minimum of various values $N[i-1,t']$.

(a) A simpler recurrence relation only requires computing the minimum of two values. Write out such a recurrence relation, which is based on whether $N[i,t]$ includes at least one coin of denomination d_i, or (alternatively) no coins of denomination d_i.

(b) Develop a dynamic programming algorithm based on the modified recurrence relation.

(c) What is the complexity of the dynamic programming algorithm from part (b)?

Chapter 7

Graph Algorithms

Graphs and graph algorithms are of fundamental importance in computer science. In this chapter, we study two basic ways of exploring graphs, namely breadth-first search and depth-first search. Then, we present various graph algorithms, many of which are based on these search techniques, including algorithms to find minimum spanning trees, shortest paths and Eulerian circuits, among others.

7.1 Graphs

7.1.1 Definitions

A ***graph*** is defined mathematically to be a pair $G = (V, E)$. V is a set whose elements are called ***vertices*** and E is a set whose elements are called ***edges***. Each edge joins two distinct vertices. An edge can be represented as a set of two vertices, e.g., $\{u, v\}$, where $u \neq v$. We may also write this edge as uv or vu.

The edge uv is ***incident*** with the two vertices u and v. Sometimes we might say that u and v are the endpoints of the edge uv. We also say that u and v are ***adjacent vertices***, or ***neighbors***, if uv is an edge.

We often denote the number of vertices in a graph by n and the number of edges by m. Clearly $m \leq \binom{n}{2}$, since there are $\binom{n}{2}$ pairs of vertices, each of which may or may not be adjacent. If $m = \binom{n}{2}$, we have a ***complete graph***, which is denoted by K_n. In a complete graph, every pair of vertices is joined by an edge.

We note that complexity of graph algorithms is often expressed in terms of both m and n.

Let $G = (V, E)$ be a graph. For a vertex $v \in V$, the ***degree*** of v, denoted $\mathsf{deg}(v)$, is the number of edges $e \in E$ such that e is incident with v. The following basic lemma relates the number of edges in a graph to the degrees of its vertices.

LEMMA 7.1 *Suppose $G = (V, E)$ is a graph having m edges and n vertices. Then*

$$\sum_{v \in V} \mathsf{deg}(v) = 2m.$$

PROOF Let $\mathcal{I} = \{(v, e) : v \in V, e \in E, v \in e\}$. It is clear that $|\mathcal{I}| = 2m$, because every edge contains two vertices. On the other hand,

$$|\mathcal{I}| = \sum_{v \in V} \mathsf{deg}(v),$$

DOI: 10.1201/9780429277412-7

because a vertex v is incident with $\deg(v)$ edges. The result follows. ∎

A ***path*** in a graph $G = (V, E)$ is a sequence of vertices $v_1, v_2, \ldots, v_k \in V$ such that $\{v_i, v_{i+1}\} \in E$ for $1 \leq i \leq k-1$. Sometimes this path will be denoted as a (v_1, v_k)-path. The ***length*** of this path is $k-1$ (i.e., the number of edges that it contains). If all the vertices in a path are distinct, the path is a ***simple path***.

LEMMA 7.2 *If there is (u, v)-path in a graph G, then there is a simple (u, v)-path in G.*

PROOF Let

$$v_1, v_2, \ldots, v_k$$

be a (u, v)-path in G having the minimum length (note that $u = v_1$ and $v = v_k$). We claim that this path is simple. Suppose not; then $v_i = v_j$ for some i, j with $1 \leq i < j \leq k$. Then

$$v_1, v_2, \ldots, v_i, v_{j+1}, \ldots, v_k$$

is also a (u, v)-path, but it has fewer edges than the assumed minimum-length path, which is impossible. This contradiction proves the desired result. ∎

A graph $G = (V, E)$ is a ***connected graph*** if there is a path joining any two distinct vertices. A graph that is not connected consists of vertex-disjoint ***connected components***, where a connected component consists of a maximal subset of vertices, any two of which are joined by a path.

A path $v_1, v_2, \ldots, v_k$ in which $v_1 = v_k$ (i.e., one in which the starting and ending vertices are the same) is called a ***circuit***. Thus, a circuit starts and ends at the same vertex. A ***cycle*** of length $k \geq 3$ in a graph $G = (V, E)$ is a circuit $v_1, v_2, \ldots, v_k = v_1 \in V$ such that $v_1, v_2, \ldots, v_{k-1}$ are distinct. Thus, every cycle is automatically a circuit, but not vice versa. This is because a circuit is allowed to visit a vertex more than once (see Example 7.1). Also, for an edge uv, note that we do not consider u, v, u to be a cycle of length two, because it does not consist of distinct edges (the edge uv is the same as the edge vu).

Example 7.1 Consider a graph G where $V = \{1, 2, 3, 4, 5, 6, 7, 8\}$ and $E = \{12, 13, 23, 24, 25, 35, 37, 38, 56, 78\}$.

Here is a pictorial representation of G:

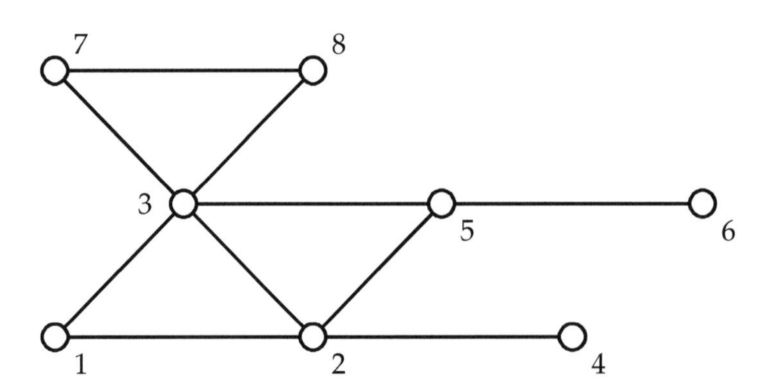

The graph G is connected and it has $n = 8$ vertices and $m = 10$ edges. The vertices have degrees as follows:

$$\begin{array}{llll} \mathsf{deg}(1) = 2, & \mathsf{deg}(2) = 4, & \mathsf{deg}(3) = 5, & \mathsf{deg}(4) = 1, \\ \mathsf{deg}(5) = 3, & \mathsf{deg}(6) = 1, & \mathsf{deg}(7) = 2, & \mathsf{deg}(8) = 2. \end{array}$$

$1,3,2,5,6$ is a $(1,6)$-path of length four, and $1,3,5,2,1$ is a cycle of length four. $1,3,7,8,3,5,2,1$ is a circuit that is not a cycle. □

A graph is an ***acyclic graph*** if it does not contain any cycles. A ***tree*** is a connected, acyclic graph. We record the following basic results without proof.

THEOREM 7.3 *Suppose $G = (V, E)$ is a connected graph with m edges and n vertices. Then $m \geq n - 1$. Further, $m = n - 1$ if and only if G is a tree.*

7.1.2 Data Structures for Graphs

There are two main data structures that are used to represent graphs: an ***adjacency matrix*** and a set of ***adjacency lists***. We discuss these now.

Let $G = (V, E)$ be a graph with $|V| = n$ and $|E| = m$. The ***adjacency matrix*** of G is an n by n matrix $A = (a_{u,v})$, whose rows and columns are indexed by V, such that

$$a_{u,v} = \begin{cases} 1 & \text{if } \{u, v\} \in E \\ 0 & \text{otherwise.} \end{cases}$$

Observe that there are exactly $2m$ entries of A equal to 1 and all the diagonal entries of A are zeroes.

Example 7.2 The adjacency matrix of the graph G from Example 7.1 is

$$\begin{pmatrix} 0 & 1 & 1 & 0 & 0 & 0 & 0 & 0 \\ 1 & 0 & 1 & 1 & 1 & 0 & 0 & 0 \\ 1 & 1 & 0 & 0 & 1 & 0 & 1 & 1 \\ 0 & 1 & 0 & 0 & 0 & 0 & 0 & 0 \\ 0 & 1 & 1 & 0 & 0 & 1 & 0 & 0 \\ 0 & 0 & 0 & 0 & 1 & 0 & 0 & 0 \\ 0 & 0 & 1 & 0 & 0 & 0 & 0 & 1 \\ 0 & 0 & 1 & 0 & 0 & 0 & 1 & 0 \end{pmatrix}.$$

□

Let $G = (V, E)$ be a graph with $|V| = n$ and $|E| = m$. An ***adjacency list representation*** of G consists of n linked lists. For every $u \in V$, there is a linked list (called an ***adjacency list***) which is named $Adj[u]$. For every $v \in V$ such that $uv \in E$, there is a node in $Adj[u]$ labeled v. The order of the nodes in an adjacency list is immaterial. However, for convenience, the nodes will typically be given in increasing order.

In an undirected graph, every edge uv corresponds to nodes in two adjacency lists: there is a node v in $Adj[u]$ and a node u in $Adj[v]$.

Example 7.3 We return to the graph from Example 7.2. The adjacency lists for this graph are

$$\begin{aligned}
Adj[1] &: \; 2 \rightarrow 3 \\
Adj[2] &: \; 1 \rightarrow 3 \rightarrow 4 \rightarrow 5 \\
Adj[3] &: \; 1 \rightarrow 2 \rightarrow 5 \rightarrow 7 \rightarrow 8 \\
Adj[4] &: \; 2 \\
Adj[5] &: \; 2 \rightarrow 3 \rightarrow 6 \\
Adj[6] &: \; 5 \\
Adj[7] &: \; 3 \rightarrow 8 \\
Adj[8] &: \; 3 \rightarrow 7.
\end{aligned}$$

□

7.2 Breadth-first Search

Breadth-first search is a fundamental way to "explore" a graph. We begin by discussing breadth-first search in the setting of undirected graphs. A breadth-first search of an undirected graph begins at an arbitrary vertex s. Here are the basic features of a breadth-first search, expressed informally.

- Breadth-first search makes use of the adjacency list representation of a graph.
- The vertex s is the ***source vertex***.
- The search "spreads out" from s, proceeding in ***layers***.
- First, all the neighbors of s are ***explored***.
- Next, the neighbors of those neighbors are explored.
- This process continues until all vertices have been explored.
- A ***queue*** is used to keep track of the vertices to be explored.

More formally, a breadth-first search (or BFS) for a graph G, starting at source vertex s, is described in Algorithm 7.1.

Algorithm 7.1 assigns colors to vertices. Vertices are initially colored *white*, then they are colored *red*, and finally they become *black*. In more detail, a vertex is *white* if it is ***undiscovered***. A vertex is *red* if it has been ***discovered***, but we are still processing its adjacent vertices. A vertex becomes *black* when all of its adjacent vertices have been processed. If G is connected, then every vertex eventually is colored black. If G is not connected, then the black vertices (at the termination of the algorithm) are precisely the vertices in the same connected component as the source s.

As well, Algorithm 7.1 records the ***predecessors*** of the vertices. For a connected graph, every vertex has a unique predecessor (except for the source, which has no

Algorithm 7.1: BFS(G, s)

```
comment: G is a graph represented by adjacency lists
for each v ∈ V(G)
    do { color[v] ← white
         π[v] ← ∅
color[s] ← red
INITIALIZEQUEUE(Q)
ENQUEUE(Q, s)
while not EMPTY(Q)
    do { u ← DEQUEUE(Q)
         for each v ∈ Adj[u]
             do { if color[v] = white
                      then { color[v] = red
                             π[v] ← u
                             ENQUEUE(Q, v)
         color[u] ← black
```

predecessor). The predecessor of vertex v is the vertex from which v is discovered; it is denoted by $\pi[v]$ in Algorithm 7.1.

When a vertex v is discovered, we color it *red*, record its predecessor, and place it at the end of the queue Q. At any point in time, Q contains all the red vertices. When we are done processing the adjacency list $Adj[u]$, the vertex u is assigned the color *black*.

Suppose we explore an edge $\{u, v\}$ starting from u. Then

- if v is *white*, then uv is a ***tree edge*** and $\pi[v] = u$ is the ***predecessor*** of v in the ***BFS tree***, where the BFS tree consists of all the tree edges.
- otherwise, uv is a ***cross edge***.

Example 7.4 We run BFS with $s = 1$ on the graph from Example 7.1.

$color[1] \leftarrow red, \quad Q = [1]$

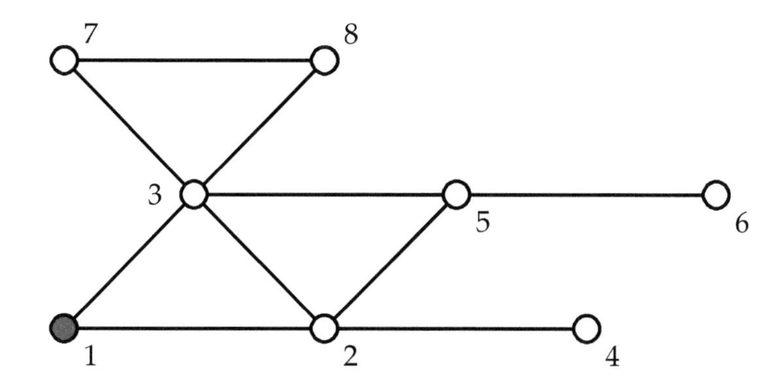

$u \leftarrow 1, Q = []$

$v \leftarrow 2, \quad color[2] \leftarrow red, \quad \pi[2] \leftarrow 1, \quad Q = [2]$

$v \leftarrow 3, \quad color[3] \leftarrow red, \quad \pi[3] \leftarrow 1, \quad Q = [2, 3]$

$color[1] \leftarrow black$

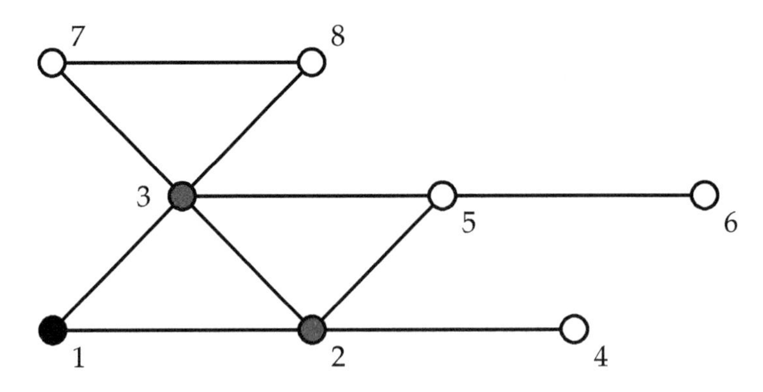

$u \leftarrow 2, \quad Q = [3]$

$v \leftarrow 4, \quad color[4] \leftarrow red, \quad \pi[4] \leftarrow 2, \quad Q = [3, 4]$

$v \leftarrow 5, \quad color[5] \leftarrow red, \quad \pi[5] \leftarrow 2, \quad Q = [3, 4, 5]$

$color[2] \leftarrow black$

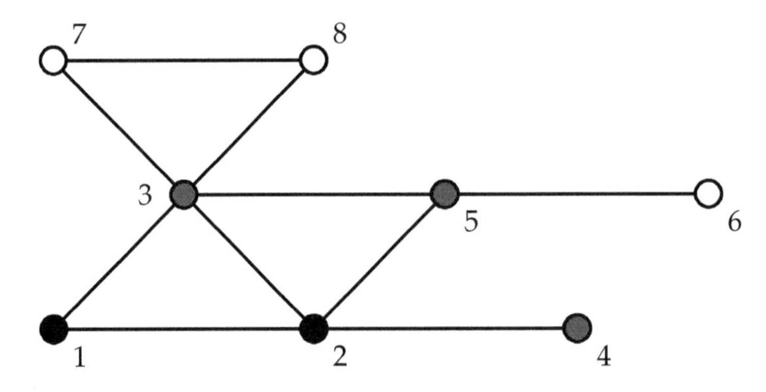

$u \leftarrow 3, \quad Q = [4, 5]$

$v \leftarrow 7, \quad color[7] \leftarrow red, \quad \pi[7] \leftarrow 3, \quad Q = [4, 5, 7]$

$v \leftarrow 8, \quad color[8] \leftarrow red, \quad \pi[8] \leftarrow 3, \quad Q = [4, 5, 7, 8]$

$color[3] \leftarrow black$

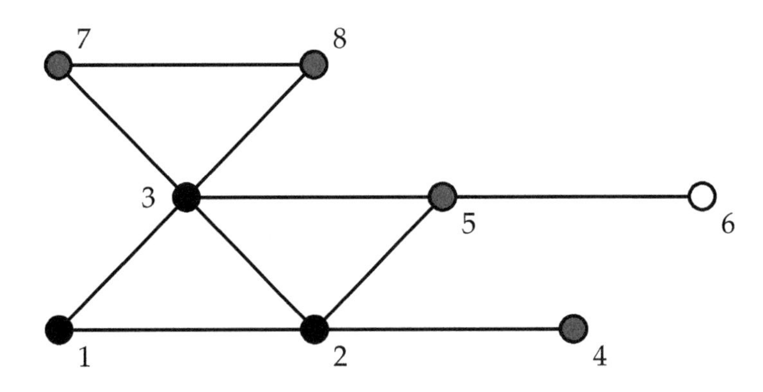

$u \leftarrow 4, \quad Q = [5,7,8]$
$color[4] \leftarrow black$

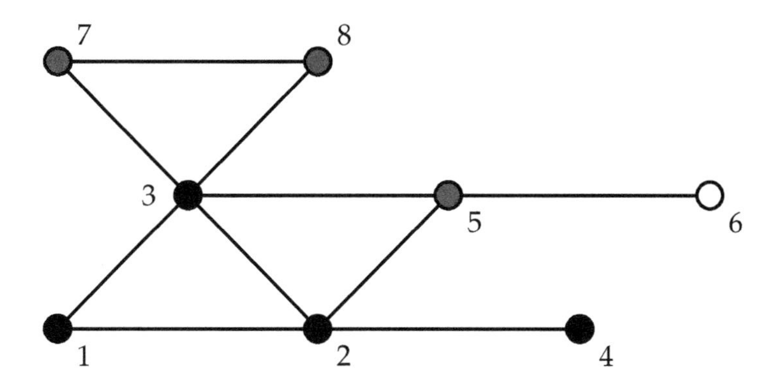

$u \leftarrow 5, \quad Q = [7,8]$
$\quad v \leftarrow 6, \quad color[6] \leftarrow red, \quad \pi[6] \leftarrow 5, \quad Q = [7,8,6]$
$color[5] \leftarrow black$

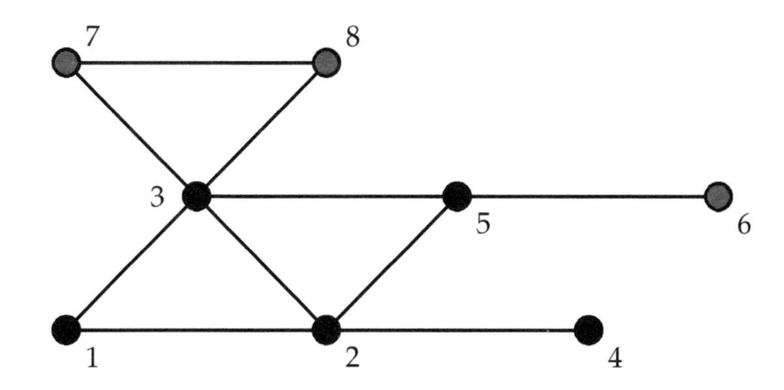

$u \leftarrow 7, \quad Q = [8,6]$
$color[7] \leftarrow black$

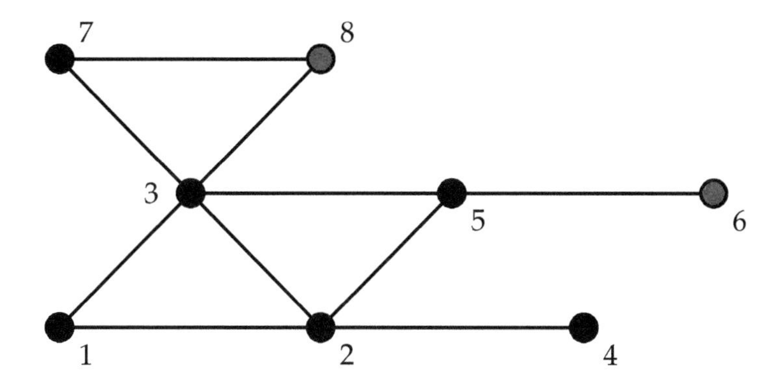

$u \leftarrow 8, \quad Q = [6]$
$color[8] \leftarrow black$

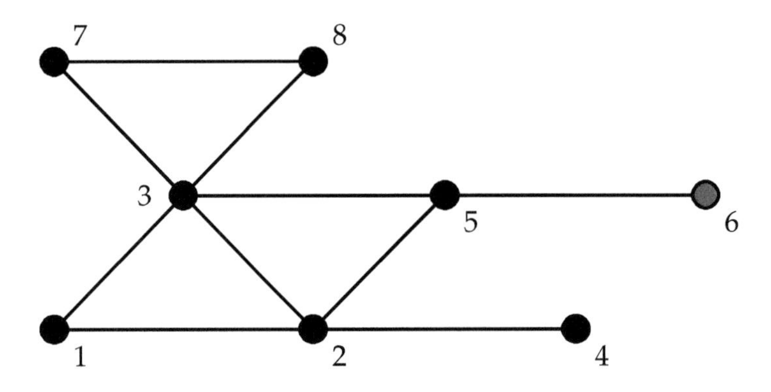

$u \leftarrow 6, \quad Q = []$
$color[6] \leftarrow black$

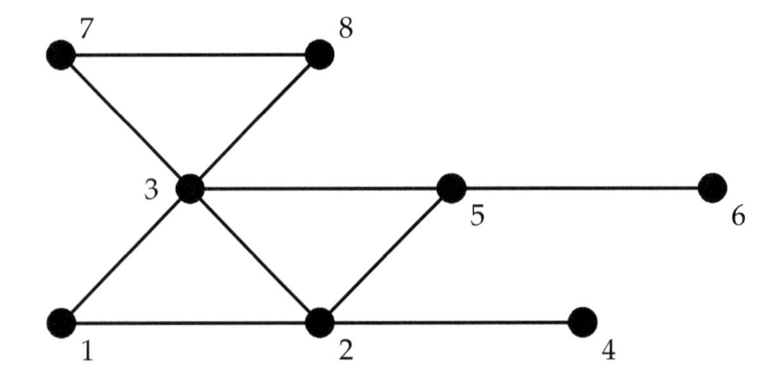

The tree edges are $12, 13, 24, 25, 37, 38, 56$.

□

The complexity of a breadth-first search of a connected graph is written as $\Theta(n + m)$. Every one of the n adjacency lists is traversed once during the search. There are n header nodes and a total of $2m$ other nodes in the adjacency lists, which simplifies to give $\Theta(n + m)$.

7.2.1 Shortest Paths

One important application of breadth-first search is to compute shortest paths from a specified source. With a few modifications, we can adapt Algorithm 7.1 so it computes the length of a ***shortest path*** from a specified source vertex s to v, for every vertex v. This quantity will be denoted by $dist[v]$. If the graph is not connected and the vertex v is in a different connected component than s, then we define $dist[v] = \infty$. As we have already mentioned, the length of a path is just the number of edges in the path.

Algorithm 7.2 is the modified algorithm. The three red statements in Algorithm 7.2 are the only changes from Algorithm 7.1. The basic idea is that, whenever we encounter a tree edge uv, we set the distance of v (i.e., the length of the shortest path from the source s to v) to be equal to $1+$ the distance of u.

Algorithm 7.2: BFS SHORTEST PATHS(G, s)

```
comment: G is a graph represented by adjacency lists
for each v ∈ V(G)
       ⎧ color[v] ← white
    do ⎨ π[v] ← ∅
       ⎩ dist[v] ← ∞
color[s] ← red
dist[s] ← 0
INITIALIZEQUEUE(Q)
ENQUEUE(Q, s)
while not EMPTY(Q)
       ⎧ u ← DEQUEUE(Q)
       ⎪ for each v ∈ Adj[u]
       ⎪         ⎧ if color[v] = white
    do ⎨         ⎪             ⎧ color[v] = red
       ⎪      do ⎨        then ⎨ π[v] ← u
       ⎪         ⎪             ⎪ ENQUEUE(Q, v)
       ⎪         ⎩             ⎩ dist[v] ← dist[u] + 1
       ⎩ color[u] ← black
```

Example 7.5 We use Algorithm 7.2 to compute distances from source 1 in the graph from Example 7.1:

$$
\begin{aligned}
dist[1] &= 0 \\
dist[2] &= dist[1] + 1 = 1 \\
dist[3] &= dist[1] + 1 = 1 \\
dist[4] &= dist[2] + 1 = 2 \\
dist[5] &= dist[2] + 1 = 2 \\
dist[7] &= dist[3] + 1 = 2 \\
dist[8] &= dist[3] + 1 = 2 \\
dist[6] &= dist[5] + 1 = 3
\end{aligned}
$$

□

A formal mathematical proof of correctness of Algorithm 7.2 is surprisingly complicated. Before doing this, we give an intuitive justification that this algorithm correctly computes distances from the source s. The distance of the source from itself is 0. Algorithm 7.2 then traverses the adjacency list $Adj[s]$. The vertices in $Adj[s]$ are the neighbors of s, so they all have distance one from s. Now, in Algorithm 7.2, after each of these vertices is assigned distance one, it is placed on the queue, Q. These vertices constitute the first "layer" of vertices that are explored during Algorithm 7.2.

The neighbors of s will be the first vertices (other than s) that are removed from the queue Q. They will be removed from Q in the same order that they were added to Q because a queue is "first-in-first-out." After each of neighbor u of s is removed from Q, we process the adjacency list $Adj[u]$. So we examine the neighbors of u one by one. A neighbor v of u might have been previously discovered. For example, s is a neighbor of u, and there might also be neighbors of u that are also neighbors of s. However, if v is a neighbor of u that has not been previously discovered, then it has distance two from s. These vertices comprise the second layer of vertices encountered in the search.

This process is continued: the previously undiscovered neighbors of the vertices that have distance two will be assigned distance three, etc. Eventually we assign a distance to all the vertices that are in the same connected component as v.

Referring to Example 7.5, the vertices 2 and 3 have distance one. When we remove vertex 2 from the queue, we traverse $Adj[s]$, which contains vertices 1, 3, 4 and 5. Vertices 1 and 3 have already been discovered, so they are ignored. However, vertices 4 and 5 have not yet been discovered, so their distances (namely, two) are now assigned and they are added to the queue.

If we want to determine the vertices in a shortest (s,u)-path, then we just use the predecessors that have been defined in Algorithm 7.1 and Algorithm 7.2. Starting from u, we compute

$$\pi(u), \pi(\pi(u)), \dots,$$

until we reach s. This gives us the shortest path from s to u in reverse order.

For example, to find the shortest $(1,6)$-path in the graph from Example 7.1, we can use the predecessors as they have been computed in Example 7.4. We have $\pi(6) = 5$, $\pi(5) = 2$ and $\pi(2) = 1$. Thus the shortest $(1,6)$-path is $1,2,5,6$.

Now we present a formal correctness proof. The values $dist[u]$ are integers computed by Algorithm 7.2. In Lemma 7.4 and Lemma 7.5, we establish some properties concerning these values. At this point, we have not shown that these are the "correct" distances from s, but we will eventually do so.

LEMMA 7.4 *If u is discovered before v, then $dist[u] \leq dist[v]$.*

PROOF The proof is by contradiction. Let v be the first vertex such that

$$dist[u] > dist[v]$$

for some u that was discovered before v. Denote

$$d = dist[v];$$

then

$$dist[u] \geq d + 1.$$

Let

$$\pi[v] = v_1 \quad \text{and} \quad \pi[u] = u_1;$$

then

$$dist[v_1] = d - 1 \quad \text{and} \quad dist[u_1] \geq d.$$

Note that v_1 was discovered before u_1 since v is the first "out-of-order" vertex. So, in order of discovery, we have

$$v_1, u_1, u, v.$$

Vertex v was discovered while processing $Adj[v_1]$ and vertex u was discovered while processing $Adj[u_1]$. But this means that v was discovered before u, a contradiction. ∎

LEMMA 7.5 *If $\{u, v\}$ is any edge, then $|dist[u] - dist[v]| \leq 1$.*

PROOF Without loss of generality, suppose u is discovered before v. Therefore we explore uv in the direction $u \to v$.

We identify three cases:

case (1)

Suppose v is white when we process $Adj[u]$. Then

$$dist[v] = dist[u] + 1.$$

case (2)

Suppose v is red when we process $Adj[u]$. Let

$$\pi[v] = v_1;$$

then v was discovered when $Adj[v_1]$ was being processed. So v_1 was discovered before u. By Lemma 7.4,

$$dist[v_1] \leq dist[u].$$

Also,

$$dist[v] = dist[v_1] + 1,$$

so

$$dist[u] \geq dist[v] - 1.$$

Since u was discovered before v, we have

$$dist[u] \leq dist[v]$$

by Lemma 7.4. Therefore,

$$dist[u] \leq dist[v] \leq dist[u] + 1.$$

case (3)

Suppose v is black when we process $Adj[u]$. Then $Adj[v]$ has been completely processed and we would already have discovered u from v, which is a contradiction.

Since these three cases exhaust all the possibilities, the proof is complete. ∎

THEOREM 7.6 *The value $dist[v]$ is the length of the shortest path from s to v.*

PROOF Let $\delta(v)$ denote the true length of the shortest path from s to v. We want to show that $\delta(v) = dist[v]$. Consider the path

$$v \quad \pi[v] \quad \pi[\pi[v]] \quad \cdots \quad s.$$

This path is a path from v to s having length $dist[v]$, so

$$\delta(v) \leq dist[v].$$

To complete the proof, we need to show that $\delta(v) \geq dist[v]$; we will prove this by induction on $\delta(v)$.

The base case is $\delta(v) = 0$. Then $v = s$ and

$$dist[v] = 0 = \delta(v).$$

As an induction assumption, we assume that

$$\delta(v) \geq dist[v]$$

if $\delta(v) \leq d - 1$. Now suppose $\delta(v) = d$. Let

$$s \quad v_1 \quad v_2 \quad \cdots \quad v_{d-1} \quad v_d = v$$

be a shortest path (having length d). Then

$$\delta(v_{d-1}) = d - 1 = dist[v_{d-1}]$$

by induction. We have that

$$dist[v] \leq dist[v_{d-1}] + 1$$

from Lemma 7.5. But

$$dist[v_{d-1}] = d - 1,$$

so

$$dist[v] \leq d = \delta(v)$$

and we're done. ∎

7.2.2 Bipartite Graphs

A graph $G = (V, E)$ is a ***bipartite graph*** if the vertex set V can be partitioned as $V = X \cup Y$ in such a way that all edges have one endpoint in X and one endpoint in Y. Equivalently, a graph is bipartite if and only if it is possible to color the vertices using two colors (red and black, say) in such a way that no edge joins two vertices of the same color.

THEOREM 7.7 *A graph is bipartite if and only if it contains no cycle of odd length.*

PROOF Suppose G contains an odd cycle,

$$v_1, v_2, \ldots, v_{2k+1}, v_1.$$

Without loss of generality, color v_1 red. Then we are forced to color v_2 black, v_3 red, $\ldots$, and eventually we color v_{2k+1} red. Then the edge $v_{2k+1}v_1$ joins two red vertices, so G is not bipartite.

The idea is illustrated in the following diagram:

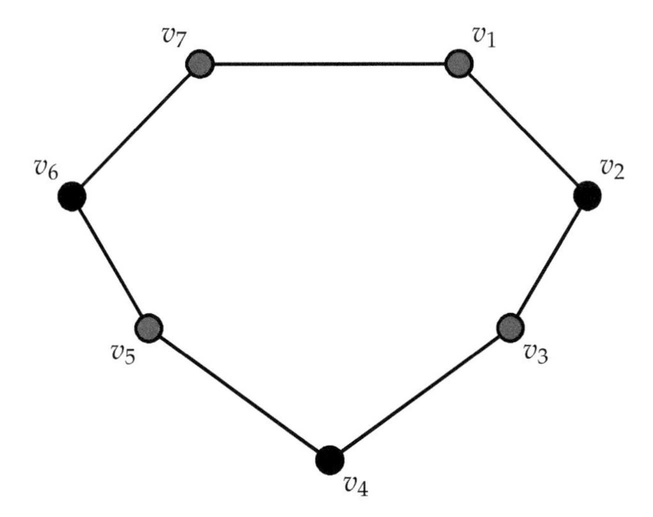

Conversely, suppose G is not bipartite. Without loss of generality, assume G is connected. Let s be any vertex. Define

$$X = \{v : dist[v] \text{ is even}\}$$

and

$$Y = \{v : dist[v] \text{ is odd}\}.$$

Since G is not bipartite, there is an edge uv where $u, v \in X$ or $u, v \in Y$. Therefore, $dist[u]$ and $dist[v]$ are both even or both odd. We showed in Lemma 7.5 that

$$|dist[u] - dist[v]| \leq 1$$

for any edge uv. Therefore, it follows that $dist[u] = dist[v]$.

Denote

$$d = dist[u] = dist[v].$$

We have two shortest paths (a (u,s)-path and a (v,s)-path) of length d in G:

$$u, u_1 = \pi[u], \ldots, u_d = \pi[u_{d-1}] = s$$

and

$$v, v_1 = \pi[v], \ldots, v_d = \pi[v_{d-1}] = s.$$

Let

$$j = \min\{i : u_i = v_i\}.$$

Note that $j \leq d$ since $u_d = v_d = s$. Then

$$v_i, \ldots, v_1, v, u, u_1, \ldots, u_i = v_i$$

is a cycle of odd length.

The following diagram illustrates how an odd length cycle could be obtained if the two paths have length seven and $j = 3$. The red edges form a cycle of length 7.

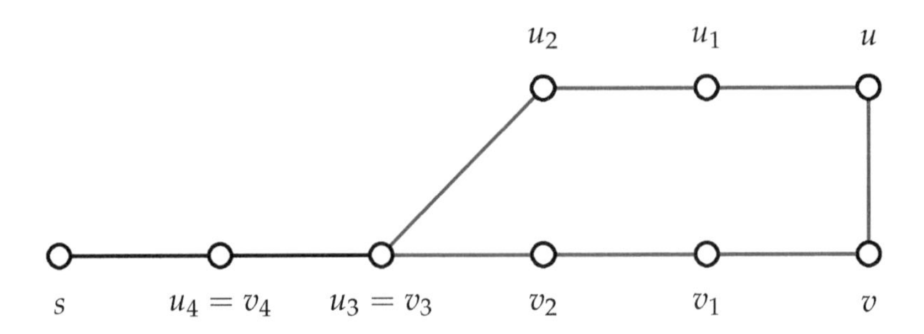

∎

From the proof of Theorem 7.7, it is easy to see that a breadth-first search can be used to test if a connected graph is bipartite. The basic idea is straightforward. If we encounter an edge $\{u,v\}$ with $dist[u] = dist[v]$, then G is not bipartite. On the other hand, if no such edge is found, then we define

$$X = \{u : dist[u] \text{ is even}\}$$

and

$$Y = \{u : dist[u] \text{ is odd}\}.$$

There will be no edge from a vertex in X to a vertex in Y, so (X,Y) forms a bipartition.

Algorithm 7.3 modifies Algorithm 7.2 so that it determines if a connected graph G is bipartite or not. It would be straightforward, using the approach described in the proof of Theorem 7.7, to extend this algorithm so that

- it constructs a bipartition (in the case where G is bipartite) or
- it finds an odd length cycle (in the case where G is not bipartite).

Also, in the case where G is not connected, it is easy to see that G is bipartite if and only if every connected component is bipartite. We leave the details for the reader to check.

```
Algorithm 7.3: BIPARTITE(G, s)

comment: G is a graph represented by adjacency lists
precondition: we assume G is connected
for each v ∈ V(G)
     do { color[v] ← white
          π[v] ← ∅
          dist[v] ← ∞
color[s] ← red
dist[s] ← 0
INITIALIZEQUEUE(Q)
ENQUEUE(Q, s)
while not EMPTY(Q)
     do { u ← DEQUEUE(Q)
          for each v ∈ Adj[u]
               do { if color[v] = white
                      then { color[v] = red
                             π[v] ← u
                             ENQUEUE(Q, v)
                             dist[v] ← dist[u] + 1
                      else { if dist[v] = dist[u]
                               then return (G is not bipartite)
          color[u] ← black
return (G is bipartite)
```

7.3 Directed Graphs

A ***directed graph*** or ***digraph*** is a pair $G = (V, E)$. V is a set of vertices (as was the case in a graph). In a digraph, however, the elements of E are called ***directed edges*** or ***arcs***. Each arc joins two vertices, where an arc is represented as an ordered pair, e.g., (u, v). The arc (u, v) is directed from u (the ***tail***) to v (the ***head***), and we allow $u = v$. Note that there might be an arc uv as well as an arc vu in a given digraph.

If we denote the number of vertices in a digraph by n and we denote the number of arcs by m, then $m \leq n^2$.

Let $G = (V, E)$ be a digraph. For a vertex $v \in V$, the ***indegree*** of v, denoted $\mathsf{deg}^-(v)$, is the number of arcs $(u, v) \in E$ (i.e., the number of arcs for which v is the head). The ***outdegree*** of v, denoted $\mathsf{deg}^+(v)$, is the number of arcs $(v, u) \in E$ (i.e., the number of arcs for which v is the tail). The following lemma is similar to Lemma 7.1. We leave the proof as an exercise for the reader.

LEMMA 7.8 *Let $G = (V, E)$ be a digraph. Then*

$$\sum_{v \in V} \mathsf{deg}^-(v) = \sum_{v \in V} \mathsf{deg}^+(v) = m.$$

If G is a digraph, then its adjacency matrix $A = (a_{u,v})$ is defined as follows:

$$a_{u,v} = \begin{cases} 1 & \text{if } (u,v) \in E \\ 0 & \text{otherwise.} \end{cases}$$

For a digraph, there are exactly m entries of A equal to 1.

Adjacency lists are defined for directed graphs exactly the same as they are defined for graphs. However, in a directed graph, every edge corresponds to a node in only one adjacency list.

A ***directed path*** in a directed graph $G = (V, E)$ is a sequence of vertices $v_1, v_2, \ldots, v_k \in V$ such that $(v_i, v_{i+1}) \in E$ for $1 \leq i \leq k - 1$. Sometimes this path will be denoted as a (v_1, v_k)-directed path. The ***length*** of this path is $k - 1$ (i.e., the number of edges that it contains). If all the vertices in a path are distinct, the path is a ***simple directed path***. When we have a directed graph, the terms "path" and "simple path" can be assumed to mean "directed path" and "simple directed path," respectively.

A directed path $v_1, v_2, \ldots, v_k$ in which $v_1 = v_k$ (i.e., one in which the starting and ending vertices are the same) is called a ***directed circuit***. A ***directed cycle*** of length $k \geq 2$ in a directed graph $G = (V, E)$ is a directed circuit $v_1, v_2, \ldots, v_k = v_1 \in V$ such that $v_1, v_2, \ldots, v_{k-1}$ are distinct. Thus, every directed cycle is automatically a directed circuit, but not vice versa, because a directed circuit is allowed to visit a vertex more than once. Note that, if u and v are distinct vertices, then u, v, u is a directed cycle of length two provided that uv and vu are both arcs in the digraph. (This is different from an undirected graph, which does not have any cycles of length two.)

When we have a directed graph, the term "cycle" can be assumed to mean "directed cycle."

7.4 Depth-first Search

We next describe another basic graph search algorithm. This algorithm is known as DEPTH-FIRST SEARCH, or DFS. We present the algorithm as it is carried out on a directed graph. The basic depth-first search is presented as Algorithms 7.4 and 7.5.

Here are a few relevant points about the depth-first search algorithm:

- A ***depth-first search*** uses a ***stack*** (or, equivalently, ***recursion***) instead of a queue.
- We define predecessors and color vertices in the same manner as in a breadth-first search.
- It is also useful for some applications to specify a ***discovery time*** $d[v]$ and a ***finishing time*** $f[v]$ for every vertex v.
- Therefore, we increment a ***time counter*** every time a value $d[v]$ or $f[v]$ is assigned.
- We will eventually visit all the vertices, and in doing so, the algorithm constructs a ***depth-first forest***.
- The complexity of DEPTH-FIRST SEARCH is $\Theta(n+m)$, just like a breadth-first search.

Algorithm 7.4: DFS(G)

comment: G is a directed graph represented by adjacency lists
for each $v \in V(G)$
  do $\begin{cases} color[v] \leftarrow white \\ \pi[v] \leftarrow \varnothing \end{cases}$
time $\leftarrow 0$
for each $v \in V(G)$
  do $\begin{cases} \textbf{if } color[v] = white \\ \quad \textbf{then } \text{DFSVISIT}(G, v) \end{cases}$

We provide a small example of a depth-first search.

Example 7.6 Consider the directed graph on vertex set $\{1,2,3,4,5,6\}$ with the following adjacency lists:

$$Adj[1] : 2 \rightarrow 3 \qquad Adj[2] : 3$$
$$Adj[3] : 4 \qquad Adj[4] : 2$$
$$Adj[5] : 4 \rightarrow 6 \qquad Adj[6] :$$

The digraph can be drawn as follows:

Algorithm 7.5: DFSVISIT(G, v)

global $color, d, f, \pi, time$
$color[v] \leftarrow red$
$time \leftarrow time + 1$
$d[v] \leftarrow time$
comment: $d[v]$ is the discovery time for vertex v
for each $w \in Adj[v]$
do { **if** $color[w] = white$
  then { $\pi[w] \leftarrow v$
    DFSVISIT(G, w) } }
$color[v] \leftarrow black$
$time \leftarrow time + 1$
$f[v] \leftarrow time$
comment: $f[v]$ is the finishing time for vertex v

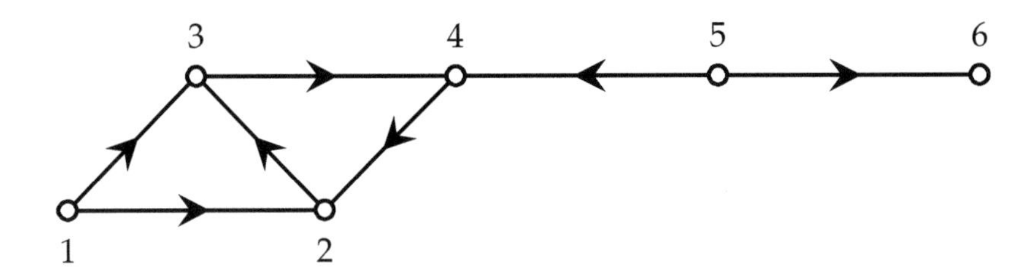

The initial call to DFSVISIT from DFS is to DFSVISIT$(G, 1)$. This leads to recursive calls DFSVISIT$(G, 2)$, DFSVISIT$(G, 3)$, and DFSVISIT$(G, 4)$. Then DFS calls DFSVISIT$(G, 5)$, which generates the recursive call DFSVISIT$(G, 6)$.

The depth-first forest consists of two trees. One tree has arcs $12, 23, 34$ (resulting from the call to DFSVISIT$(G, 1)$) and the other tree has arc 56 (resulting from the call to DFSVISIT$(G, 5)$). In the following diagram, the tree edges are colored red.

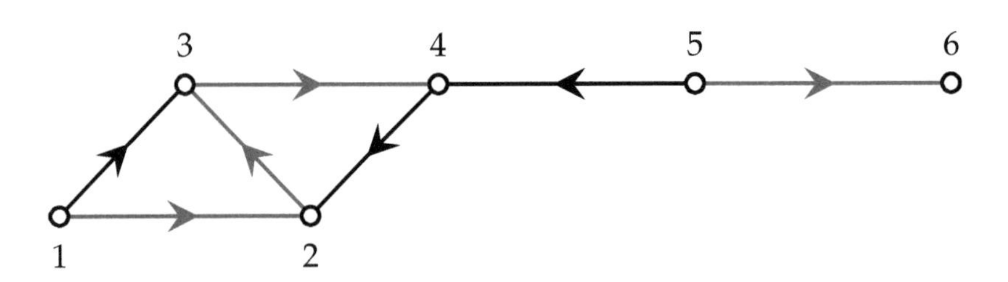

The discovery/finishing times and predecessors are as follows:

v	$d[v]$	$f[v]$	$\pi[v]$
1	1	8	–
2	2	7	1
3	3	6	2
4	4	5	3
5	9	12	–
6	10	11	5

□

We next classify the edges of a directed graph in a depth-first search. We define four types of edges:

- An arc uv is a ***tree edge*** if $u = \pi[v]$.
- An arc uv is a ***forward edge*** if it is not a tree edge, and v is a descendant of u in a tree in the depth-first forest.
- An arc uv is a ***back edge*** if u is a descendant of v in a tree in the depth-first forest.
- Any other arc is a ***cross edge***. A cross edge can join two different trees, or two branches of the same tree.

In Example 7.6, the arc 13 is a forward edge, because 3 is a descendant of 1 in the tree having vertices $1, 2, 3, 4$. The arc 42 is a back edge, because 4 is a descendant of 2 in the tree having vertices $1, 2, 3, 4$. The arc 54 is a cross edge; it joins vertices in two different trees. See the following diagram.

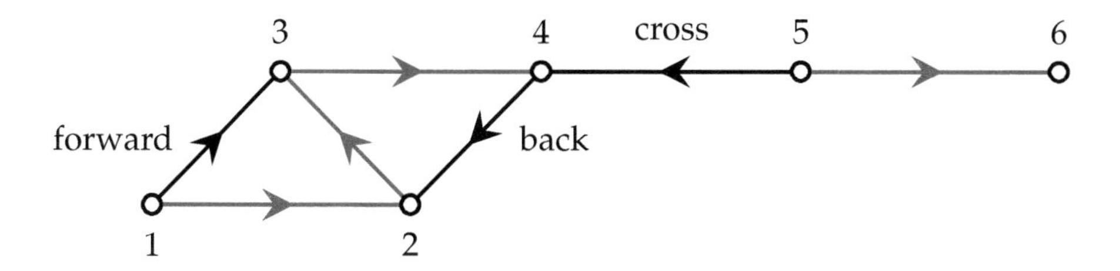

We now record some properties of edges in a depth-first search. In Table 7.1, we indicate the color of a vertex v when an edge uv is discovered, and the relation between the start and finishing times of u and v, for each possible type of edge uv. Observe that two intervals $(d[u], f[u])$ and $(d[v], f[v])$ never overlap. Any two distinct intervals are either disjoint or nested. This important fact is sometimes called the *parenthesis theorem*.

TABLE 7.1: Edge types in a depth-first search.

edge type	color of v	discovery/finish times
tree edge	*white*	$d[u] < d[v] < f[v] < f[u]$
forward edge	*black*	$d[u] < d[v] < f[v] < f[u]$
back edge	*red*	$d[v] < d[u] < f[u] < f[v]$
cross edge	*black*	$d[v] < f[v] < d[u] < f[u]$

If we perform a depth-first search of an undirected graph, we only encounter tree edges and back edges.

We also provide some diagrams to illustrate how forward, back and cross edges can arise during a depth-first search. In Figure 7.1, the red edges are tree edges and the black edges are the ones that are being classified.

7.4.1 Topological Ordering

We begin with a couple of definitions. A directed graph G is a ***directed acyclic graph***, or ***DAG***, if G contains no directed cycle.

A directed graph $G = (V, E)$ has a ***topological ordering***, or ***topological sort***, if there is a linear ordering of all the vertices in V, say "$<$", such that $u < v$ whenever $uv \in E$.

Here is a small example.

Example 7.7 Consider the following digraph:

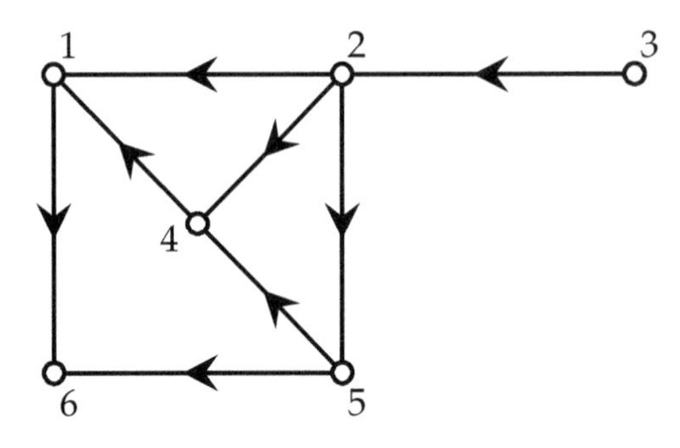

The ordering

$$3 < 2 < 5 < 4 < 1 < 6$$

is in fact a topological ordering. Thus we can redraw the graph using the topological ordering so that all arcs are directed from left to right:

forward edge uv

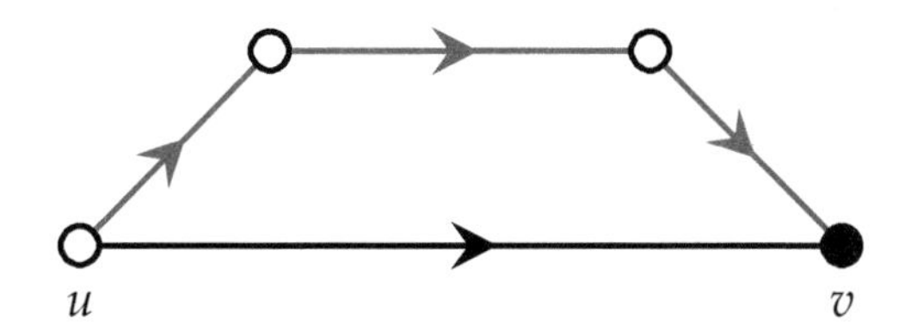

v is black when uv is processed.

back edge uv

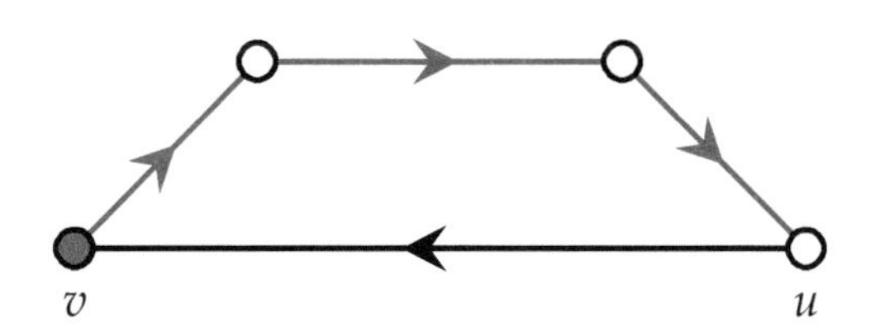

v is red when uv is processed.

cross edge uv

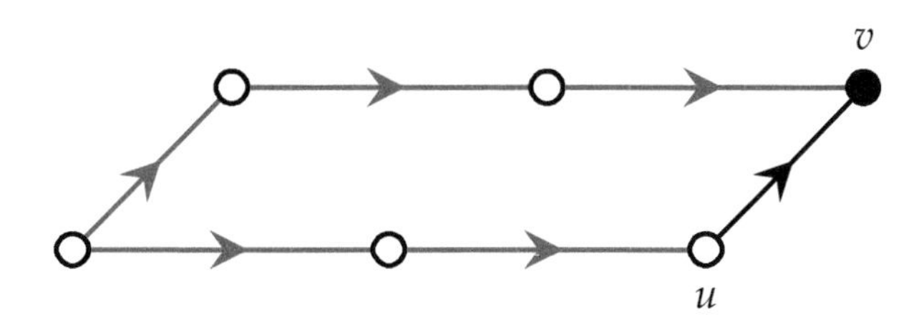

If the top path is followed first, then v is black when uv is processed.

FIGURE 7.1: Properties of edges in a depth-first search.

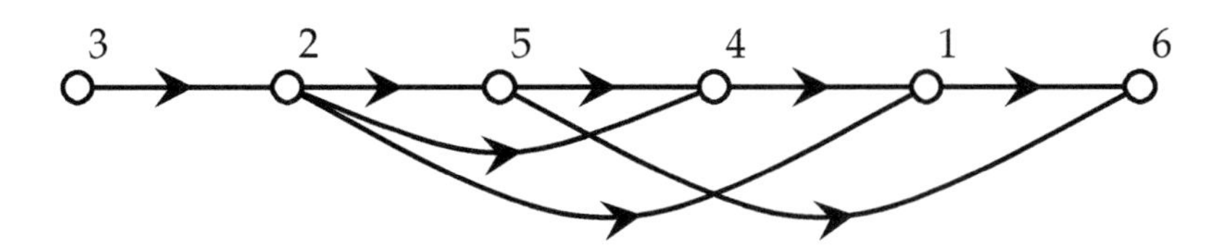

□

LEMMA 7.9 *A directed acyclic graph contains a vertex v such that* $\mathsf{deg}^-(v) = 0$ *(i.e., a vertex having indegree 0).*

PROOF Suppose we have a directed graph in which $\mathsf{deg}^-(v) > 0$ for all v (i.e., every vertex has positive indegree). Let v_1 be any vertex. For every $i \geq 1$, let $v_{i+1}v_i$ be an arc. In the sequence $v_1, v_2, v_3, \ldots,$ consider the first repeated vertex, $v_i = v_j$ where $j > i$. Then $v_j, v_{j-1}, \ldots, v_i, v_j$ is a directed cycle. ∎

We illustrate the idea of the preceding proof. Suppose, for example, that $j = i + 7$. Then we obtain a directed cycle as follows.

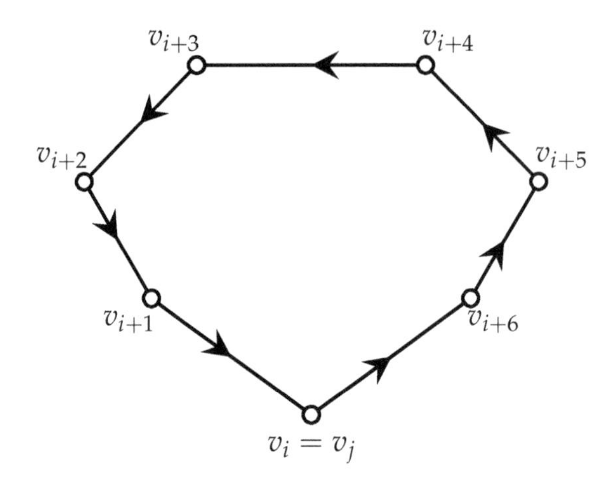

THEOREM 7.10 *A directed graph D has a topological sort if and only if it is a directed acyclic graph.*

PROOF Suppose D has a directed cycle $v_1, v_2, \ldots, v_j, v_1$. Then

$$v_1 < v_2 < \cdots < v_j < v_1,$$

so a topological ordering does not exist.

Conversely, suppose D is a directed acyclic graph. Then Algorithm 7.6 constructs a topological ordering. ∎

Algorithm 7.6: TOPORDERING(D)

```
precondition: D is a directed acyclic graph
D_1 ← D
for i ← 1 to n
  do { let v_i be a vertex in D_i such that deg^-(v_i) = 0
     { construct D_{i+1} from D_i by deleting v_i and all arcs v_i v_j
return (v_1, v_2, ..., v_n)
```

We leave it as an exercise for the reader to prove that Algorithm 7.6 is correct. Unfortunately, this algorithm is not very efficient (we also leave the determination of its complexity as an exercise). However, a related algorithm, known as KAHN'S ALGORITHM, that was published in 1962, can be implemented to run in time $\Theta(n + m)$. This algorithm uses an adjacency list representation and a queue. See Algorithm 7.7.

Algorithm 7.7: KAHN(D)

```
precondition: D is a directed acyclic graph
compute deg^-(v) for all vertices v
INITIALIZEQUEUE(Q)
for all v
  do { if deg^-(v) = 0
     {   then ENQUEUE(Q, v)
for i ← 1 to n
  do { if EMPTY(Q)
     {   then return ()
     {   else { v ← DEQUEUE(Q)
     {        { output (v)
     {        { for all w ∈ Adj(v)
     {        {   do { deg^-(w) ← deg^-(w) − 1
     {        {      { if deg^-(w) = 0
     {        {      {   then ENQUEUE(Q, w)
```

We now develop an algorithm to find a topological ordering of a DAG that is based on DEPTH-FIRST SEARCH. This algorithm is generally attributed to Tarjan.

LEMMA 7.11 *A directed graph is a directed acyclic graph if and only if a depth-first search encounters no back edges.*

PROOF Suppose that a depth-first search encounters a back edge. It is easy to see that the existence of a back edge immediately creates a directed cycle.

Conversely, suppose $C = v_1, v_2, \ldots, v_\ell$ is a directed cycle. We will show that there is a back edge in the depth-first search. Without loss of generality, suppose that v_1 is the vertex in C having the lowest discovery time. Consider the arc $v_\ell v_1$. We will prove that $v_\ell v_1$ is a back edge. First, since $d[v_\ell] > d[v_1]$, this arc must be a cross edge or a back edge (see Table 7.1). Suppose $v_\ell v_1$ is a cross edge. Then v_1 is black and v_ℓ is red when the arc $v_\ell v_1$ is processed. But v_1 is not colored black until all vertices reachable from v_1 are black. This is a contradiction, and hence $v_\ell v_1$ is a back edge. ∎

We can easily identify a back edge uv when we process it, because v must be red (see Table 7.1).

LEMMA 7.12 *Suppose D is a directed acyclic graph. Then $f[v] < f[u]$ for every arc uv.*

PROOF Look at the classification on Table 7.1. In a DAG, there are no back edges. For any other type of arc uv, it holds that $f[v] < f[u]$. ∎

Therefore, if D is a DAG and we order the vertices in reverse order of finishing time, then we get a topological ordering.

So, to find a topological ordering of a DAG, we simply run DEPTH-FIRST SEARCH and use a *stack* to output the vertices in reverse order of finishing time. If we encounter a back edge at any time the search, then we can quit—the graph is not a DAG. Also, it is easy to recognize back edges as we encounter them: vw is a back edge if and only if $color[w] = red$.

Algorithm 7.8: DFS-TOPORD(G)

```
comment: G is a directed graph which may or may not be a DAG
INITIALIZESTACK(S)
DAG ← true
for each v ∈ V(G)
  do { color[v] ← white
     { π[v] ← ∅
time ← 0
for each v ∈ V(G)
  do { if color[v] = white
     {   then DFSVISIT(G, v)
if DAG
  then return (S)
  else return (DAG)
```

Algorithms 7.8 and 7.9 are the modified algorithms. The red statements are the only changes from a basic depth-first search. At the termination of Algorithm 7.8, if we output a stack S, then we can POP the stack n times to obtain the vertices in topological order.

Algorithm 7.9: DFSVISIT-TOPORD(G, v)

```
global S, color, d, f, π, time
color[v] ← red
time ← time + 1
d[v] ← time
comment: d[v] is the discovery time for vertex v
for each w ∈ Adj[v]
     ⎧ if color[w] = white
     ⎪   then ⎧ π[w] ← v
  do ⎨        ⎩ DFSVISIT-TOPORD(G, w)
     ⎪ if color[w] = red
     ⎩   then DAG ← false
color[v] ← black
PUSH(S, v)
time ← time + 1
f[v] ← time
comment: f[v] is the finishing time for vertex v
```

Example 7.8 We consider the graph from Example 7.7. It has the following adjacency lists:

$$\begin{array}{llll} Adj[1]: & 6 & Adj[4]: & 1 \\ Adj[2]: & 1 \to 4 \to 5 & Adj[5]: & 4 \to 6 \\ Adj[3]: & 2 & Adj[6]: & \end{array}$$

The initial calls are DFSVISIT-TOPORD$(G, 1)$, DFSVISIT-TOPORD$(G, 2)$, and DFSVISIT-TOPORD$(G, 3)$.

The discovery/finishing times are as follows:

v	$d[v]$	$f[v]$
1	1	4
2	5	9
3	11	12
4	6	7
5	8	9
6	2	3

The topological ordering is $3, 2, 5, 4, 1, 6$ (reverse order of the finishing times). ☐

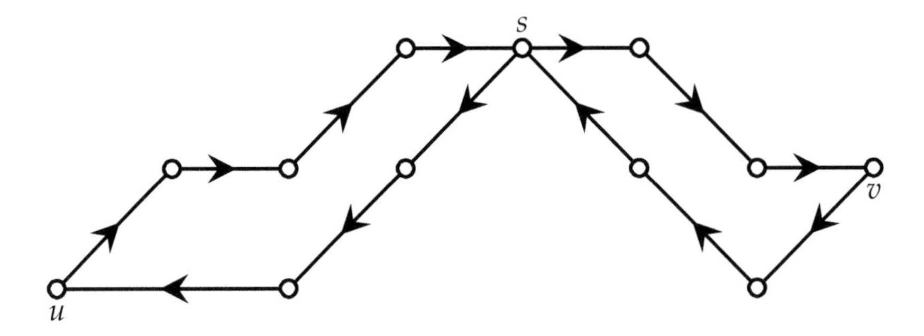

FIGURE 7.2: Reachability in a directed graph.

7.5 Strongly Connected Components

7.5.1 Connectivity and Strong Connectivity

Recall that an undirected graph is connected if there is a path between any two vertices. A graph is connected if all vertices are colored black after a breadth-first search starting from an arbitrary source vertex, or if every vertex is colored black after one initial call to DFSVISIT.

We have defined a directed path in a digraph $G = (V, E)$ to be a sequence of vertices $v_1, v_2, \ldots, v_k \in V$ such that $(v_i, v_{i+1}) \in E$ for $1 \leq i \leq k-1$. Sometimes this path will be denoted as a (v_1, v_k)-directed path. In a directed graph, a vertex v is ***reachable*** from a vertex w if there is a (v, w)-directed path.

A directed graph G is a ***strongly connected digraph*** if any vertex is reachable from any other vertex.

LEMMA 7.13 *A directed graph G is strongly connected if all vertices are reachable from s and s is reachable from all vertices, for an arbitrary (fixed) vertex s.*

PROOF Let u and v be any two vertices. Since s is reachable from u and v is reachable from s, it follows that v is reachable from u by concatenating the two paths. See Figure 7.2. Similarly, u is reachable from v. ∎

Lemma 7.13 leads to a simple way to determine if a directed graph is strongly connected. But we require one important observation first.

Testing to see if every vertex is reachable from a given source s is straightforward. As in the undirected case, we make one initial call to DFSVISIT(s) and then check to see that all vertices are colored black.

It is perhaps less obvious to see how we can easily determine if s is reachable from every other vertex. Certainly we do not want to run DFSVISIT from all the other vertices! It turns out that this can be done much more easily by considering the ***reverse digraph***. The reverse digraph of $G = (V, E)$, which is denoted $G^R =$

(V, E^R), is constructed from G by reversing the direction of every arc in E. That is,

$$(u,v) \in E^R \Leftrightarrow (v,u) \in E.$$

Our algorithm will make use of the following simple lemma.

LEMMA 7.14 *Let $G = (V,E)$ be a digraph and let G^R be its reverse. Then s is reachable from u in G if and only if u is reachable from s in G^R.*

PROOF If

$$s, v_2, \ldots, v_{k-1}, u$$

is a directed path in G, then

$$u, v_{k-1}, \ldots, v_2, s$$

is a directed path in G^R, and conversely. ∎

As a result of Lemma 7.14, to test if a digraph is strongly connected, it suffices to check if every vertex is reachable from s in both G and G^R. This only requires us to run DFSVISIT once in G and once in G^R. See Algorithm 7.10.

Algorithm 7.10: STRONGLYCONNECTED(G, s)

```
external DFSVISIT
comment: s is any vertex in G
DFSVISIT(G, s)
if there exists a non-black vertex in G
  then return (G is not strongly connected)
reverse the direction of all edges in G to construct G^R
DFSVISIT(G^R, s)
if there exists a non-black vertex in G^R
  then return (G is not strongly connected)
  else return (G is strongly connected)
```

Analyzing the complexity of Algorithm 7.10 is straightforward. Both calls to DFSVISIT have complexity $O(n+m)$. Checking to see if there exists a white vertex takes time $O(n)$. Finally, constructing the adjacency lists for G^R can be done in time $\Theta(n+m)$ (we leave it as an exercise for the reader to fill in the details). Therefore the complexity of Algorithm 7.10 is $\Theta(n+m)$.

7.5.2 Sharir's Algorithm

If an undirected graph is not connected, then we can partition the set of vertices into connected components. Suppose G is an undirected graph. Define $x \sim y$ if $x = y$; or if $x \neq y$ and there exists a path joining x and y. The ***connected components*** of G are the equivalence classes under this equivalence relation.

In an undirected graph, each initial call to DFSVISIT from DFS explores one

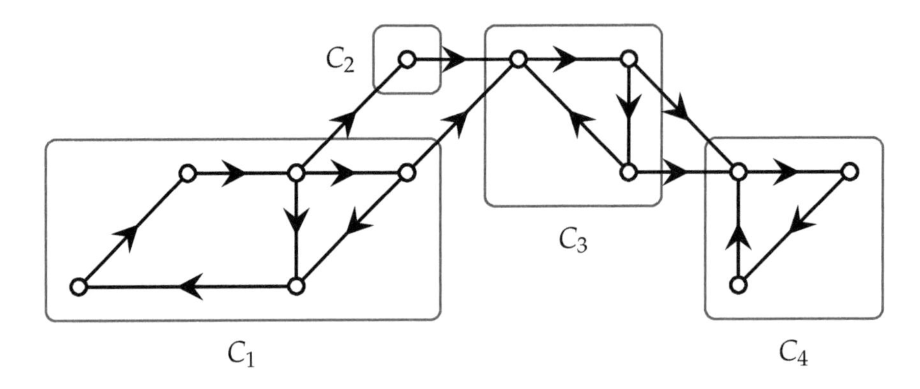

FIGURE 7.3: A directed graph and its strongly connected components.

connected component. It is a simple matter to maintain a counter that is incremented each time an initial call to DFSVISIT is made. In this way, every vertex will be labeled with an integer indicating which component it belongs to.

We seek to do something similar in the setting of directed graphs. Suppose $G = (V, E)$ is a directed graph. For two vertices x and y of G, define $x \sim y$ if $x = y$; or if $x \neq y$ and there exist directed paths from x to y *and* from y to x (i.e., x is reachable from y and vice versa).

The following lemma is straightforward; we leave the proof details for the reader to fill in.

LEMMA 7.15 *The relation $\sim$ is an **equivalence relation** on the set of vertices of a directed graph.*

The ***strongly connected components*** of the directed graph G are defined to be the equivalence classes of vertices defined by the relation $\sim$. It is not hard to see that a strongly connected component of a digraph G is a maximal strongly connected subgraph of G.

As an example, the directed graph in Figure 7.3 has four strongly connected components, which are denoted C_1, C_2, C_3 and C_4.

The ***component graph*** of a directed graph G is a directed graph whose vertices are the strongly connected components of G. There is an arc from C_i to C_j if and only if there is an arc in G from some vertex of C_i to some vertex of C_j.

To illustrate, the component graph of the directed graph in Figure 7.3 is given in the following diagram:

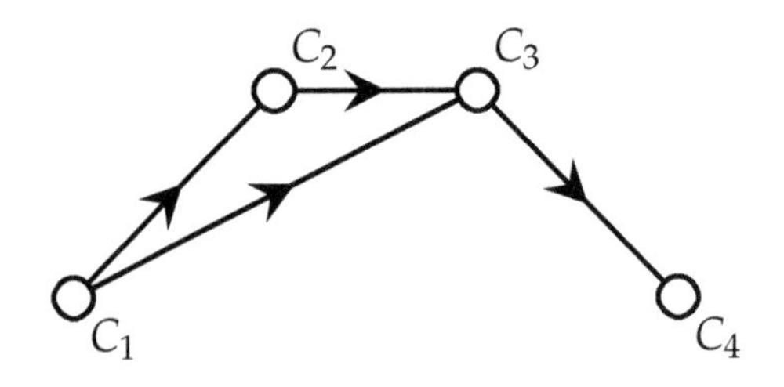

It can be proven that the component graph of any directed graph G is a DAG (directed acyclic graph). This is not hard to see, as a directed cycle in the component graph of G would mean that the union of the strongly connected components in the directed cycle would actually comprise a single strongly connected component.

We next describe an algorithm, based on DEPTH-FIRST SEARCH, that finds the strongly connected components of a directed graph G. The algorithm is commonly known as SHARIR'S ALGORITHM, but it is sometimes referred to as KOSARAJU'S ALGORITHM. In fact, Kosaraju discovered the algorithm in 1978 but did not publish it. Sharir rediscovered the algorithm and published it in 1981.

A high-level description of the main steps in SHARIR'S ALGORITHM is given in Algorithm 7.11.

Algorithm 7.11: SHARIR(G)

1. Perform a depth-first search of G, recording the finishing times $f[v]$ for all vertices v.
2. Construct the reverse graph G^R of G.
3. Perform a depth-first search of G^R, performing initial calls to DFSVISIT in decreasing order of the values $f[v]$ computed in step 1.
4. The strongly connected components of G are the trees in the depth-first forest constructed in step 3.

Example 7.9 The following directed graph G is the same graph as in Figure 7.3.

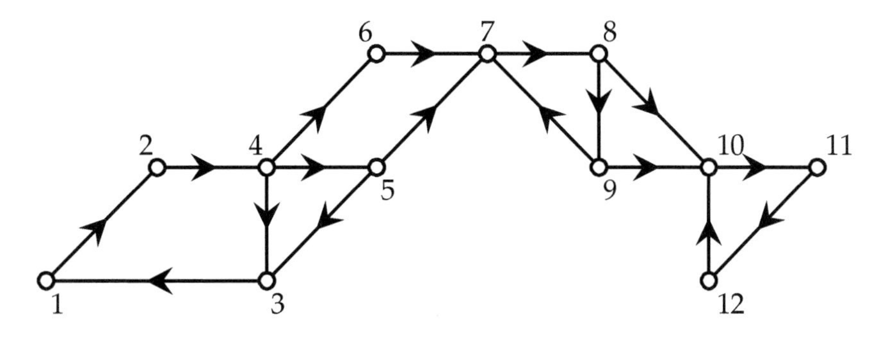

Here are the discovery and finish times for each vertex:

v	$d[v]$	$f[v]$
1	1	24
2	2	23
3	19	20
4	3	22
5	18	21
6	4	17

v	$d[v]$	$f[v]$
7	5	16
8	6	15
9	7	14
10	8	13
11	9	12
12	10	11

The vertices in decreasing order of finishing time are

$$1,2,4,5,3,6,7,8,9,10,11,12.$$

Now we construct the reverse graph, G^R:

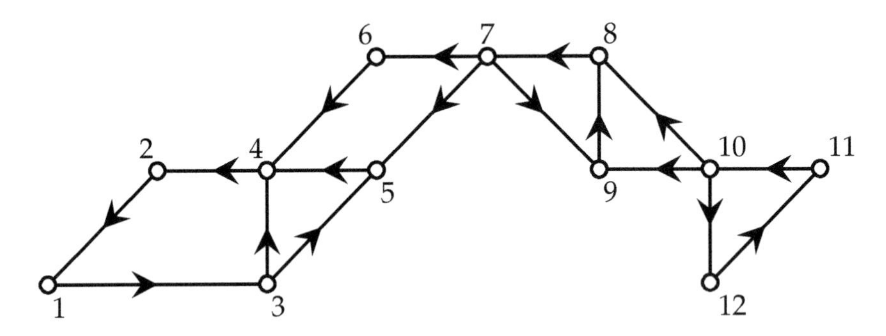

We now process the vertices in the order

$$1,2,4,5,3,6,7,8,9,10,11,12.$$

The depth-first search of G^R proceeds as follows:

1. An initial call to DFSVISIT from vertex 1 explores the strongly connected component having vertices $\{1,2,3,4,5\}$.
2. An initial call to DFSVISIT from vertex 6 explores the strongly connected component having vertices $\{6\}$.
3. An initial call to DFSVISIT from vertex 7 explores the strongly connected component having vertices $\{7,8,9\}$.
4. An initial call to DFSVISIT from vertex 10 explores the strongly connected component having vertices $\{10,11,12\}$.

□

In a bit more detail, we describe in Algorithms 7.12 and 7.13 how the depth-first search of G^R is carried out (this is step 3 of Algorithm 7.11). We augment a basic depth-first search by keeping track of the strongly connected component that each vertex belongs to. In order to do this, we maintain a counter, *scc*, that is incremented every time an initial call is made to DFSVISIT-SCC. This counter is the label assigned to the strongly connected component that is being processed at any given time. Also, for any vertex v, $comp[v]$ is the strongly connected component containing v.

We now sketch a proof of correctness of SHARIR'S ALGORITHM. We begin by stating and proving a useful fact about edges in the component graph. In step 1 of Algorithm 7.11, we carry out a depth-first search of G, obtaining discovery and finishing times for each vertex. For a strongly connected component C, define

$$f[C] = \max\{f[v] : v \in C\}$$

Algorithm 7.12: DFS-SCC(G^R)

```
precondition: f[v_{i_1}] > f[v_{i_2}] > ··· > f[v_{i_n}]
for j ← 1 to n
  do color[v_{i_j}] ← white
scc ← 0
for j ← 1 to n
     { if color[v_{i_j}] = white
  do {       { scc ← scc + 1
     {  then { DFSVISIT-SCC(G^R, v_{i_j}, scc)
return (comp)
```

Algorithm 7.13: DFSVISIT-SCC(G^R, v, scc)

```
global color, comp, scc
color[v] ← red
comp[v] ← scc
for each w ∈ Adj[v]
  do { if color[w] = white
     {   then DFSVISIT-SCC(H, w, scc)
color[v] ← black
```

and

$$d[C] = \min\{d[v] : v \in C\}.$$

LEMMA 7.16 *If C_i, C_j are strongly connected components, and there is an arc from C_i to C_j in the component graph, then $f[C_i] > f[C_j]$.*

PROOF Suppose $d(C_i) < d(C_j)$. Let $u \in C_i$ be the first discovered vertex. All vertices in $C_i \cup C_j$ are reachable from u, so they are descendants of u in the DFS tree. Hence $f(v) < f(u)$ for all $v \in C_i \cup C_j$, $v \neq u$. Therefore $f(C_i) > f(C_j)$.

Suppose $d(C_i) > d(C_j)$. In this case, no vertices in C_i are reachable from C_j, so $f(C_j) < d(C_i) < f(C_i)$. ∎

We will make use of the contrapositive of Lemma 7.16: If $f[C_i] < f[C_j]$, then there is no arc from C_i to C_j in the component graph.

Now, note that G and G^R have the same strongly connected components. Let $u = v_{i_1}$ be the first vertex visited in step 3 of Algorithm 7.11. Let C be the strongly connected component containing u and let C' be any other strongly connected component.

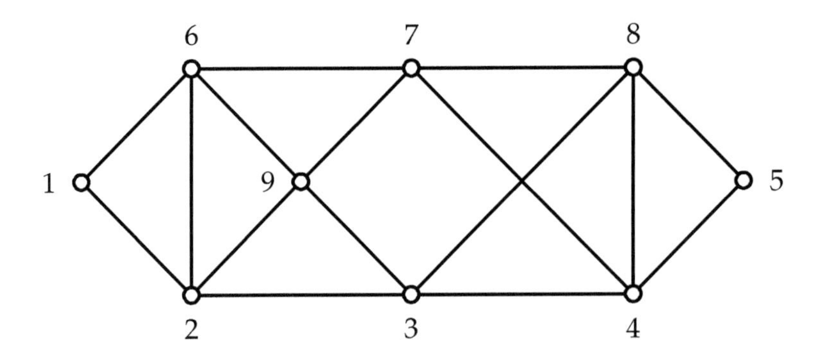

FIGURE 7.4: An Eulerian graph.

We have $f(C) > f(C')$, so there is no edge from C' to C in G (by Lemma 7.16). Therefore there is no edge from C to C' in G^R and hence no vertex in C' is reachable from u in G^R. It follows that the call to DFSVISIT-SCC from vertex u explores the vertices in C (and only those vertices); this forms one DFS tree in G^r.

The next initial call to DFSVISIT-SCC, say from vertex u', explores the vertices in the strongly connected component containing u'. This process continues: every time we make an initial call to DFSVISIT-SCC, we are exploring a new strongly connected component.

7.6 Eulerian Circuits

One of the oldest graph-theoretic problems is to find an Eulerian circuit in a given graph. An ***Eulerian circuit*** is a circuit that includes every edge in the graph exactly once. Informally, this means that it is possible to start at any vertex v and trace all the edges in the graph in a unicursal fashion (i.e., without lifting a pencil from the paper), finishing at v. A graph that has an Eulerian circuit is termed an ***Eulerian graph***.

Example 7.10 Consider the graph in Figure 7.4. It is relatively easy to find an Eulerian circuit of this graph by trial and error. Here is one Eulerian circuit:

$$1 \quad 2 \quad 3 \quad 4 \quad 5 \quad 6 \quad 4 \quad 7 \quad 6 \quad 3 \quad 9 \quad 7 \quad 8 \quad 9 \quad 2 \quad 8 \quad 1$$

□

Leonhard Euler first posed this problem in 1736. He observed that not every graph has an Eulerian circuit. The following theorem gives necessary and sufficient conditions for a graph to be Eulerian.

THEOREM 7.17 *A graph $G = (V, E)$ has an Eulerian circuit if and only if the following two conditions are satisfied:*

1. *G is connected, and*
2. *for every vertex $v \in V$, the degree of v is even.*

As a result of Theorem 7.17, it is straightforward to determine if a graph G is Eulerian. We can check to see if G is connected using a depth-first search or a breadth-first search. It is also easy to compute the degrees of all the vertices and verify if they are all even. For example, the graph in Figure 7.4 is a connected graph in which every vertex has degree two or four. Therefore, the graph is Eulerian from Theorem 7.17.

Theorem 7.17 is often termed ***Euler's Theorem***, but the first complete proof of it is due to Hierholzer in 1873. The necessary conditions are fairly obvious. It is clear that an Eulerian graph must be connected, so we examine the second condition. Let us think of the edges in an Eulerian circuit to be directed. So we write the edges in a circuit $v_1, v_2, v_3, \ldots$ as directed edges, $v_1v_2, v_2v_3, \ldots$. The key observation is that every time an edge enters a vertex v, another edge is used to immediately leave v. So the edges incident with a vertex v are used up two at a time. Since we eventually need to use up all the edges incident with each vertex, it follows immediately that every vertex has even degree.

Establishing the sufficiency of the two conditions is more difficult. It is possible to give an inductive proof that any graph satisfying conditions 1 and 2 of Theorem 7.17 is Eulerian. The basic idea of the proof is to start at a given vertex s and follow an arbitrary sequence of edges until we return to s. We are guaranteed that this will happen because every vertex has even degree. This creates a circuit, say C_1; however, C_1 might not use up all the edges in the graph.

Suppose there are edges that are not included in C_1. Then, there must be a vertex in C_1 that has incident edges that are not in C_1 (this follows because the graph G is assumed to be connected). Now we can inductively find an Eulerian circuit C' of the graph that is obtained by deleting the edges of C_1 from E. If we combine C' and C_1, then we obtain an Eulerian circuit of the original graph.

We now turn the above proof into an efficient recursive algorithm, having complexity $O(n^2 + m)$, to construct an Eulerian circuit. The resulting algorithm, which was first published in 1873, is often termed HIERHOLZER'S ALGORITHM.

As described above, we start at a vertex s and follow an arbitrary sequence of edges until we return to s, thus creating a circuit C_1. The vertices in C_1 are placed (in order) in a queue Q. Then we process Q, by deleting vertices from Q, adding them to the Eulerian circuit *Eul* that we are constructing. If we encounter a vertex v that still has incident edges that have not been used yet, then we recursively call our procedure from v. After we return from the recursive call, we continue to process vertices in Q.

The recursive call that is made from v creates another queue and it also generates another circuit, say C_2, that begins and ends at v. The circuit C_2 will be added to *Eul* after the initial portion of C_1 (from s to v) and before the rest of C_1 is processed. Various recursive calls will be generated; the result is that each circuit is spliced into the circuit at the previous level of the recursion. The end result is a single Eulerian circuit.

Algorithm 7.14 uses this approach to find an Eulerian circuit in an Eulerian graph G. We assume the vertex set is $V = \{1, \ldots, n\}$ and adjacency lists for the graph have been constructed. This algorithm constructs an Eulerian circuit *Eul*.

Here are a few details about Algorithm 7.14:

- As we search for edges to include in *Eul*, we proceed through every adjacency list once from beginning to end. These adjacency lists are global.

- The value $drem[v]$ indicates the number of edges incident with v that have not been included in the Eulerian circuit *Eul* at a given point in time. The value $UsedEdge[w, v]$ is **true** if and only if vw is an edge of the graph that has not yet been added to *Eul*.

- One slightly subtle point is that, when we include an edge vw in the Eulerian circuit, say after finding $w \in Adj[v]$, we have to set

  $$UsedEdge[v, w] \leftarrow \textbf{false} \quad \text{and} \quad UsedEdge[w, v] \leftarrow \textbf{false}.$$

  This is because we may later encounter the vertex v in $Adj[w]$ and we want to ensure we do not re-use this edge. We also update

  $$drem[v] \leftarrow drem[v] - 1 \quad \text{and} \quad drem[w] \leftarrow drem[w] - 1$$

  to reflect the fact that the number of available edges incident with v and the number of available edges incident with w are both decreased by one.

Example 7.11 Let us see how Algorithm 7.14 would execute on the graph presented in Figure 7.4.

1. The initial call EULER(1) creates a queue Q_1:

   $$1 \quad 2 \quad 3 \quad 4 \quad 5 \quad 6 \quad 3 \quad 9 \quad 2 \quad 8 \quad 1.$$

2. When we process the queue Q_1, we add the vertices $1, 2, 3$ to *Eul*. Then we find $drem[4] > 0$, so we call EULER(4).

3. EULER(4) creates a queue Q_2:

   $$4 \quad 6 \quad 7 \quad 4.$$

   We add the vertices $4, 6$ to *Eul*. Then $drem[7] > 0$, so we call EULER(7).

Algorithm 7.14: EULER(s)

```
precondition: G is connected and every vertex has even degree
global drem, Adj, UsedEdge, Eul, len
v ← s
INITIALIZEQUEUE(Q)
ENQUEUE(Q, v)
repeat
  repeat
    w ← the next vertex in Adj[v]
  until not UsedEdge[v, w]
  ENQUEUE(Q, w)
  UsedEdge[v, w] ← true
  UsedEdge[w, v] ← true
  drem[v] ← drem[v] − 1
  drem[w] ← drem[w] − 1
  v ← w
until w = s
while not EMPTY(Q)
  do { v ← DEQUEUE(Q)
       if drem[v] = 0
         then { len ← len + 1
                Eul[len] ← v
         else EULER(v)
main
  len ← 0
  for v ← 1 to n
    do { drem[v] ← deg(v)
         for w ← 1 to n
           do UsedEdge[v, w] ← true
  for all vw ∈ E
    do UsedEdge[v, w] ← false
  EULER(1)
```

4. EULER(7) creates the queue Q_3:

$$7 \quad 8 \quad 9 \quad 7.$$

   We add all the vertices $7, 8, 9, 7$ to *Eul*. Then we return to the previous level of the recursion.

5. We continue to process the rest of the queue Q_2, adding the vertex 4 to *Eul*. Then we return to the previous level of the recursion.

6. We continue to process the rest of the queue Q_1, adding the vertices $5, 6, 3, 9, 2, 8, 1$ to *Eul*.

7. The resulting Eulerian circuit *Eul* is

$$1 \quad 2 \quad 3 \quad 4 \quad 6 \quad 7 \quad 8 \quad 9 \quad 7 \quad 4 \quad 5 \quad 6 \quad 3 \quad 9 \quad 2 \quad 8 \quad 1.$$

□

The complexity of Algorithm 7.14 is easy to analyze. The initialization of *UsedEdge* takes time $O(n^2)$. The main part of the algorithm takes time $O(n+m)$ because we proceed through every adjacency list once from beginning to end, and every edge enters and leaves a queue exactly once.

7.7 Minimum Spanning Trees

A ***spanning tree*** in a connected, undirected graph $G = (V, E)$ is a subgraph T that is a tree containing every vertex of V. T is a spanning tree of G if and only if T is an acyclic subgraph of G that has $n-1$ edges (where $n = |V|$). The **Minimum Spanning Tree** problem, or **MST**, is presented as Problem 7.1.

Problem 7.1: Minimum Spanning Tree
Instance: A connected, undirected graph $G = (V, E)$ and a ***weight function*** $w : E \rightarrow \mathbb{R}$.
Find: A spanning tree T of G such that $\sum_{e \in T} w(e)$ is minimized (this is called a ***minimum spanning tree***, or ***MST***).

We will study *greedy* algorithms to solve the MST problem. KRUSKAL'S ALGORITHM and PRIM'S ALGORITHM are two popular algorithms that we will discuss. However, we first record some useful facts about trees that we will require in the subsequent discussion in this section. These facts are all rather obvious, but we include complete proofs, as the proof details are surprisingly tricky.

LEMMA 7.18 *Every tree contains a vertex of degree one.*

PROOF The proof is reminiscent of the proof of Lemma 7.9. Suppose we have a graph G in which $\deg(v) \geq 2$ for all v. We prove that G contains a cycle, so it cannot be a tree.

Let v_1 be any vertex and let v_1v_2 be any edge. For every $i \geq 2$, let $\{v_i, v_{i+1}\}$ be an edge such that $v_{i+1} \neq v_{i-1}$ for all $i \geq 2$ (this is possible because every vertex has degree at least two). In the infinite sequence $v_1, v_2, v_3, \ldots$, consider the first repeated vertex, $v_i = v_j$ where $j > i$. Note that $j \geq i + 3$. Then $v_i, v_{i+1}, \ldots, v_j$ is a cycle. ∎

LEMMA 7.19 *A tree on n vertices has $n - 1$ edges.*

PROOF The proof is by induction on the number of vertices, n. A tree on $n = 1$ vertex has $n - 1 = 0$ edges, so the lemma is valid for $n = 1$.

As an induction assumption, suppose the lemma is valid for trees having $n - 1$ vertices, for some $n \geq 2$. Now assume T is a tree having n vertices. By Lemma 7.18, T has a vertex v having degree one. If we delete v and its unique incident edge, we obtain a tree T' having $n - 1$ vertices. By induction, T' has $n - 2$ edges. Since we deleted one edge from T to form T', it follows that T has $n - 1$ edges. Therefore the lemma is valid for all $n \geq 1$ by induction. ∎

LEMMA 7.20 *There is a unique simple path between any two distinct vertices in a tree.*

PROOF Let u and v be two distinct vertices in a tree T. A tree is connected, so there is at least one simple path between u and v, say P_1, consisting of vertices

$$v_1, v_2, \ldots, v_{k-1}, v_k,$$

where $u = v_1$ and $v = v_k$. (Note that $k = 2$ is allowed, if uv is an edge in T.) Suppose that there is another simple (u, v)-path, say P_2, consisting of vertices

$$w_1, w_2, \ldots, w_{j-1}, w_j,$$

where $u = w_1$ and $v = w_j$. We will eventually obtain a contradiction.

The two paths P_1 and P_2 are not identical, so there exists at least one index i such that $v_i \neq w_i$. Choose the minimum such i; we observe that $i \geq 2$. We also note that $v_{i-1} = w_{i-1}$.

Now choose the minimum $j > i$ such that $v_j \in P_2$. Note that $j \leq k$ because $v_k \in P_2$. Suppose that $v_j = w_\ell$. We claim that

$$v_{i-1}, v_i, \ldots, v_j, w_{\ell-1}, w_{\ell-2}, \ldots, w_{i-1}$$

is a cycle. In words, this cycle is obtained by taking the portion of P_1 from v_{i-1} to v_j, say P_1', and then joining it (in reverse order) to the portion of P_2 from w_{i-1} to w_ℓ, say P_2'.

No interior vertex of P_1' occurs in P_2, because of the way j was chosen. It is possible that an interior vertex of P_2' occurs in P_1, but no such vertex occurs in P_1', again by the way j was chosen. So we have a set of distinct vertices, which is a requirement of a cycle. Next, it is clear that every pair of consecutive vertices is an edge. The last thing to check is that the cycle has length at least three. If its length is equal to two, then each of P_1' and P_2' consists of the single edge $v_{i-1}v_i$. But this is not consistent with the way that i was chosen.

We conclude that we have constructed a cycle in T, which contradicts the fact that T is acyclic. ∎

LEMMA 7.21 *If T is a tree and an edge $e \notin T$ is added to T, then the resulting graph contains a unique cycle C. Further, if $e' \in C$, then $T \cup \{e\} \setminus \{e'\}$ is a tree.*

PROOF Suppose $e = \{u, v\}$. From Lemma 7.20, there is a unique simple (u, v)-path in T, which has length at least two. This path, together with the edge e, forms a cycle, say C, in $T \cup \{e\}$. We need to show that C is the unique cycle in $T \cup \{e\}$. Suppose there is another cycle, say D, in $T \cup \{e\}$. Clearly D must contain the edge e because T is acyclic. Now, if we delete e from C and D, we obtain two simple (u, v)-paths in T. This contradicts Lemma 7.20. Therefore we conclude that C is the unique cycle in $T \cup \{e\}$.

Now suppose we delete an edge $e' \in C$ from $T \cup \{e\}$. We need to show that the resulting graph $T \cup \{e\} \setminus \{e'\}$ is connected and acyclic. We leave these verifications as an exercise for the reader. ∎

7.7.1 Kruskal's Algorithm

We describe KRUSKAL'S ALGORITHM, which was published in 1956, in this section.

As a preprocessing step, we will sort and relabel the edges so

$$w(e_1) \leq w(e_2) \leq \cdots \leq w(e_m),$$

where $m = |E|$. Assuming we can obtain a list of all the edges in time $\Theta(m)$, e.g., using an adjacency list representation of the graph, this will take time $\Theta(m \log m)$.

Algorithm 7.15 is a high-level description of KRUSKAL'S ALGORITHM.

```
Algorithm 7.15: KRUSKAL(G, w)

comment: we assume w(e_1) ≤ w(e_2) ≤ ··· ≤ w(e_m)
precondition: G is a connected graph
A ← ∅
for j ← 1 to m
   do { if A ∪ {e_j} does not contain a cycle
      {   then A ← A ∪ {e_j}
return (A)
```

KRUSKAL'S ALGORITHM is a greedy algorithm. It examines the edges in increasing order of weight, and adds an edge to the spanning tree A that is "under construction" provided that no cycle is contained in the set A of chosen edges.

At any time during the execution of KRUSKAL'S ALGORITHM, we have a ***forest*** of vertex-disjoint trees (a forest is a graph in which every connected component is a tree). Initially, we have n one-vertex trees, and eventually we end up with a single tree on all n vertices (assuming the graph is connected).

Consider an edge $e_j = uv$. Addition of the edge e_j to the set A will create a cycle if and only if u and v are in the same tree in the current forest. Thus, an edge e_j will be added to A if and only if it joins two different trees together. Therefore, the number of trees is reduced by one at each iteration.

We present a small example to illustrate KRUSKAL'S ALGORITHM.

Example 7.12 We start with the following graph.

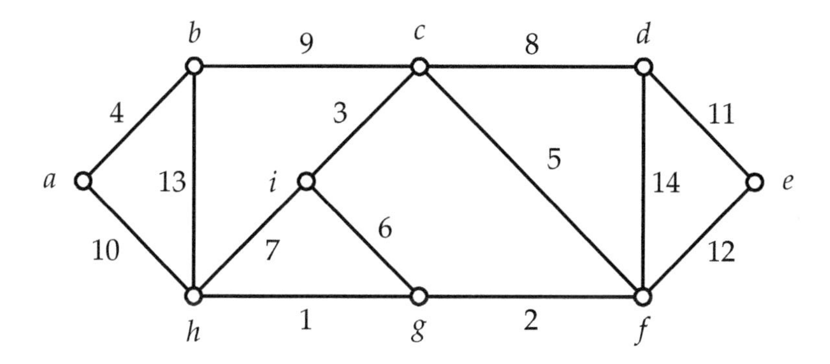

We begin by choosing (in order) the edges gh, fg, ci, ab, and cf (which have weights $1, 2, 3, 4$, and 5, respectively). None of these edges create any cycles. Therefore, at this point, the partial spanning tree consists of the red edges in the following graph:

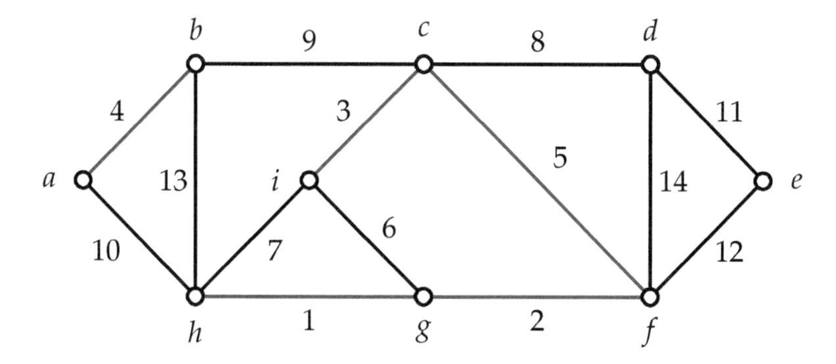

We next examine the edge gi, which has weight 6. We cannot include this edge, because we have already chosen the edges ic, cf, and fg, and if we also included the edge gi, we would have a cycle i, c, f, g, i.

The next edge, hi, has weight 7. This edge cannot be included in the spanning tree, because it would create a cycle h, g, f, c, i, h.

The next edge, cd, has weight 8. It can be included in the spanning tree, as can the edge bc that has weight 9. At this point, we have chosen the following red edges:

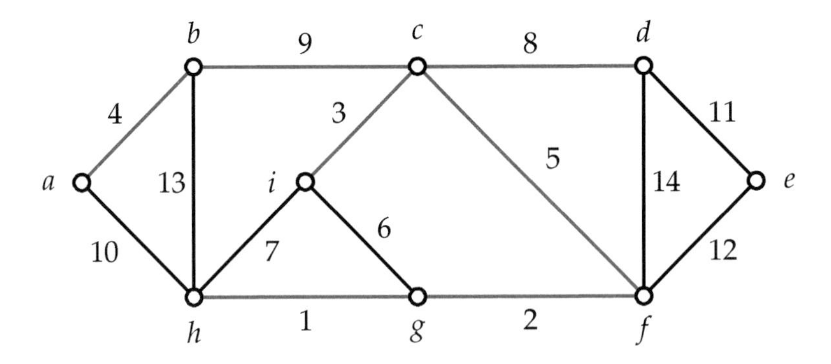

The next edge is ah, which has weight 10. It cannot be included in the spanning tree because it would create a cycle a, b, c, f, g, h, a.

The next edge, de, has weight 11 and it can be added to the spanning tree. At this point, we have selected a sufficient number of edges to form a spanning tree, so we are done. We end up with the following set of red edges, which are the edges in the minimum spanning tree.

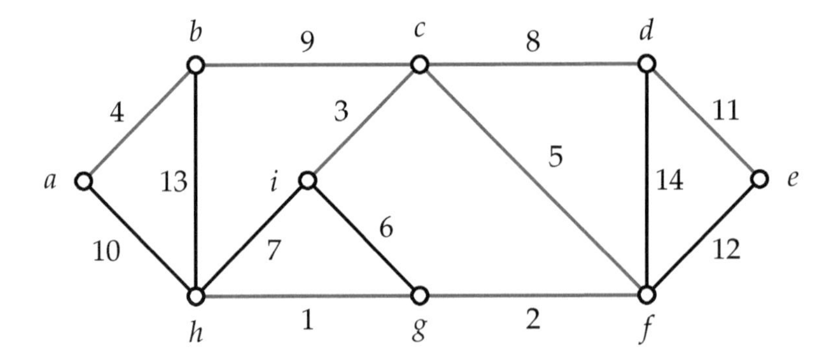

□

We next discuss some implementation details for KRUSKAL'S ALGORITHM. As we have already mentioned, at any given point in time, we want to determine if an edge uv creates a cycle if it is added to the set of edges that have been selected previously. One obvious way to test this condition would be to perform a depth-first search on the set of selected edges together with the edge uv and verify that there are no back edges. This approach would have complexity $\Theta(n)$ each time it is done. Since it is not very fast, we desire a more efficient test.

Given that the set of selected edges is always a forest, we only need to check if the edge uv has endpoints in the same tree. (If the endpoints are in different trees, then we can add this edge to A.) Observe that we have a partition of the set of vertices into trees, which defines an equivalence relation on the set of vertices of the graph. The equivalence classes are just the vertex sets of the trees in the forest. Thus, given the edge uv, we seek to determine if u and v are in the same equivalence class. This can be checked if we have an efficient way of identifying a unique ***leader vertex*** that acts as an unambiguous "label" for the equivalence class (i.e., tree) it lies in.

We also need a mechanism to update the equivalence relation when we decide to add a new edge to A. This has the effect of merging two trees into one. That is, when we add the edge uv, two trees in the forest become one new tree, whose set of vertices is the union of the sets of vertices of the two trees that are being merged.

Using the language of abstract data types (ADTs), we need a data structure that supports the two operations FIND and UNION. Not surprisingly, this is called a ***union-find data structure***. Two vertices u and v will be in the same tree if and only if $\text{FIND}(u) = \text{FIND}(v)$, and the UNION operation merges two trees.

Here is a simple way to define a leader vertex for each tree so that the FIND operation is straightforward to define. We make use of an auxiliary array L. A leader vertex is any vertex w such that $L[w] = w$. If v is not a leader vertex, then we construct the sequence of vertices

$$v, L[v], L[L[v]], \cdots, w \tag{7.1}$$

until we reach a leader vertex, w. Then we define $w = \text{FIND}(v)$; the vertex w is the leader vertex for the tree containing v.

Of course we need to ensure that each tree has a unique leader vertex. Initially, there are n one-vertex trees and $L[v] = v$ for all v (i.e., every vertex is a leader vertex). When we use an edge uv to merge two trees, we perform the following UNION operation:

1. $u' \leftarrow Find(u)$
2. $v' \leftarrow Find(v)$
3. $L[u'] \leftarrow v'$.

The result is that the new merged tree has the unique leader vertex v'.

The array L implicitly defines a set of ***directed trees*** (we will call these the ***union-find trees***). For each vertex v such that $L[v] \neq v$, construct a directed edge $(v, L[v])$. These directed edges form a set of trees, each of which has edges directed to a unique leader vertex. Note that the union-find trees contain the same vertices as the trees in A, but *not* the same edges.

How efficient are these operations? We can concentrate on FIND, because a UNION operation basically consists of two FIND operations. The cost of a FIND operation depends on the length of the sequence (7.1) that must be followed to identify a leader vertex. If we are not careful, we could end up with sequences of length $\Omega(n)$ as various trees are merged. However, a simple optimization will ensure that the sequences (7.1) are always of length $O(\log n)$.

Define the ***depth*** of a union-find tree to be the maximum number of edges that must be followed until the leader of that tree is found. Suppose we also keep track of the depth of each union-find tree. For example, whenever w is a leader vertex, *depth*$[w]$ denotes the depth of the tree for which w is the leader vertex (if w is not a leader vertex, *depth*$[w]$ is not defined). Initially, every vertex is a leader vertex and *depth*$[w] = 0$ for all vertices w.

When we use the UNION operation to merge two union-find trees, say T_1 and T_2, we will ensure that the leader of the new tree is the leader of the tree having larger depth (if T_1 and T_2 have the same depth, then we can choose either leader to be the leader of the new tree). If we merge two trees of depth d, then we get a tree of depth $d+1$. If we merge a tree of depth d and one of depth less than d, we end up with a tree of depth d.

We now give pseudocode descriptions of FIND and UNION, as well as the appropriate initialization of the arrays L and *depth*. See Algorithms 7.16, 7.17 and 7.18.

Algorithm 7.16: INITIALIZEUNIONFIND(V)

```
comment: V = {1,...,n} is the set of vertices in a graph
for w ← 1 to n
  do { L[w] ← w
     { depth[w] ← 0
return (L, depth)
```

Algorithm 7.17: FIND(v, L)

```
w ← v
while L[w] ≠ w
  do w ← L[w]
return (w)
```

With these optimizations, UNION and FIND each take $O(\log n)$ time. This follows from the following property concerning the depths of union-find trees.

LEMMA 7.22 *Suppose that two union-find trees are merged using Algorithm 7.18 and suppose T is a union-find tree of depth d. Then T contains at least 2^d vertices.*

PROOF Initially, every union-find tree consists of one node and has depth 0. Let's consider what happens when two trees T_1 and T_2 are merged using Algorithm 7.18. For $i = 1, 2$, let d_i denote the depth of T_i. We can assume inductively that T_i contains at least 2^{d_i} vertices, for $i = 1, 2$. Let T be the new tree obtained by merging T_1 and T_2.

First, suppose that $d_1 = d_2 = d$, say. Then T has depth $d+1$. The number of vertices in T is at least $2^d + 2^d = 2^{d+1}$, as desired.

If $d_1 < d_2 = d$, then T has depth d. The tree T_2 has at least 2^d vertices, so T also has at least 2^d vertices. If $d_1 > d_2$, the proof is similar. ∎

Algorithm 7.18: UNION($u, v, L, depth$)

external FIND
$u' \leftarrow$ FIND(u)
$v' \leftarrow$ FIND(v)
$d_1 \leftarrow depth[u']$
$d_2 \leftarrow depth[v']$
if $d_1 < d_2$
  then $\begin{cases} L[u'] \leftarrow v' \\ depth[u'] \leftarrow \text{undefined} \end{cases}$
else if $d_1 > d_2$
  then $\begin{cases} L[v'] \leftarrow u' \\ depth[v'] \leftarrow \text{undefined} \end{cases}$
  else $\begin{cases} L[u'] \leftarrow v' \\ depth[u'] \leftarrow \text{undefined} \\ depth[v'] \leftarrow depth[v'] + 1 \end{cases}$
return $(L, depth)$

Lemma 7.22 establishes that a UNION or FIND operation can always be carried out in time $O(\log n)$. This is because, at any point in time, any union-find tree T has at most n nodes, and therefore the depth of T is at most $\log_2 n$.

This leads to an implementation of KRUSKAL'S ALGORITHM having complexity $O(m \log n)$. The pre-sort has complexity $O(m \log m)$, which is the same as $O(m \log n)$ because a connected graph has $n - 1 \leq m \leq \binom{n}{2}$ and hence $\log m \in \Theta(\log n)$. The iterative part of the algorithm looks at each of the m edges in turn. For each edge, $O(1)$ FIND operations are carried out, each of which has complexity $O(\log n)$. So the iterative part of the algorithm has complexity $O(m \log n)$, as does the entire algorithm.

Example 7.13 We look again at the graph in Example 7.12 to illustrate how the union-find data structure can be used in KRUSKAL'S ALGORITHM. Initially, the union-find trees consist of isolated vertices and every vertex is a leader vertex. (Leader vertices will be colored red in this example.)

$a \quad b \quad d \quad e \quad c \quad i \quad h \quad g \quad f$

Considering edge gh, we have FIND(g) $= g$, FIND(h) $= h$. The two trees have depth 0. Suppose we direct $g \rightarrow h$; then the tree $\{g, h\}$ has depth 1.

$a \quad b \quad d \quad e \quad c \quad i \quad h \leftarrow g \quad f$

Considering edge fg, we have FIND$(f) = f$, FIND$(g) = h$. The two trees have depth 0 and 1, resp. Thus we *must* direct $f \rightarrow h$ (see the previous slide). Then the resulting tree $\{f, g, h\}$ has depth 1.

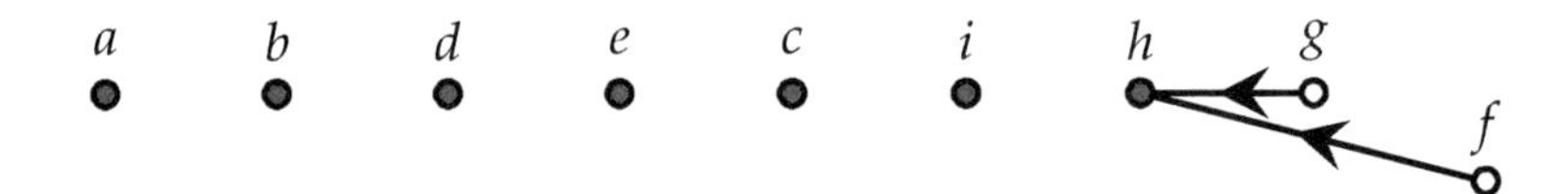

Considering edge ci, we have FIND$(c) = c$, FIND$(i) = i$. The two trees have depth 0. Suppose we direct $c \rightarrow i$; then the tree $\{c, i\}$ has depth 1.

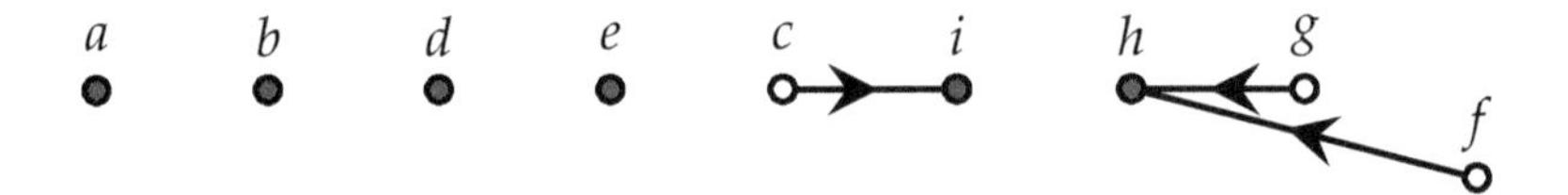

Considering edge ab, we have FIND$(a) = a$, FIND$(b) = b$. The two trees have depth 0. Say we direct $a \rightarrow b$; then the tree $\{a, b\}$ has depth 1.

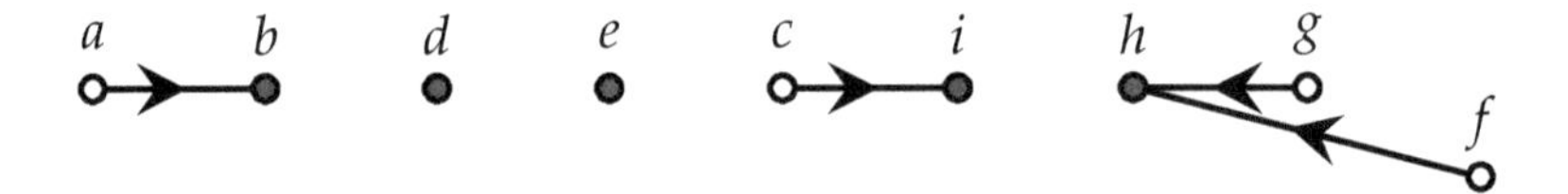

Considering edge cf, we have FIND$(c) = i$, FIND$(f) = h$. The two trees have depth 1. Suppose we direct $i \rightarrow h$; then the tree $\{c, f, g, h, i\}$ has depth 2.

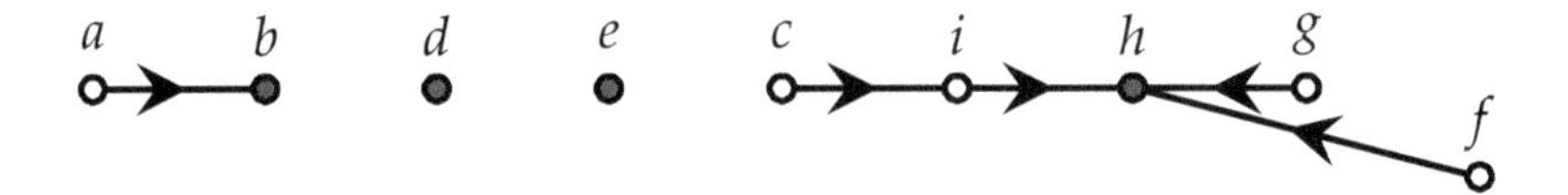

Edges gi and hi are not added to the MST because FIND(g) = FIND(i) and FIND(h) = FIND(i). Considering edge cd, we have FIND$(c) = h$, FIND$(d) = d$. The two trees have depth 0 and 2, resp. Thus we *must* direct $d \rightarrow h$. Then the resulting tree containing vertices $\{c, d, f, g, h, i\}$ has depth 2.

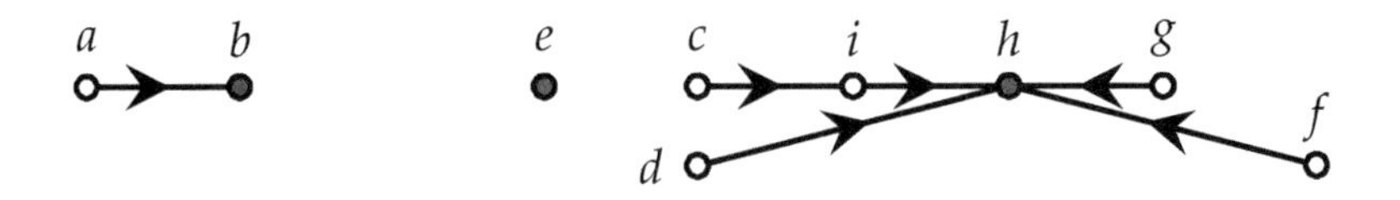

There are two more edges that will be added to the tree, namely *bc* and *de*. The reader can check the details.

□

We now give a proof of correctness of KRUSKAL'S ALGORITHM. For simplicity, we assume that all edge weights are distinct. In this case, we actually prove that there is a unique minimum spanning tree, and it is the spanning tree constructed by KRUSKAL'S ALGORITHM.

Let A be the spanning tree constructed by KRUSKAL'S ALGORITHM and let A' be any minimum spanning tree. Suppose the edges in A are named $f_1, f_2, \ldots, f_{n-1}$, where

$$w(f_1) < w(f_2) \cdots < w(f_{n-1}).$$

The proof will be a proof by contradiction. Suppose $A \neq A'$ and let f_j be the first edge in $A \setminus A'$. Then $A' \cup \{f_j\}$ contains a unique cycle, say C, from Lemma 7.21. Let e' be the first (i.e., lowest weight) edge of C that is not in A (such an edge exists because $C \nsubseteq A$). Define

$$A'' = A' \cup \{f_j\} \setminus \{e'\}.$$

A'' is also a spanning tree (from Lemma 7.21). See Figure 7.5 for a diagram.

Then

$$w(A'') = w(A') + w(f_j) - w(e').$$

Since A' is a minimum spanning tree, we must have $w(A'') \geq w(A')$. Therefore, $w(f_j) \geq w(e')$. However, we are assuming that the edge weights are all distinct, so $w(f_j) > w(e')$.

What happened when KRUSKAL'S ALGORITHM considered the edge e'? This occurred before it considered the edge f_j, because $w(f_j) > w(e')$. Because KRUSKAL'S ALGORITHM rejected the edge e', the set of edges $\{f_1, \ldots, f_{j-1}, e'\}$ must contain a cycle. However, A' contains all these edges and A' is a tree, so we have a contradiction. Therefore $A = A'$ and A is the unique minimum spanning tree. This completes the correctness proof.

Assuming the edge weights are distinct simplifies the proof given above. The result is still true even when there are "repeated" edge weights, though the MST might not be unique in this case.

7.7.2 Prim's Algorithm

PRIM'S ALGORITHM is another popular algorithm to find minimum spanning trees. It is interesting to note that Jarnik actually discovered the algorithm in 1930, while Prim rediscovered it in 1957.

Here are the basic ideas in the algorithm:

- We initially choose an arbitrary vertex u_0.
- Define $V_A = \{u_0\}$ and $A = \{e\}$, where e is the *minimum-weight* edge incident with u_0.

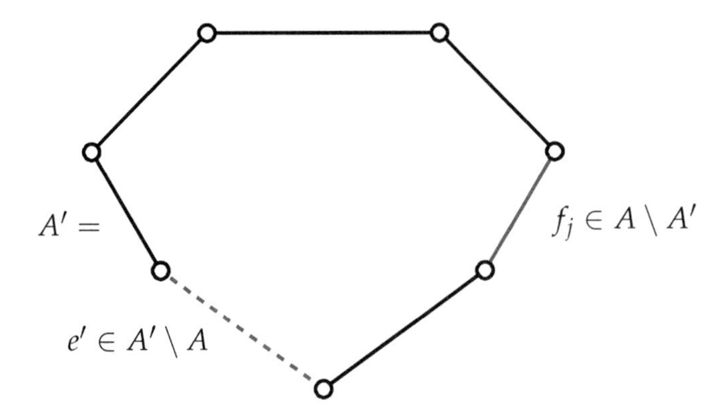

FIGURE 7.5: Correctness of KRUSKAL'S ALGORITHM.

- As we proceed, A is always a *single tree* and V_A will denote the set of vertices in A.
- At each step, we select the minimum-weight edge that joins a vertex $u \in V_A$ to a vertex $v \notin V_A$.
- We add v to V_A and we add the edge uv to A.
- Repeat these operations until A is a spanning tree.

We illustrate the execution of PRIM'S ALGORITHM with an example.

Example 7.14 We run PRIM'S ALGORITHM on the graph that we used in Example 7.12. Suppose we start at a; then $V_A = \{a\}$. In the following, vertices in V_A and edges in A are colored red.

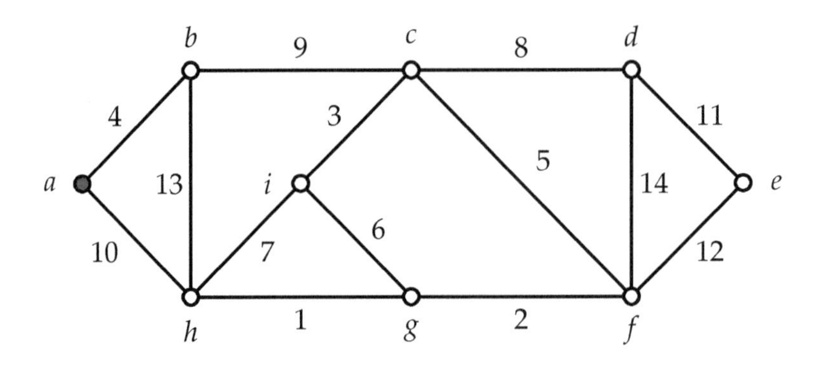

We consider possible edges ab and ah; the minimum edge is ab; then $V_A = \{a, b\}$.

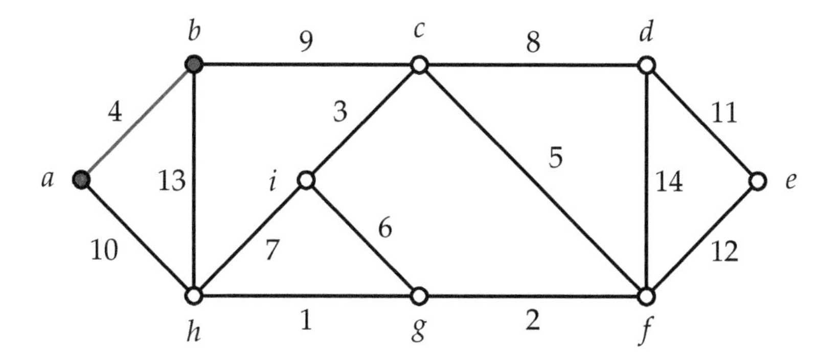

We next consider possible edges ah, bh and bc; the minimum edge is bc; then $V_A = \{a, b, c\}$.

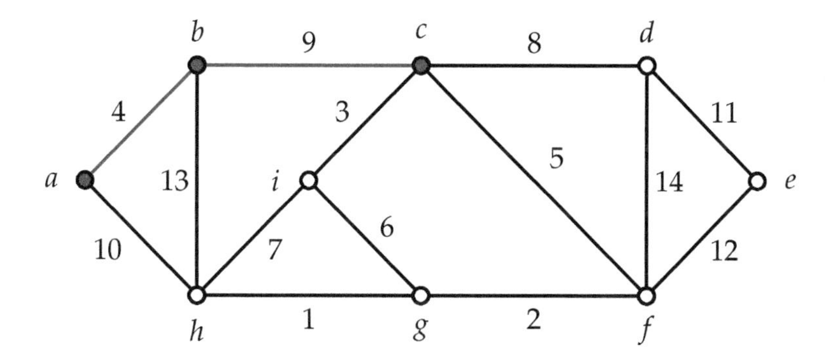

We next consider possible edges ah, bh, hi, cd, cf and ci; the minimum edge is ci; then $V_A = \{a, b, c, i\}$.

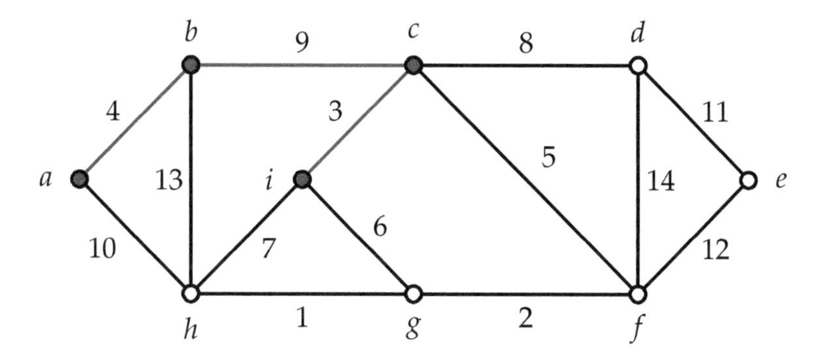

We next consider possible edges $ah, bh, hi, gi, cf andcd$; the minimum edge is cf; then $V_A = \{a, b, c, f, i\}$.

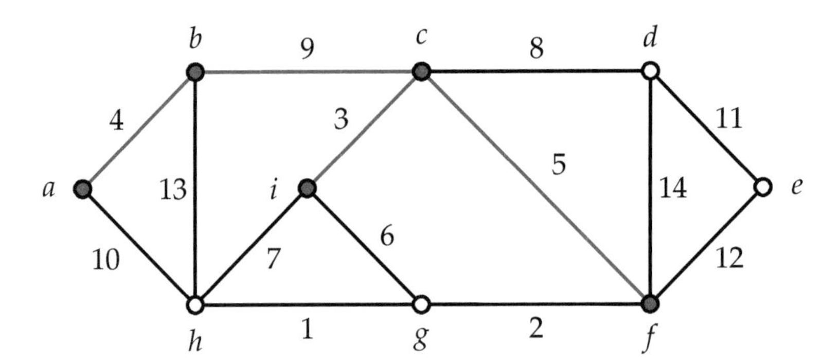

We next consider possible edges *ah*, *bh*, *hi*, *gi*, *fg*, *cd*, *df* and *ef*; the minimum edge is *fg*; then $V_A = \{a, b, c, f, g, i\}$.

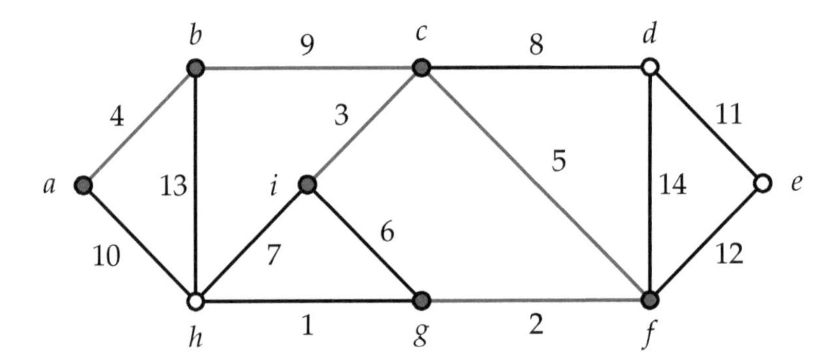

We next consider possible edges *ah*, *bh*, *hi*, *gh*, *cd*, *df* and *ef*; the minimum edge is *gh*; then $V_A = \{a, b, c, f, g, h, i\}$.

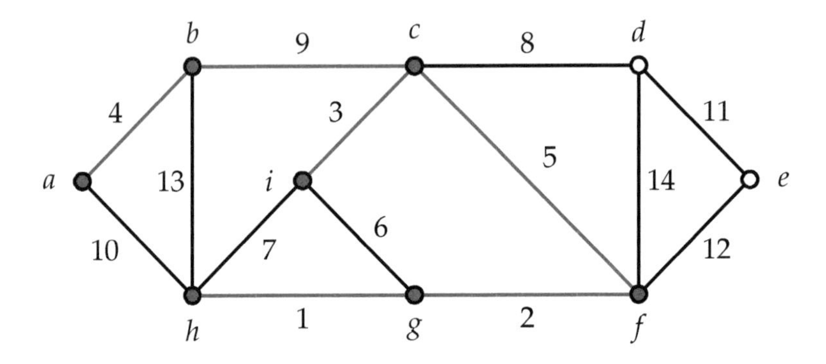

We next consider possible edges *cd*, *df* and *ef*; the minimum edge is *cd*; then $V_A = \{a, b, c, d, f, g, h, i\}$.

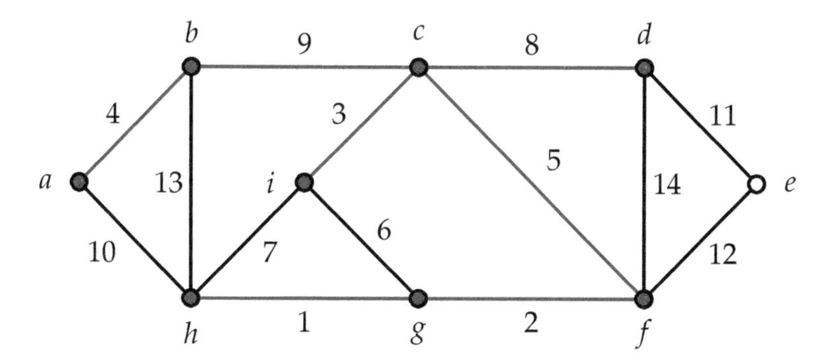

Finally, we consider possible edges de and ef; the minimum edge is de; $V_A = \{a, b, c, d, e, f, g, h, i\}$. We are done!

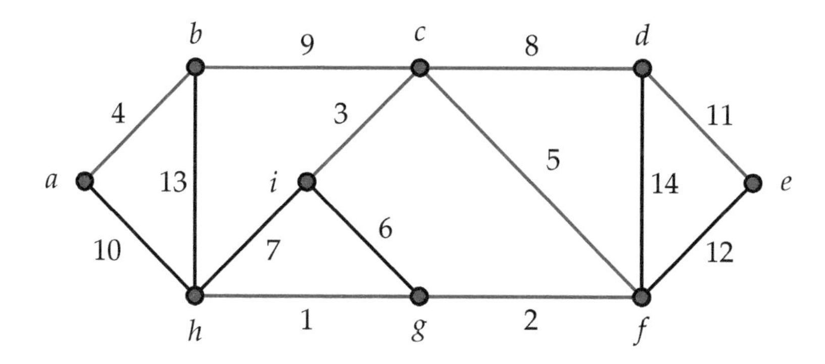

□

We now present an efficient way to determine, at each stage of the algorithm, the edge that should be added to A (and the vertex that should be added to V_A). In what follows, we assume for convenience that $w(u, v) = \infty$ if $\{u, v\} \notin E$.

For a vertex $v \notin V_A$, define

$$N[v] = u, \text{ where } \{u, v\} \text{ is a minimum-weight edge such that } u \in V_A$$
$$W[v] = w(N[v], v).$$

Initially, we choose a starting vertex u_0. Then

$$N[v] = u_0 \quad \text{and}$$
$$W[v] = w(u_0, v)$$

for all v.

At any given time, the vertex that is added to V_A is the vertex $v \notin V_A$ with the minimum W-value. The edge to be added to A is $\{v, N(v)\}$. However, after adding a new vertex to V_A and a new edge to A, we have to update the values $W[v']$ and $N[v']$.

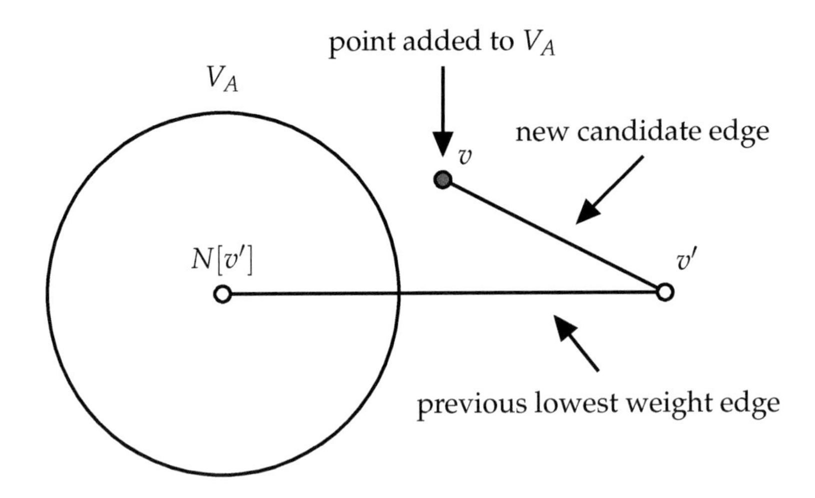

FIGURE 7.6: Updating $N[v']$ and $W[v']$ in PRIM'S ALGORITHM.

When we add the vertex v to V_A, the edge $\{v', v\}$ is a new "candidate edge" that might have weight that is less than the current minimum weight $W[v']$. So we update $W[v']$ using the formula

$$W[v'] = \min\{W[v'], w(v, v')\}.$$

If $W[v']$ is reduced in value, then we also must update the value of $N[w']$ to v. See Figure 7.6.

We present PRIM'S ALGORITHM as Algorithm 7.19. There are a few implementation details that we need to consider. First, we can test if a vertex v is in the set V_A by simply keeping track of the elements in V_A using a bit-vector. Suppose $V = \{v_1, \ldots, v_n\}$. Then we define $S[i] = 1$ if $v_i \in V_A$, and $S[i] = 0$, otherwise. In this way, we can easily test membership in V_A in $O(1)$ time.

In the **for all** loop that is executed in each iteration of the **while** loop, we could iterate through all the vertices and update W- and N-values of vertices not in V_A, as needed. We can analyze this approach as follows:

- There are $n - 1$ iterations of the **while** loop.
- Finding v takes time $O(n)$.
- Updating W-values takes time $O(n)$.
- Therefore the algorithm has complexity $O(n^2)$.

An alternative approach is to use a priority queue, implemented as a min-heap, to store the W-values. (Heaps were discussed in Section 3.4.1.) We can analyze this strategy as follows:

Algorithm 7.19: PRIM(G, w)

precondition: G is a connected graph
$A \leftarrow \varnothing$
$V_A \leftarrow \{u_0\}$, where u_0 is arbitrary
for all $v \in V \backslash \{u_0\}$
  do $\begin{cases} W[v] \leftarrow w(u_0, v) \\ N[v] \leftarrow u_0 \end{cases}$
while $|A| < n - 1$
  do {
    choose $v \in V \backslash V_A$ such that $W[v]$ is minimized
    $V_A \leftarrow V_A \cup \{v\}$
    $u \leftarrow N[v]$
    $A \leftarrow A \cup \{uv\}$
    for all $v' \in V \backslash V_A$
      do {
        if $w(v, v') < W[v']$
          then $\begin{cases} W[v'] \leftarrow w(v, v') \\ N[v'] \leftarrow v \end{cases}$
return (A)

- Initializing the priority queue is done using a BUILDHEAP operation that takes time $O(n)$.
- A DELETEMIN operation can be used to find v; it takes time $O(\log n)$.
- Each update of a W-value takes time $O(\log n)$. This is a basically a modification of a HEAPINSERT operation that uses a similar bubble-up process.
- In total, we perform n DELETEMIN operations.
- Suppose we use an adjacency list representation of the graph. When we add v to V_A, we only update W-values of vertices in $Adj[v]$. So the total number of updates that are performed is at most m.
- It follows that the algorithm has complexity $O(m \log n)$. This turns out to be worse than the previous simple implementation if $m \in \Omega(n^2)$, but it is better for sparse graphs, e.g., graphs where the number of edges is $O(n)$.

We do not give a proof of correctness for PRIM'S ALGORITHM at this time. However, the correctness of PRIM'S ALGORITHM will follow from a correctness proof we give for a more general algorithm in the next section.

7.7.3 A General Algorithm

In this section, we present a "general" algorithm to find a minimum spanning tree. This algorithm yields both KRUSKAL'S ALGORITHM and PRIM'S ALGORITHM as special cases.

We begin with some relevant definitions. Let $G = (V, E)$ be a graph. A ***cut*** is a partition of V into two non-empty (disjoint) sets, i.e., a pair $(S, V \backslash S)$, where $S \subseteq V$ and $1 \leq |S| \leq n - 1$.

Let $(S, V \backslash S)$ be a cut in a graph $G = (V, E)$. An edge $e \in E$ is a ***crossing edge*** with respect to the cut $(S, V \backslash S)$ if e has one endpoint in S and one endpoint in $V \backslash S$.

Let $A \subseteq E$ (i.e., A is a subset of the edges in E). We say that a cut $(S, V \backslash S)$ ***respects*** the set of edges A provided that no edge in A is a crossing edge.

In the general algorithm, at any point in time in the construction of a minimum spanning tree, we will have a set A of vertex-disjoint trees (possibly including trivial trees consisting of isolated vertices). In this situation, a cut $(S, V \setminus S)$ respects A if and only if S consists of the union of the vertex sets one or more of the trees in A.

The general minimum spanning tree algorithm is also a greedy algorithm. We present it as Algorithm 7.20.

Algorithm 7.20: GENERALGREEDYMST(G, w)

precondition: G is a connected graph
$A \leftarrow \varnothing$
while $|A| < n - 1$
do $\begin{cases} \text{let } (S, V \backslash S) \text{ be a cut that respects } A \\ \text{let } e \text{ be a minimum-weight crossing edge} \\ A \leftarrow A \cup \{e\} \end{cases}$
return (A)

We show that KRUSKAL'S ALGORITHM and PRIM'S ALGORITHM are both special cases of Algorithm 7.20:

- In PRIM'S ALGORITHM, the edge uv added at any stage of the algorithm is a minimum-weight edge respecting the cut $(V_A, V \setminus V_A)$.
- In KRUSKAL'S ALGORITHM, the edge uv added at any stage of the algorithm is a minimum-weight edge respecting the cut $(S, V \setminus S)$, where S is the tree containing u.

Thus, once we prove the correctness of Algorithm 7.20, we obtain proofs of correctness of KRUSKAL'S ALGORITHM and PRIM'S ALGORITHM for free.

Let $G = (V, E)$ be a graph on n vertices. We prove that the spanning tree A constructed by Algorithm 7.20 is a minimum spanning tree, assuming all edge weights are distinct. (The algorithm is correct even if the edge weights are not all distinct, but the proof is slightly more complicated.)

Let $e_1, \ldots, e_{n-1}$ be the edges in A, in the order that they are added to A. We will prove by induction on j that $\{e_1, \ldots, e_j\}$ is contained in a minimum spanning tree, for all j, $0 \leq j \leq n - 1$.

We can take $j = 0$ as a trivial base case, if we interpret $\{e_1, \ldots, e_j\}$ as being the empty set when $j = 0$.

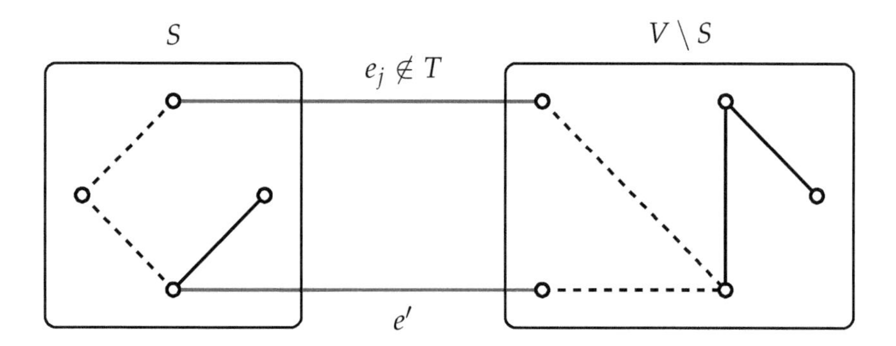

FIGURE 7.7: Proof of the general greedy algorithm to find a minimum spanning tree.

As an induction assumption, suppose that

$$A_{j-1} = \{e_1, \dots, e_{j-1}\} \subseteq T,$$

where T is a minimum spanning tree of the graph G, and consider e_j. If $e_j \in T$, we're done, so assume $e_j \notin T$. We will obtain a contradiction in this case.

There is a cut $(S, V \setminus S)$ respecting A_{j-1} for which e_j is the minimum crossing edge. $T \cup \{e_j\}$ contains a unique cycle C, from Lemma 7.21. There is an edge $e' \neq e_j$ such that $e' \in C$ and e' is a crossing edge for the cut $(S, V \setminus S)$.

Let

$$T' = T \cup \{e_j\} \setminus \{e'\}.$$

T' is also a spanning tree (from Lemma 7.21). Also, we have

$$w(T') = w(T) + w(e_j) - w(e').$$

Because T' is a spanning tree and T is a minimum spanning tree, we have $w(T') \geq w(T)$; hence $w(e_j) \geq w(e')$. But e_j is the minimum-weight crossing edge, so $w(e_j) < w(e')$, which is a contradiction. See Figure 7.7 for a diagram illustrating this situation. In this diagram, the edges in C are the dashed edges and the two (red) crossing edges e' and e_j.

7.8 Single Source Shortest Paths

We study shortest path problems in this section. First we look at a variant called the **Single Source Shortest Paths** problem; see Problem 7.2.

In Problem 7.2, if there is no directed path from u_0 to v, then the shortest (u_0, v)-path would have weight infinity.

Problem 7.2: Single Source Shortest Paths

Instance: A directed graph $G = (V, E)$, a non-negative ***weight function*** $w : E \rightarrow \mathbb{R}^+ \cup \{0\}$, and a ***source vertex*** $u_0 \in V$.

Find: For every vertex $v \in V$, a directed path P from u_0 to v such that

$$w(P) = \sum_{e \in P} w(e)$$

is minimized.

It is important to recognize that the term ***shortest path***, in the context of a weighted graph, really means ***minimum-weight path***. That is, we are interested in the *weight* of the path rather than the *number of edges* in the path.

Note also that Problem 7.2 requires us to find n different shortest paths with the same source, one for each vertex $v \in V$. We will consider other variations of this problem a bit later, most notably, the **All Pairs Shortest Paths** problem, where we are asked to find the shortest (i.e., minimum-weight) (u, v)-path, for all u and v.

If we have a graph in which all edges have weight 1 (or weight w for some fixed $w > 0$), then we can just use a breadth-first search to solve the **Single Source Shortest Paths** problem.

7.8.1 Dijkstra's Algorithm

DIJKSTRA'S ALGORITHM, which was discovered in 1956, solves the **Single Source Shortest Paths** problem for graphs that have no negative edge weights. It works for both directed and undirected graphs.

Here are the main ideas in the algorithm:

- At any given time, S is a subset of vertices such that the shortest paths from u_0 to all vertices in S are known; initially, $S = \{u_0\}$.
- For all vertices $v \in S$, $D[v]$ is the weight of the shortest path P_v from u_0 to v, and all vertices on P_v are in the set S.
- For all vertices $v \notin S$, $D[v]$ is the weight of the shortest path P_v from u_0 to v in which *all interior vertices are in* S.
- For $v \neq u_0$, $\pi[v]$ is the ***predecessor*** of v on the path P_v.
- At each stage of the algorithm, we choose $v \in V \backslash S$ so that $D[v]$ is minimized, and then we add v to S (see Lemma 7.23).
- Then the arrays D and π are updated appropriately.

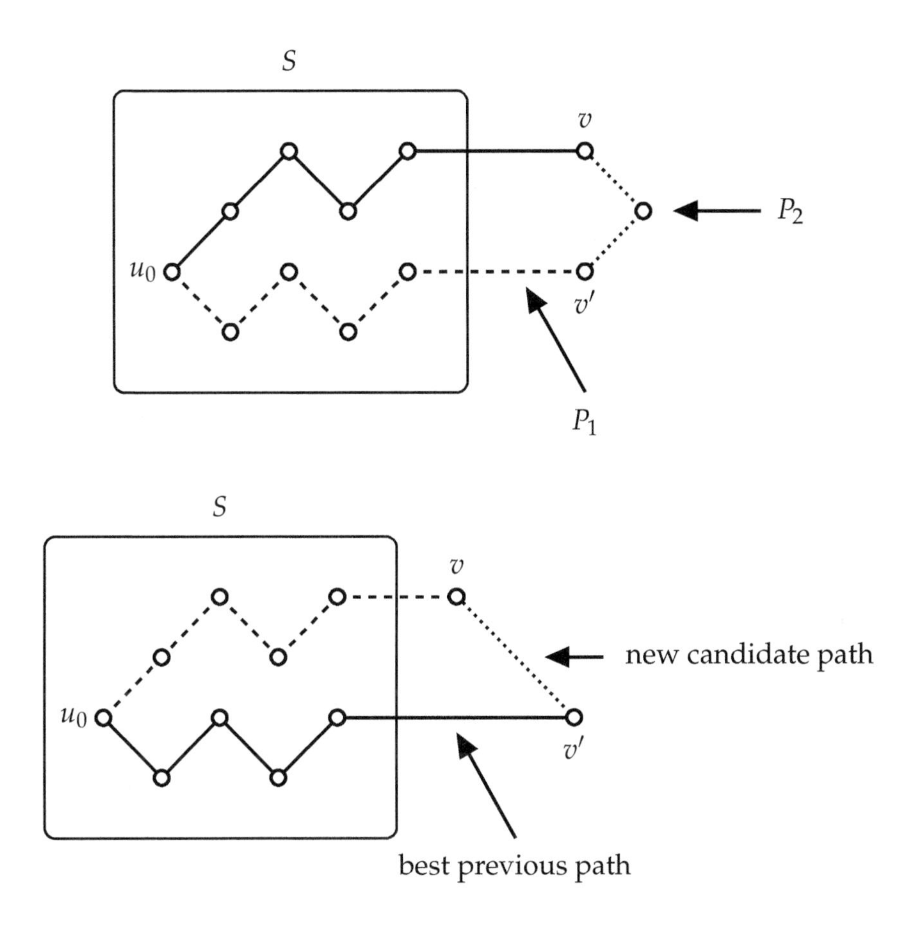

FIGURE 7.8: Adding a vertex to S and updating D-values.

LEMMA 7.23 *Suppose v has the smallest D-value of any vertex not in S. Then $D[v]$ equals the weight of the shortest (u_0, v)-path.*

PROOF Let P be a (u_0, v)-path having weight $D[v]$ and having all interior vertices in S. Suppose there is a (u_0, v)-path P' with weight less than $D[v]$. Let v' be the first vertex of P' not in S. Observe that $v' \neq v$. Decompose P' into two paths: a (u_0, v')-path P_1 and a (v', v)- path P_2 (see Figure 7.8, where P_1 consists of the dashed edges and P_2 consists of the dotted edges). We have

$$\begin{aligned} w(P') &= w(P_1) + w(P_2) \\ &\geq D[v'] + w(P_2) \\ &\geq D[v] \qquad (w(P_2) \geq 0 \text{ because all edge weights are } \geq 0). \end{aligned}$$

This is a contradiction because we assumed $w(P') < D[v]$. ∎

Lemma 7.23 says we can add v to S. Then, to update a value $D[v']$ for $v' \notin S$, we consider the new "candidate" path consisting of the shortest (u_0, v)-path together with the edge vv' (see Figure 7.8, where the (u_0, v)-path is shown with dashed edges and vv' is a dotted edge). If this path is shorter than the current best (u_0, v')-path, then update. Note that updating is only required for vertices $v' \in Adj[v]$.

Algorithm 7.21: DIJKSTRA(G, w, u_0)

```
S ← {u0}
D[u0] ← 0
for all v ∈ V\{u0}
   do { D[v] ← w(u0, v)
        π[v] ← u0
while |S| < n
   do { choose v ∈ V\S such that D[v] is minimized
        S ← S ∪ {v}
        for all v' ∈ V\S
           do { if D[v] + w(v, v') < D[v']
                  then { D[v'] ← D[v] + w(v, v')
                         π[v'] ← v
return (D, π)
```

Algorithm 7.22: FINDPATH(u_0, π, v)

```
comment: π is computed by Algorithm 7.21
path ← v
u ← v
while u ≠ u0
   do { u ← π[u]
        path ← u || path
return (path)
```

We present DIJKSTRA'S ALGORITHM as Algorithm 7.21. The predecessor array π allows us to determine the shortest paths as indicated in Algorithm 7.22. The edges $\{\pi[v], v\}$ actually comprise a spanning tree, called a ***shortest path spanning tree***, which is rooted at u_0.

We present an example to show the execution of DIJKSTRA'S ALGORITHM.

Example 7.15 Consider the graph depicted in Figure 7.9. The set S and the D-values are computed as shown in Figure 7.10.

□

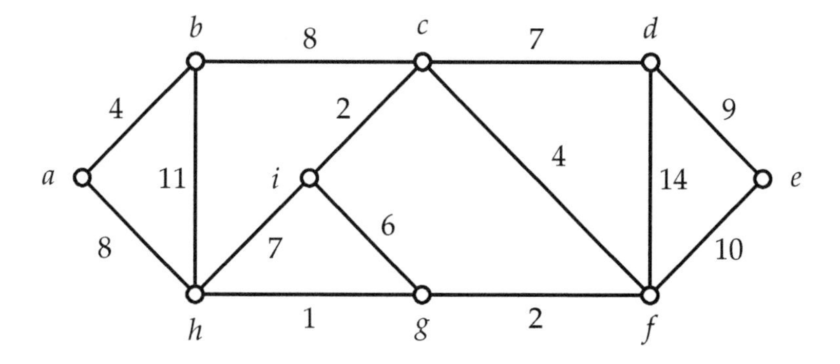

FIGURE 7.9: An example of DIJKSTRA'S ALGORITHM.

S	D-values b	c	d	e	f	g	h	i
$\{a\}$	4	∞	∞	∞	∞	∞	8	∞
$\{a,b\}$	4	12	∞	∞	∞	∞	8	∞
$\{a,b,h\}$	4	12	∞	∞	∞	9	8	15
$\{a,b,g,h\}$	4	12	∞	∞	11	9	8	15
$\{a,b,f,g,h\}$	4	12	25	21	11	9	8	15
$\{a,b,c,f,g,h\}$	4	12	19	21	11	9	8	14
$\{a,b,c,f,g,h,i\}$	4	12	19	21	11	9	8	14
$\{a,b,c,d,f,g,h,i\}$	4	12	19	21	11	9	8	14
$\{a,b,c,d,e,f,g,h,i\}$	4	12	19	21	11	9	8	14

FIGURE 7.10: Computing the set S and the D-values.

It is interesting to observe that the structure of DIJKSTRA'S ALGORITHM (for shortest paths) is very similar to that of PRIM'S ALGORITHM (for minimum spanning trees). The only difference is the updating step:

PRIM	DIJKSTRA
if $w(v,v') < W[v']$ **then** $W[v'] \leftarrow w(v,v')$	**if** $D[v] + w(v,v') < D[v']$ **then** $D[v'] \leftarrow D[v] + w(v,v')$

It follows that the complexity of DIJKSTRA'S ALGORITHM is the same as the complexity of PRIM'S ALGORITHM, since they are pretty much the same algorithm.

Subsequent shortest path algorithms we will be studying will solve shortest path problems in directed graphs that have no negative-weight (directed) cycles. First, we note that, if there are negative-weight directed cycles in a directed graph, then we cannot even define shortest paths in a sensible way, as we demonstrate in the following example.

Example 7.16 Consider the following graph.

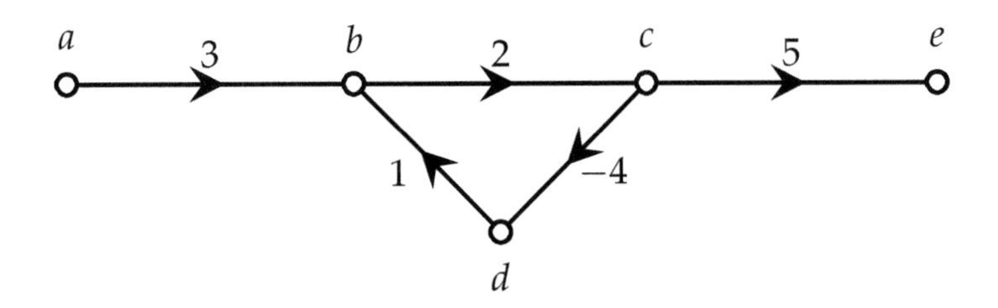

It is clear that the shortest *simple* (a, e)-directed path is a, b, c, e, which has weight 10. However, there is a negative-weight directed cycle, namely, b, c, d, b, which has weight -1. For a positive integer r, consider the directed path

$$a, b, c, \underbrace{d, b, c}_{r}, e,$$

where the segment b, c, d is repeated r times. This (non-simple) directed path has weight $10 - r$. Since the integer r can be arbitrarily large, we can construct non-simple directed paths with arbitrarily small negative weight. So there is no shortest (a, e)-directed path. □

In a directed graph having negative-weight cycles, there are shortest *simple* paths. Unfortunately, there are apparently no known efficient algorithms to find the shortest simple paths in graphs containing negative-weight cycles.

If there are no negative-weight cycles in a directed graph G, then, for any two vertices u and v, there is a shortest (u, v)-path that is simple (a shortest path could contain cycles of weight 0, but removal of all such cycles would yield a simple shortest path having the same weight).

If we considered the analogous situation for undirected graphs, we would replace every edge $\{u, v\}$, by two arcs, (u, v) and (v, u) to obtain an "equivalent" directed graph. However, if we want this directed graph to contain no negative-weight directed cycles, then the original undirected graph should contain no negative-weight edges. This is because a negative-weight edge in the undirected graph would give rise to a negative-weight cycle (consisting of two edges) in the associated directed graph.

7.8.2 Shortest Paths in Directed Acyclic Graphs

A directed acyclic graph (or DAG) obviously cannot contain a negative-weight directed cycle because it contains no directed cycles at all. There is a nice way to solve the **Single Source Shortest Paths** problem in a DAG that is based in a topological sort. First, we find a topological ordering of the vertices, which can be done using Algorithm 7.8. As we have discussed, this algorithm returns a stack S. If we perform n POP operations on X, we obtain a topological ordering, say $v_1, \ldots, v_n$. Then we can easily find all the shortest paths in G with source v_1 using

Algorithm 7.23: DAG SHORTEST PATHS(G, w)

comment: assume G is a directed acyclic graph on n vertices
$S \leftarrow$ DFS-TOPORD(G)
comment: S is a stack
obtain the ordering $v_1, \ldots, v_n$ from S
comment: this is done by popping the stack n times
for $j \leftarrow 1$ **to** n
  do $\begin{cases} D[v_1] \leftarrow \infty \\ \pi[v_j] \leftarrow \text{undefined} \end{cases}$
$D[v_1] \leftarrow 0$
for $j \leftarrow 1$ **to** $n - 1$
  do { **for all** $v' \in Adj[v_j]$
    do { **if** $D[v_j] + w(v_j, v') < D[v']$
      then $\begin{cases} D[v'] \leftarrow D[v_j] + w(v_j, v') \\ \pi[v'] \leftarrow v_j \end{cases}$
return (D, π)

Algorithm 7.23. Observe that the updating step in this algorithm is the same as in other algorithms we have considered previously.

Algorithm 7.23 is correct even if there are negative-weight edges in the DAG. The correctness of this algorithm follows from the following loop invariant:

> After the jth iteration (of the main **for** loop of Algorithm 7.23), $D[v]$ is the shortest (v_1, v)-path having interior vertices in $\{v_1, \ldots, v_j\}$.

This loop invariant can be proven by induction on j. Then, if we take $j = n$, the correctness of the algorithm is proven.

Recall that the complexity of Algorithm 7.8 is $O(m + n)$ since it is based on a depth-first search. It is not hard to see that Algorithm 7.23 also has complexity $O(m + n)$.

Algorithm 7.23 can be modified in an obvious way to find shortest paths from any given source (not just the "first" vertex v_1). We leave this as an exercise.

Example 7.17 Here is a directed acyclic graph, where all edges are directed from left to right and the vertices $v_1, \ldots, v_6$ are already in topological order:

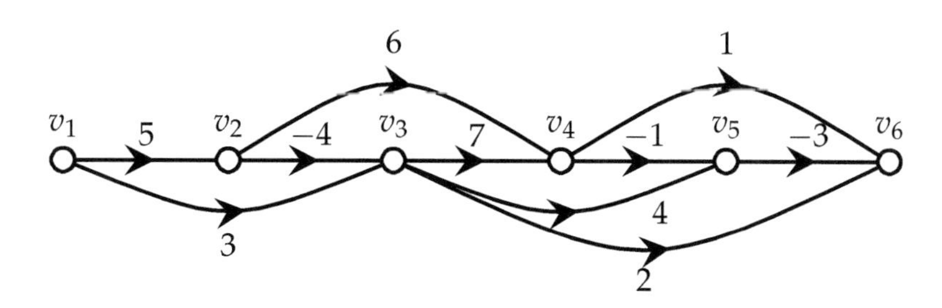

TABLE 7.2: Example of Algorithm 7.23.

j	$D[v_1]$	$D[v_2]$	$D[v_3]$	$D[v_4]$	$D[v_5]$	$D[v_6]$
0	0	∞	∞	∞	∞	∞
1	0	5	3	∞	∞	∞
2	0	5	1	11	∞	∞
3	0	5	1	8	5	3
4	0	5	1	8	5	3
5	0	5	1	8	5	2

It is clear that there are no directed cycles, since there are no edges going from right to left. Table 7.2 indicates the values in the array after each iteration of the main **for** loop (i.e., the loop on j, from 1 to $n-1$). The last line of the table contains the lengths of all the shortest (v_1, v_i)-paths. □

7.8.3 Bellman-Ford Algorithm

The BELLMAN-FORD algorithm solves the single source shortest path problem in any directed graph without negative-weight cycles. As with various other algorithms we have studied, the algorithm was named after the first person to discover it. The so-called BELLMAN-FORD algorithm was discovered by Shimbel in 1955; it was later published by Ford (1956) and independently by Bellman (1958). We present the BELLMAN-FORD algorithm as Algorithm 7.24.

Algorithm 7.24: BELLMAN-FORD(G, w, u_0)

```
comment: assume E = {e_1, ..., e_m}
for all u
  do { D[u] ← ∞
       π[u] ← undefined
D[u_0] ← 0
for i ← 1 to n − 1
  do { for j ← 1 to m
         do { denote e_j = (u, v)
              if D[u] + w(u, v) < D[v]
                then { D[v] ← D[u] + w(u, v)
                       π[v] ← u
return (D, π)
```

Algorithm 7.24. makes use of the same updating step that we previously employed in DIJKSTRA'S ALGORITHM as well as in Algorithm 7.23. To be precise, the updating step replaces $D[v]$ by $D[u] + w(u, v)$ if $D[u] + w(u, v) < D[v]$. This updating step is often referred to as ***relaxing*** the edge (u, v).

Algorithm 7.24 can be described succinctly as follows: we relax every edge in the graph, and repeat this over and over again, until every edge has been relaxed $n-1$ times. Also, as usual, the values $\pi[v]$ are predecessors on shortest paths.

DIJKSTRA'S ALGORITHM has complexity $O(m \log n)$ (using priority queues) whereas the BELLMAN-FORD algorithm has complexity $O(mn)$. However, DIJKSTRA'S ALGORITHM requires that the graph contain no negative-weight edges, whereas the BELLMAN-FORD algorithm only has the weaker requirement that there are no negative-weight cycles in the graph.

It will be useful to keep track of how the values $D[v]$ change as Algorithm 7.24 is executed. Therefore, for $i = 0, 1, 2, \ldots$, we let $D_i[v]$ denote the value of $D[v]$ after the ith iteration of the outer **for** loop.

In each iteration of the outer **for** loop, we relax the edges of the graph in some specified order that basically depends on how the edges are named. Before discussing the correctness, it will be useful to clarify how the execution of Algorithm 7.24 depends on this ordering of the edges.

Example 7.18 Consider the following graph:

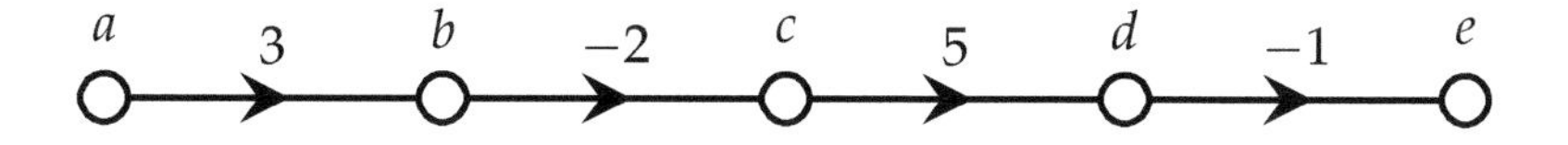

Suppose the edges are named $e_1 = (a,b)$, $e_2 = (b,c)$, $e_3 = (c,d)$ and $e_4 = (d,e)$. Then the following D-values will be computed:

$$\begin{aligned}D_1[a] &= 0\\ D_1[b] &= 3\\ D_1[c] &= 1\\ D_1[d] &= 6\\ D_1[e] &= 5.\end{aligned}$$

Thus the lengths of all the shortest paths with source a have been computed after one iteration of the outer **for** loop.

Let us contrast this with the execution of the algorithm when the edges are named $e_4 = (a,b)$, $e_3 = (b,c)$, $e_2 = (c,d)$ and $e_1 = (d,e)$:

i	$D_i[a]$	$D_i[b]$	$D_i[c]$	$D_i[d]$	$D_i[e]$
0	0	∞	∞	∞	∞
1	0	3	∞	∞	∞
2	0	3	1	∞	∞
3	0	3	1	6	∞
4	0	3	1	6	5

The same D-values are eventually computed, but it takes four iterations instead of one iteration to compute all of them. □

Our correctness proofs will not make any assumptions about how efficient the edge orderings turn out to be. That is, we consider the "worst case" in establishing properties of Algorithm 7.24.

For $i = 0, 1, \ldots$, and for any vertex v, define $\delta_i[v]$ to be the weight of the shortest (u_0, v)-path that has at most i edges (if at least one such path exists), and $\delta_i[v] = \infty$, otherwise. The following lemma, which describes the behavior of Algorithm 7.24, is valid for any directed graph.

LEMMA 7.24 *Let $G = (V, E)$ be a directed graph, let $u_0 \in V$ be a designated source vertex and let $D_i[v]$ be defined as above, for $i = 0, 1, \ldots$. Then the following properties hold.*

1. *If there is a directed path from u_0 to v that contains at most i edges, then*

$$D_i[v] \leq \delta_i[v].$$

2. *If there is no directed path from u_0 to v that contains at most i edges, then*

$$D_i[v] = \infty.$$

PROOF The proof is by induction on i. The base case is $i = 0$. By definition, $D_i[u_0] = 0$ and $D_i[v] = \infty$ for $v \neq u_0$. Since the only path with source u_0 containing no arc is the empty path, the two desired properties hold when $i = 0$.

As an induction assumption, assume that 1 and 2 hold for $i = j - 1 \geq 0$ and suppose that $i = j$. Suppose that there is at least one directed path from u_0 to v that contains at most j edges. Let P be the minimum-weight path of this type, so the weight of P is $\delta_j[v]$. Let v' be the predecessor of v on the path P. The portion of P from u_0 to v' is a directed path P' from u_0 to v' that contains at most $j - 1$ edges. By induction, $D_{j-1}[v'] \leq \delta_{j-1}[v']$. Now,

$$\begin{aligned} D_j[v] &\leq D_{j-1}[v'] + w(u, v) \\ &\leq \delta_{j-1}[v'] + w(v', v) \\ &= w(P) \\ &= \delta_j[v]. \end{aligned}$$

This proves the desired result by induction. Note that, in the above, we are using the fact that P must consist of a minimum-weight (u_0, v')-path containing at most $i - 1$ edges, namely P', together with the arc $v'v$. ∎

Now, assuming that G has no negative-weight directed cycles, we can prove that Algorithm 7.24 is correct.

THEOREM 7.25 *Let $G = (V, E)$ be a directed graph that contains no negative-weight directed cycles. Let $u_0 \in V$ be a designated source vertex and let $D_i[v]$ be defined as above, for $i = 0, 1, \ldots$.*

1. *If there is a directed path from* u_0 *to* v*, then* $D_{n-1}[v]$ *is equal to the length of the minimum-weight* (u_0, v)*-path.*

2. *If there is no directed path from* u_0 *to* v*, then* $D_{n-1}[v] = \infty$*.*

PROOF Since there are no negative-weight cycles in G, it follows that there exists a simple minimum-weight path (which has the same weight, of course). A simple path contains at most $n-1$ edges. The stated results now follow immediately from Lemma 7.24. ∎

We can run Algorithm 7.24 on any directed graph G. However, it may return incorrect values if G contains a negative-weight cycle. Fortunately, the BELLMAN-FORD algorithm can also detect the presence of a negative-weight cycle in G. The idea is to perform an additional (i.e., an nth) iteration of the **for** loop. The graph G has a negative-weight cycle if and only if there is at least one "improvement" in this extra iteration.

LEMMA 7.26 *The directed graph* G *contains a negative-weight cycle if and only if*

$$D_{n-1}[u] + w(u,v) < D_{n-1}[v]$$

for some arc (u, v)*.*

PROOF Suppose that $v_1, v_2, \ldots, v_k, v_1$ is a negative-weight cycle. For convenience, denote $v_{k+1} = v_1$. Then

$$\sum_{i=1}^{k} w(v_i, v_{i+1}) < 0. \tag{7.2}$$

Suppose that

$$D_{n-1}[u] + w(u,v) \geq D_{n-1}[v]$$

for every arc (u, v). Then, considering the arcs (v_i, v_{i+1}) for $1 \leq i \leq k$, we obtain

$$\sum_{i=1}^{k} (D_{n-1}[v_i] + w(v_i, v_{i+1})) \geq \sum_{i=1}^{k} D_{n-1}[v_{i+1}].$$

Clearly

$$\sum_{i=1}^{k} D_{n-1}[v_i] = \sum_{i=1}^{k} D_{n-1}[v_{i+1}]$$

because $v_1 = v_{k+1}$. Hence

$$\sum_{i=1}^{k} w(v_i, v_{i+1}) \geq 0,$$

which contradicts (7.2).

Conversely, suppose that

$$D_{n-1}[u] + w(u,v) < D_{n-1}[v]$$

for some arc (u, v). This means that there is a (u_0, v)-directed path having at most n vertices whose weight is less than the optimal (u_0, v)-directed path that has at most $n-1$ edges. If G contains no negative-weight directed cycles, this is impossible, because there is a minimum-weight (u_0, v)-directed path that has at most $n-1$ edges. ∎

The enhancement provided by Lemma 7.26 is included in Algorithm 7.25; it is the most commonly used version of the BELLMAN-FORD algorithm.

```
Algorithm 7.25: BELLMAN-FORD2(G, w, u0)

comment: assume E = {e1, ..., em}
for all u
   do { D[u] ← ∞
        π[u] ← undefined
D[u0] ← 0
for i ← 1 to n − 1
   do { for j ← 1 to m
           do { denote ej = (u, v)
                if D[u] + w(u, v) < D[v]
                  then { D[v] ← D[u] + w(u, v)
                         π[v] ← u
for j ← 1 to m
   do { denote ej = (u, v)
        if D[u] + w(u, v) < D[v]
          then return (G contains a negative-weight cycle)
return (D, π)
```

The BELLMAN-FORD algorithm can also be modified to include an "early stopping" condition. If there are no changes to any D-values in iteration $i \leq n-1$, then we can stop, because there would then be no changes in any subsequent iterations.

7.9 All-Pairs Shortest Paths

We define the **All-Pairs Shortest Paths** problem, which is a variation of the **Single Source Shortest Paths** problem, as Problem 7.3.

We will study algorithms to solve the **All-Pairs Shortest Paths** problem for directed graphs that contain no negative-weight directed cycles (negative-weight edges are allowed, however). This is the same requirement that was used in the last section in reference to the BELLMAN-FORD algorithm, and it ensures that all the desired minimum-weight paths exist.

Problem 7.3: All-Pairs Shortest Paths

Instance: A directed graph $G = (V, E)$, and a ***weight matrix*** W, where $W[i, j]$ denotes the weight of edge ij, for all $i, j \in V, i \neq j$.

Find: For all pairs of vertices $u, v \in V$, $u \neq v$, a directed path P from u to v such that

$$w(P) = \sum_{ij \in P} W[i, j]$$

is minimized.

We use the following conventions for the weight matrix W:

$$W[i, j] = \begin{cases} w_{ij} & \text{if } (i, j) \in E \\ 0 & \text{if } i = j \\ \infty & \text{otherwise.} \end{cases}$$

One possible solution to Problem 7.3 is to run the BELLMAN-FORD algorithm n times, once for every possible source. The complexity of this approach is $O(n^2 m)$. We now investigate some other, hopefully more efficient, algorithms to solve this problem.

7.9.1 A Dynamic Programming Approach

Dynamic programming is one possible approach that can be used to solve Problem 7.3. We follow the standard steps to develop a dynamic programming algorithm.

optimal structure

Let k be the predecessor of j on the minimum-weight (i, j)-path P having at most m edges. Then the portion of P from i to k, say P', is a minimum-weight (i, k)-path having at most $m - 1$ edges.

subproblems

Let $L_m[i, j]$ denote the minimum-weight (i, j)-path having at most m edges. Each L_m is an n by n matrix. We want to compute the matrix L_{n-1}.

recurrence relation

We begin by initializing $L_1 = W$. The general form of the recurrence relation is as follows: For $m \geq 2$,

$$L_m[i, j] = \min\{L_{m-1}[i, k] + L_1[k, j] : 1 \leq k \leq n\}. \quad (7.3)$$

Note that $k = i$ or $k = j$ does not cause any problems in this recurrence relation.

compute optimal solutions

We simply compute $L_1, L_2, \ldots, L_{n-1}$ in turn.

The resulting dynamic programming algorithm is Algorithm 7.26. Note that each matrix L_m has n^2 entries. For $m \geq 2$, each entry of each L_m takes time $\Theta(n)$ to compute. There are $\Theta(n)$ matrices to compute, so the complexity of Algorithm 7.26 is $\Theta(n^4)$.

Algorithm 7.26: FAIRLYSLOWALLPAIRSSHORTESTPATH(W)

```
L_1 ← W
for m ← 2 to n − 1
  do { for i ← 1 to n
         do { for j ← 1 to n
                do { ℓ ← ∞
                     for k ← 1 to n
                       do ℓ ← min{ℓ, L_{m−1}[i,k] + W[k,j]}
                     L_m[i,j] ← ℓ
return (L_{n−1})
```

7.9.2 Successive Doubling

A more efficient algorithm is obtained by "successive doubling." The idea is to construct $L_1, L_2, L_4, \ldots L_{2^t}$, where t is the smallest integer such that $2^t \geq n-1$. This can be done by making use of a modified optimal structure for shortest paths.

Let k be the midpoint of j on the minimum-weight (i,j)-path P having at most $2m$ edges. Then the portion of P from i to k is a minimum-weight (i,k)-path having at most m edges, and the portion of P from k to j is a minimum-weight (k,j)-path having at most m edges. This analysis yields the following recurrence relation, which is an alternative to (7.3). For $m \geq 1$,

$$L_{2m}[i,j] = \min\{L_m[i,k] + L_m[k,j] : 1 \leq k \leq n\}. \tag{7.4}$$

The recurrence relation (7.4) leads immediately to Algorithm 7.27.

It is straightforward to see that the complexity of Algorithm 7.27 is $O(n^3 \log n)$. This is because there are $O(\log n)$ matrices that are computed, each of which takes time $O(n^3)$ to compute.

7.9.3 Floyd-Warshall Algorithm

Our third solution is known as the FLOYD-WARSHALL algorithm. There were several people involved in its discovery, including Roy (1959), Floyd (1962) and Warshall (1962). It uses a somewhat different approach than the two previous algorithms. Instead of placing an upper bound on the number of edges in the paths,

Algorithm 7.27: FASTERALLPAIRSSHORTESTPATH(W)

```
L_1 ← W
m ← 1
while m < n − 1
  do { for i ← 1 to n
         do { for j ← 1 to n
                do { ℓ ← ∞
                     for k ← 1 to n
                       do ℓ ← min{ℓ, L_m[i,k] + L_m[k,j]}
                     L_{2m}[i,j] ← ℓ
       m ← 2m
return (L_m)
```

as we did in Algorithms 7.26 and 7.27, we gradually allow more and more possible interior vertices in the paths we construct. This is similar to what was done in Algorithm 7.23.

We need the following definition. For a directed path, $P = v_1, v_2, \ldots, v_k$, the ***interior vertices*** in are $v_2, \ldots, v_{k-1}$.

We now proceed to develop a dynamic programming algorithm that makes use of the idea of successively expanding the set of interior vertices.

optimal structure

Let P be the minimum-weight (i, j)-path (P exists because the graph G contains no negative-weight cycles). If P contains n as an interior vertex, then we can decompose P into two disjoint paths:

1. an (i, n)-directed path P_1 and
2. an (n, j)-directed path P_2.

Clearly P_1 must be a minimum-weight (i, n)-path having interior vertices that are in the set $\{1, \ldots, n-1\}$, and P_1 must be a minimum-weight (n, j)-path having interior vertices that are in the set $\{1, \ldots, n-1\}$.

On the other hand, if P does not contain vertex n as an interior vertex, then P is the minimum-weight (i, j)-path having interior vertices in the set $\{1, \ldots, n-1\}$.

subproblems

The relevant subproblems will involve (i, j)-paths having interior vertices that are in the set $\{1, \ldots, m\}$, for $m = 1, 2, \ldots, n$. Therefore, we let $D_m[i, j]$ denote the minimum weight of an (i, j)-path in which all interior vertices are in the set $\{1, \ldots, m\}$.

recurrence relation

We begin by initializing $D_0 = W$. (Here the minimum-weight paths consist of single arcs.)

The general form of the recurrence relation is as follows: For $m \geq 1$,

$$D_m[i,j] = \min\{D_{m-1}[i,j], D_{m-1}[i,m] + D_{m-1}[m,j]\}.$$

This is based on essentially the same analysis that was used in "optimal structure."

compute optimal solutions

We compute $D_1, D_2, \ldots D_{n-1}$ in turn.

Algorithm 7.28 is the dynamic programming algorithm that arises from this approach. It incorporates predecessors in the usual way. The final predecessor matrix, π_n, can be used to construct the shortest paths.

Algorithm 7.28: FLOYDWARSHALL(W)

```
D_0 ← W
for i ← 1 to n
  do { for j ← 1 to n
         do { if (i ≠ j) and (W[i,j] < ∞)
                then π_0[i,j] ← i
                else π_0[i,j] ← undefined
for m ← 1 to n
  do { for i ← 1 to n
         do { for j ← 1 to n
                do { if D_{m-1}[i,m] + D_{m-1}[m,j] < D_{m-1}[i,j]
                       then { D_m[i,j] ← D_{m-1}[i,m] + D_{m-1}[m,j]
                              π_m[i,j] ← π_{m-1}[m,j]
                       else D_m[i,j] ← D_{m-1}[i,j]
return (D_n)
```

The complexity of Algorithm 7.28 is clearly $O(n^3)$. We illustrate the execution of this algorithm in a small example.

Example 7.19 Consider the graph shown in Figure 7.11. It contains negative-weight edges, but no negative-weight cycles. Algorithm 7.28 computes the matrices $D_0, \ldots, D_4$ that are presented in Figure 7.12. □

It is useful to have a detailed correctness proof for Algorithm 7.28.

THEOREM 7.27 *Suppose $G = (V, E)$ is a directed graph that contains no negative-weight directed cycles. Then Algorithm 7.28 correctly computes all minimum-weight (i, j)-directed paths.*

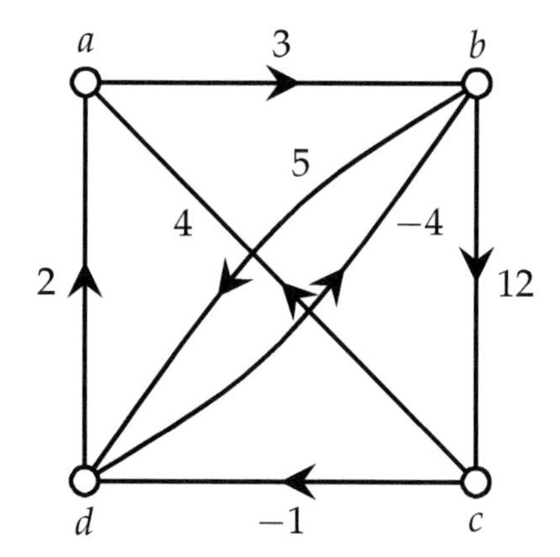

FIGURE 7.11: A graph containing negative-weight edges but no negative-weight cycles.

$$D_0 = \begin{pmatrix} 0 & 3 & \infty & \infty \\ \infty & 0 & 12 & 5 \\ 4 & \infty & 0 & -1 \\ 2 & -4 & \infty & 0 \end{pmatrix}$$

$$D_1 = \begin{pmatrix} 0 & 3 & \infty & \infty \\ \infty & 0 & 12 & 5 \\ 4 & 7 & 0 & -1 \\ 2 & -4 & \infty & 0 \end{pmatrix}$$

$$D_2 = \begin{pmatrix} 0 & 3 & 15 & 8 \\ \infty & 0 & 12 & 5 \\ 4 & 7 & 0 & -1 \\ 2 & -4 & 8 & 0 \end{pmatrix}$$

$$D_3 = \begin{pmatrix} 0 & 3 & 15 & 8 \\ 16 & 0 & 12 & 5 \\ 4 & 7 & 0 & -1 \\ 2 & -4 & 8 & 0 \end{pmatrix}$$

$$D_4 = \begin{pmatrix} 0 & 3 & 15 & 8 \\ 7 & 0 & 12 & 5 \\ 1 & -5 & 0 & -1 \\ 2 & -4 & 8 & 0 \end{pmatrix}$$

FIGURE 7.12: Computing W-matrices.

PROOF For $0 \leq m \leq n$, we prove that the minimum weight of any (i, j)-directed path having all of its interior vertices in the set $\{1, \dots, m\}$ is equal to $D_m[i, j]$. The proof is by induction on m, and the base case $m = 0$ is obvious.

As an induction hypothesis, suppose that the assertion is correct for $m = \ell - 1$. Consider $m = \ell$. Let P be a minimum-weight (i, j)-directed path such that all of its interior vertices are in the set $\{1, \dots, \ell\}$. We do not have to assume that P is a simple path. We consider two cases:

case 1

If P has all of its interior vertices in the set $\{1, \dots, \ell - 1\}$, then $w(P) = D_{\ell-1}[i, j]$, by induction.

case 2

If P contains ℓ as an interior vertex, then we can decompose P into two disjoint paths: an (i, ℓ) path P_1 and an (ℓ, j) path P_2. It does not matter if P_1 and P_2 are simple.

Clearly P_1 is a shortest (i, ℓ) path in which the interior vertices are in the set $\{1, \dots, \ell - 1\}$, and P_2 is a shortest (ℓ, j) path in which the interior vertices are in the set $\{1, \dots, \ell - 1\}$.

Hence, by induction,

$$w(P_1) = D_{\ell-1}[i, \ell] \quad \text{and} \quad w(P_2) = D_{\ell-1}[\ell, j].$$

Then,

$$\begin{aligned} w(P) &= w(P_1) + w(P_2) \\ &= D_{\ell-1}[i, \ell] + D_{\ell-1}[\ell, j]. \end{aligned}$$

Since

$$D_\ell[i, j] = \min\{D_{\ell-1}[i, \ell] + D_{\ell-1}[\ell, j], D_{\ell-1}[i, j]\},$$

the desired result is proven by induction. ∎

REMARK The two cases considered in the proof of Theorem 7.27 are illustrated in Figure 7.13. P_1 is shown with dotted edges and P_2 has dashed edges. The path P has solid edges. ∎

In the proof of Theorem 7.27, we did not need to distinguish between simple and non-simple shortest paths. However, if we use the predecessors as they are computed in Algorithm 7.28 to compute the vertices in a shortest path, we would like to have a guarantee that a simple shortest path is produced. Basically, the idea is to show that the shortest paths implied by the definition of the π_m-matrices in Algorithm 7.28 are simple paths. For $m = 0$, this is certainly true because the paths in question consist of single arcs. So let's consider what happens as we proceed from π_{m-1} to π_m.

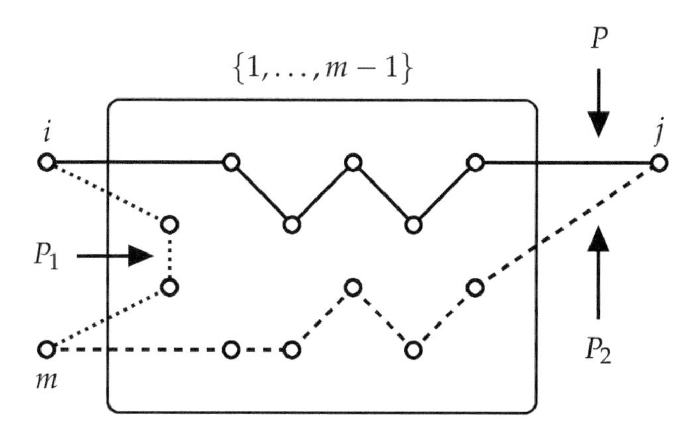

FIGURE 7.13: Updating step in the FLOYD-WARSHALL algorithm.

We examine the two cases in the proof of Theorem 7.27. If the minimum-weight path arises from case 1, there is no change in the path, so this case cannot cause any problems. However, in case 2, we are concatenating two simple paths to create a new path, so it is not immediate that the result is a simple path.

The predecessor value $\pi_m[i,j] \neq \pi_{m-1}[i,j]$ only when the shortest path from case 2 only has a *strictly* lower weight than the "previous" minimum-weight path. So, in this case, we have

$$w(P_1) + w(P_2) < D_{m-1}[i,j].$$

We claim that the concatenation of P_1 and P_2 is a simple path. Suppose this is not the case; then there is a vertex $k \neq m$ that occurs in both P_1 and P_2. The portion of P_1 from k to m, together with the portion of P_2 from m to k forms a directed cycle. Delete this directed cycle from the concatenation of P_1 and P_2. The resulting (i,j)-directed path, say P', satisfies the following properties:

- $w(P') \leq w(P_1) + w(P_2) < D_{m-1}[i,j]$, and
- P' has all its interior vertices in $\{1,\ldots,m-1\}$.

However, this is impossible, since the minimum weight of an (i,j)-directed path with all its interior vertices in $\{1,\ldots,m-1\}$ is $D_{m-1}[i,j]$. Therefore, whenever we change a predecessor, i.e., $\pi_m[i,j] \neq \pi_{m-1}[i,j]$, the "new" path determined by the predecessors is still a simple path.

Finally, it is useful to have a method to ensure negative-weight cycle detection. This turns out to be easy: If $D_m[i,i] < 0$ for any m, it means that the graph has a negative-weight cycle. It suffices to check this condition at the termination of the algorithm, since $D_n[i,i] \leq D_m[i,i]$ for any m.

If no negative-weight cycle is found, then all the D-values are correct. This is proven in the following lemma.

LEMMA 7.28 *The directed graph G contains a negative-weight cycle if and only if $D_n[i,i] < 0$ for some i.*

PROOF Any value $D_n[i,j]$ is the weight of some (i,j)-directed path in G. Suppose that $D_n[i,i] < 0$ for some i. Then there is a directed path beginning and ending at i that has negative weight. This directed path might not be a directed cycle, because some vertex other than i might be repeated. But in this case, there must be a subset of the vertices and edges in this directed path that form a negative-weight directed cycle.

Conversely, suppose that G contains a negative-weight directed cycle. Let i be any vertex in this cycle. Clearly, this cycle is an (i,i)-directed path having negative weight, so the value $D_n[i,i]$ computed by Algorithm 7.28 will be at most the weight of this cycle, so $D_n[i,i] < 0$. ∎

7.10 Notes and References

There are many introductory textbooks on graph theory. One that we particularly recommend is

- *Graph Theory*, by Bondy and Murty [10].

For a more algorithmic point of view, we suggest

- *Graphs, Algorithms and Optimization, Second Edition*, by Kocay and Kreher [43].

Our treatment of BREADTH-FIRST SEARCH and DEPTH-FIRST SEARCH is strongly influenced by Cormen, Leiserson, Rivest and Stein [16]. The basic idea of a depth-first search dates back at least to the 19th century, while using a breadth-first search strategy seems to have first appeared in the 1940s.

The algorithms for finding topological orderings that we described in Section 7.4.1 are found in Kahn [36] and Tarjan [59].

Sharir published his algorithm to find strongly connected components in 1981 in [55]. However, we should note that Tarjan published a different algorithm, also having complexity $O(m+n)$, in 1972; see [58].

There are surprisingly few detailed pseudocode descriptions of algorithms to find Eulerian circuits. One such algorithm is presented in Kocay and Kreher [43]. It is based on HIERHOLZER'S ALGORITHM and represents the circuit using a linked list. Our algorithm, Algorithm 7.14, is a recursive algorithm that employs a queue to avoid the use of linked lists. This permits a fairly succinct, easily implemented, algorithm.

It should be mentioned that the first published algorithm for the **Minimum Spanning Tree** problem was due to Borůvka in 1926. Its complexity is $O(m \log n)$, similar to PRIM'S ALGORITHM and KRUSKAL'S ALGORITHM.

Union-find data structures were introduced in 1964 by Galler and Fisher [24]. There has since been a considerable amount of work on this topic. Galil and Italiano [23] is a 1991 survey on union-find data structures.

Schrijver [53] is an informative history of the shortest path problem; it contains numerous references to various shortest path algorithms. We refer the interested reader to that paper for additional background on this topic.

Exercises

7.1 Give a proof of Lemma 7.8.

7.2 Prove that Algorithm 7.6 is correct.

7.3 Suppose T is a tree and an edge $e \notin T$ is added to T. We proved in Lemma 7.21 that the resulting graph contains a unique cycle C. Prove that, if $e' \in C$, then $T \cup \{e\} \setminus \{e'\}$ is a tree.

7.4 Describe how to modify Algorithm 7.23 to find shortest paths from any given source v_j.

7.5 Suppose that G is an undirected graph in which every edge has weight $1, 2$ or 3. Describe an algorithm to find the shortest (u, v)-path, for two specified vertices u and v, in time $O(n + m)$.

HINT First, construct a new graph from G in which edges do not have weights.

7.6 Suppose $G = (V, E)$ is a connected undirected graph. For two vertices $u, v \in V$, define $dist(u, v)$ to be the distance from u to v (i.e., the number of edges in the shortest path from u to v). Let u_0 be a specified source vertex. The *width* of G with respect to u_0, denoted $\mathsf{width}(G, u_0)$, is the maximum number of vertices whose distance from u_0 is equal to d, for some positive integer d.

As an example, suppose there are three vertices of distance 1 from u_0, six vertices of distance 2 from u_0, and four vertices of distance 3 from u_0. Then $\mathsf{width}(G, u_0) = 6$.

The *width* of G, denoted $\mathsf{width}(G)$, is defined as follows:

$$\mathsf{width}(G) = \max\{\mathsf{width}(G, u_0) : u_0 \in V\}.$$

Give a high-level description of an algorithm to compute the $\mathsf{width}(G)$. Your algorithm should be based on BREADTH-FIRST SEARCH (BFS), but you do not have to write out detailed code for BFS. You can use some simple auxiliary data structures if you wish to do so. Your algorithm should have complexity $O(n^2 + nm)$. Briefly justify the correctness of your algorithm and the complexity.

7.7 Suppose we have a set of n politicians, all of whom are from one of two parties. Suppose we are given a list of m pairs of politicians, where every pair consists of two enemies. Suppose also that this list of pairs forms a connected graph. Finally, suppose that enemies are always from different parties.

(a) Suppose we have the following set of nine pairs of enemies involving eight politicians:

$$\begin{array}{lll} \{\text{Smith, Taylor}\}, & \{\text{Taylor, Adams}\}, & \{\text{Adams, Baker}\}, \\ \{\text{Baker, Moore}\}, & \{\text{Smith, Norton}\}, & \{\text{Moore, Norton}\}, \\ \{\text{Norton, Lloyd}\}, & \{\text{Lloyd, Baker}\}, & \{\text{Moore, Taylor}\}. \end{array}$$

In this example, determine which politicians belong to the same party.

(b) Describe briefly how you would efficiently solve this problem in general.

7.8 An ***Eulerian path*** in a graph G is a path that contains every edge in G. An Eulerian path that begins and ends at the same vertex is an Eulerian circuit. In this question, we investigate Eulerian paths that begin and end at different vertices.

(a) Suppose G is a connected graph with exactly two vertices of odd degree, say u and v. Then it can be shown that G has an Eulerian path that begins at u and ends at v. Using Algorithm 7.14 as a subroutine, find such an Eulerian path. (This is an example of a reduction.)

HINT Construct H from G by adding one new vertex and two new edges in a suitable way.

(b) Suppose G is a connected graph with exactly $2s$ vertices of odd degree. Show how to efficiently find s Eulerian paths in G, such that each edge of G is contained in exactly one of these Eulerian paths.

HINT Construct H from G by adding s new vertices and $2s$ new edges in a suitable way.

7.9 Suppose that G is a directed acyclic graph (i.e., a DAG) on the vertex set $V = \{x_1, \ldots, x_n\}$ and x_1 is a vertex having indegree equal to 0. For every vertex x_i, we want to determine the number (denoted $N[i]$) of directed paths in G from x_1 to x_i.

(a) Give a short high-level description of an algorithm to solve this problem, as well as detailed pseudocode. The complexity of the algorithm should be $O(n + m)$.

(b) Illustrate the execution of your algorithm step-by-step on the DAG having the following directed edges:

$$\begin{array}{llllll} x_1x_2, & x_1x_4, & x_2x_3, & x_2x_4, & x_2x_7, & x_3x_6, \\ x_4x_5, & x_4x_6, & x_5x_6, & x_5x_7, & x_6x_8, & x_7x_8. \end{array}$$

(c) Prove that your algorithm is correct. This will be a proof by induction making use of an appropriate loop invariant.

7.10 Let $G = (V, E)$ be a connected graph with n vertices and suppose that every edge $\{u, v\} \in E$ has a weight $w(u, v) > 0$. Assume further that the edge weights are all distinct. Let T be any spanning tree of G. The *critical edge* is the edge in T with the highest weight. A spanning tree of G is a *minimum critical spanning tree* if there is no other spanning tree of G having a critical edge of lower weight. Prove that a minimum spanning tree of G is a minimum critical spanning tree.

Chapter 8

Backtracking Algorithms

In this chapter, we discuss backtracking algorithms, which are examples of exhaustive search algorithms. Backtracking algorithms can be used to generate combinatorial structures and solve optimization problems; several interesting examples are given in this chapter. Enhancements such as bounding functions and branch-and-bound are also presented.

8.1 Introduction

A ***backtracking algorithm*** is a recursive method of generating feasible solutions to a given problem one step at a time. A backtracking algorithm is an example of an ***exhaustive search***, as every possible solution is generated. In its most basic form, a backtracking algorithm is similar to a depth-first search of the ***state space tree***, which consists of all the possible partial or complete solutions to a given problem. Backtracking algorithms are often used to solve optimization problems.

We begin by presenting a "generic" backtrack algorithm. For many problems of interest, a feasible solution (or partial solution) can be represented as a list $X = [x_1, x_2, \ldots, x_\ell]$, in which each x_i is chosen from a finite ***possibility set***, $\mathcal{P}_i$. The x_i's are defined one at a time, in order, as the state space tree is traversed. Hence, the backtrack algorithm considers all members of $\mathcal{P}_1 \times \cdots \times \mathcal{P}_i$ for each $i = 1, 2, \ldots$. The value ℓ is the number of co-ordinates in the current list X. This is the same as the depth of the corresponding node in the state space tree.

Given $X = [x_1, \ldots, x_\ell]$, constraints for the problem under consideration may restrict the possible values for $x_{\ell+1}$ to a subset $\mathcal{C}_{\ell+1} \subseteq \mathcal{P}_{\ell+1}$ that we call the ***choice set***. The general backtracking algorithm is presented as Algorithm 8.1. We initially set $X = []$ (an empty list) and call BACKTRACK(0).

The first step of Algorithm 8.1 is to identify if the current list X is a "complete" solution. In some (but not all) problems, complete solutions have a fixed, pre-determined length, in which case this step just consists of checking the value of ℓ.

The operation "process it" could mean several things, e.g., save X for future use; print it out; or check to see if it is better than the best solution found so far (according to some optimality measure).

The second step is to construct the choice set $\mathcal{C}_{\ell+1}$ for the current list $X = [x_1, \ldots, x_\ell]$. Note that we do not rule out the possibility that a "complete" solution

DOI: 10.1201/9780429277412-8

Algorithm 8.1: BACKTRACK(ℓ)

```
global X
if X = [x_1, ..., x_ℓ] is a complete solution
  then process it
compute C_{ℓ+1}
for each x ∈ C_{ℓ+1}
  do { x_{ℓ+1} ← x
       BACKTRACK(ℓ + 1)
main
 X = []
 BACKTRACK(0)
```

could be contained in a larger "complete" solution. In general, a given X cannot be extended if and only if $\mathcal{C}_{\ell+1} = \emptyset$.

The third step (which is executed if $\mathcal{C}_{\ell+1} \neq \emptyset$) is to assign every possible value in $\mathcal{C}_{\ell+1}$ in turn as the next coordinate, $x_{\ell+1}$. The algorithm is called recursively each time an assignment to $x_{\ell+1}$ is made.

Many backtracking algorithms are exponential-time algorithms, for the simple reason that they are generating an exponential number of solutions. For example, it is clear that it would require $\Omega(n!)$ time to generate all $n!$ permutations, no matter what algorithm is employed. Many of the optimization problems for which backtracking algorithms are often used are not known to be solvable in polynomial time (specifically, because they are NP-hard; see Section 9.7). This does not mean that problem instances of "small" or "moderate" size cannot be solved, and, indeed, backtracking algorithms can often solve "practical" instances of interest for many problems, despite the fact that they might not be polynomial-time algorithms.

8.1.1 Permutations

To illustrate the basic ideas, we develop a simple backtracking algorithm to generate all the permutations of $\{1, \ldots, n\}$. Here, a partial or complete solution will have the form $X = [x_1, \ldots, x_\ell]$ where $x_1, \ldots, x_\ell$ are distinct elements from the set $\{1 \ldots, n\}$. X is a complete solution if $\ell = n$, and it is a partial solution if $0 \leq \ell \leq n - 1$.

Clearly, x_1 can take on any value from 1 to n, so

$$\mathcal{C}_1 = \{1, \ldots, n\}.$$

For $1 \leq \ell \leq n - 1$, $x_{\ell+1}$ must be distinct from $x_1, \ldots, x_\ell$. Therefore,

$$\mathcal{C}_{\ell+1} = \{1, \ldots, n\} \setminus \{x_1, \ldots, x_\ell\},$$

for $1 \leq \ell \leq n - 1$.

Algorithm 8.2: PERMUTATIONS(ℓ, n)

```
global X
if ℓ = n
  then X is a permutation
  else ⎧ if ℓ = 0
       ⎪   then C ← {1,...,n}
       ⎪   else C ← {1,...,n} \ {x_1,...,x_ℓ}
       ⎨ for all i ∈ C
       ⎪   do ⎧ x_{ℓ+1} ← i
       ⎩      ⎩ PERMUTATIONS(ℓ+1, n)
main
  X = []
  PERMUTATIONS(0, n)
```

One straightforward way to generate all permutations by backtracking is presented in Algorithm 8.2. In our pseudocode, we denote the "current" choice set $\mathcal{C}_{\ell+1}$ simply by $\mathcal{C}$. This is reasonable because the choice set is a local variable (i.e., there is a different choice set defined in each level of the recursion).

Note also that we are using set notation for $\mathcal{C}$. Some programming languages automatically include built-in set operations, so it is almost trivial to translate the given pseudocode into a working computer program. But it sometimes might be preferable to use some other type of data structure to store the choice sets. In such a situation, we should also consider how we would iterate through the elements in the choice sets.

Here is one possible alternative. In Section 3.2.1, we discussed how to use a boolean vector to store the elements of a set that is contained in a specified universe $\mathcal{U}$. It is often the case that all the possibility sets $\mathcal{P}_i$ are identical, say $\mathcal{P}_1 = \cdots = \mathcal{P}_n = \mathcal{U}$. Given that each choice set $\mathcal{C}_i \subseteq \mathcal{P}_i$, it is convenient to represent a choice set as a bit string of length $|\mathcal{U}|$.

For example, in the case of the PERMUTATIONS problem, we have $\mathcal{P}_1 = \cdots = \mathcal{P}_n = \{1, \ldots, n\}$, so each choice set $\mathcal{C}$ can be represented as a bit string of length n. We would iterate through the elements of a choice set $\mathcal{C}$ by checking each of the n bits in turn. If the jth bit has the value 1, then this indicates that $j \in \mathcal{C}$.

The state space tree for PERMUTATIONS when $n = 3$ is shown in Figure 8.1. Each node is labeled with the current value of X. The six permutations of $\{1, 2, 3\}$ are found in the leaf nodes. Algorithm 8.2 performs a depth-first search of this state space tree. However, here the graph being searched (i.e., the state space tree) is constructed implicitly as the algorithm is executed. In contrast, graph search algorithms studied in Chapter 7 search a graph that has been constructed "ahead of time."

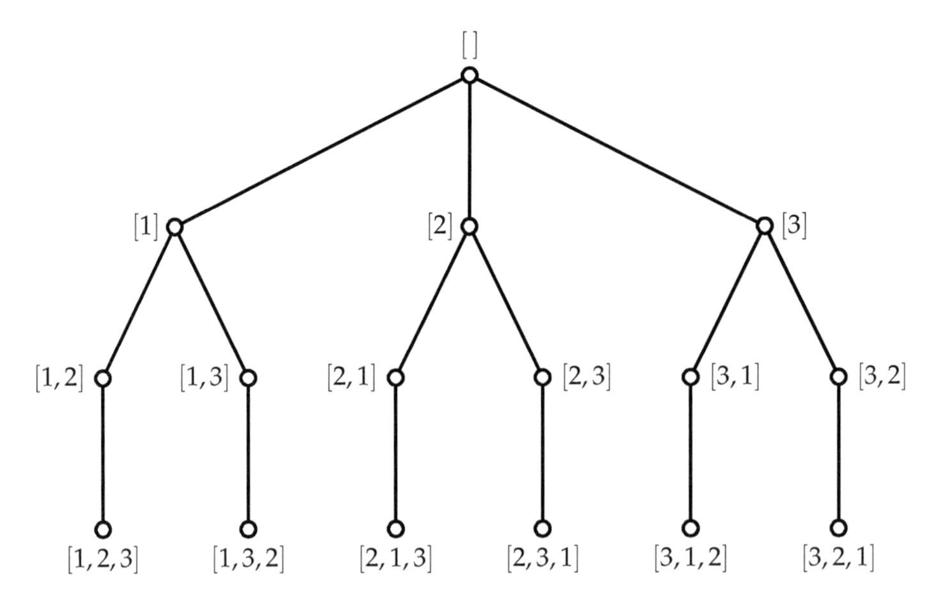

FIGURE 8.1: The state space tree for PERMUTATIONS with $n = 3$.

8.1.2 The n-queens Problem

The **8-queens** problem requires us to find all possible ways of placing eight mutually non-attacking queens on a standard 8 by 8 chessboard. More generally, we can define the n-**queens** problem, as is done in Problem 8.1.

> **Problem 8.1:** n**-queens**
>
> **Instance:** A positive integer n
>
> **Find:** All possible ways of placing n mutually non-attacking queens on an n by n chessboard, where two queens are ***non-attacking*** if they are in different rows, different columns, different diagonals and different back-diagonals.

The number of solutions $Q(n)$ to the n-**queens** problem, for $4 \leq n \leq 16$, is given in Table 8.1. One solution to the 8-queens problem is given in Figure 8.2.

Suppose we label the cells in the chessboard by (i, j), $1 \leq i \leq n$, $1 \leq j \leq n$, where i denotes a row and j denotes a column. The rows are numbered in increasing order from top to bottom and the columns are numbered in increasing order from left to right, as shown in Figure 8.2.

A solution to the n-queens problem consists of n queens, one from each row of the chessboard. Therefore we can write a solution in the form $X = [x_1, \ldots, x_n]$, where $1 \leq x_i \leq n$ for $1 \leq i \leq n$. The list X corresponds to the n cells

$$(1, x_1), (2, x_2), \ldots, (n, x_n).$$

Thus, the solution to the 8-queens problem given in Figure 8.2 would be represented by $X = [6, 4, 7, 1, 8, 2, 5, 3]$.

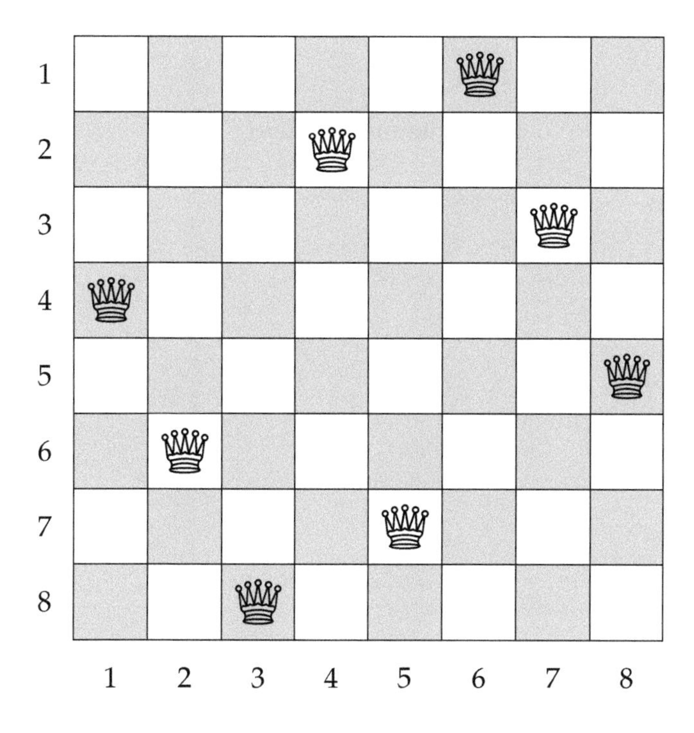

FIGURE 8.2: A solution to the 8-queens problem.

TABLE 8.1: Number of solutions to the n-**queens** problem.

n	$Q(n)$
4	2
5	10
6	4
7	40
8	92
9	352
10	724
11	2680
12	14200
13	73712
14	365596
15	2279184
16	14772512

Algorithm 8.3: n-QUEENS(ℓ, n)

```
global X
if ℓ = n
  then X is a solution to the n-queens problem
  else { if ℓ = 0
           then C ← {1,...,n}
           else { C ← {1,...,n} \ {x_1,...,x_ℓ}
                  C ← C \ {x_i − i + ℓ + 1 : 1 ≤ i ≤ ℓ}
                  C ← C \ {x_i + i − ℓ − 1 : 1 ≤ i ≤ ℓ}
         for all c ∈ C
           do { x_{ℓ+1} ← c
                n-QUEENS(ℓ + 1, n)
main
  X = []
  n-QUEENS(0, n)
```

We require that the x_i's are all distinct, since non-attacking queens cannot be in the same column of the chessboard. That is, a solution must be a permutation. In addition, we also require that no two x_i's correspond to cells in the same diagonal or back-diagonal. This suggests that we can design a simple backtracking algorithm to solve the n-**queens** problem by using Algorithm 8.2 as a starting point and modifying the choice sets in such a way that we forbid the selection of two cells in the same diagonal or back-diagonal.

It is easy to determine if two cells are in the same diagonal: cells (i_1, j_1) and (i_2, j_2) are in the same diagonal if and only if

$$j_1 - i_1 = j_2 - i_2, \tag{8.1}$$

and these two cells are in the same back-diagonal if and only if

$$j_1 + i_1 = j_2 + i_2. \tag{8.2}$$

Before defining $x_{\ell+1} \leftarrow c$ in our backtracking algorithm, we need to verify that the cell $(\ell+1, c)$ is not in the same diagonal or back diagonal as a previously chosen cell (i, x_i), where $1 \leq i \leq \ell$.

So, from (8.1) and (8.2), we see that

$$c \neq \begin{cases} x_i - i + \ell + 1 & \text{if } x_i - i + \ell + 1 \leq n \\ x_i + i - (\ell + 1) & \text{if } x_i + i - (\ell + 1) \geq 1. \end{cases} \tag{8.3}$$

Condition (8.3) must hold for all i such that $1 \leq i \leq \ell$. Thus, we remove any "bad" values in the range $1, \dots, n$ from $\mathcal{C}_{\ell+1}$.

For example, suppose $n = 8$, $\ell = 2$, $x_1 = 6$ and $x_2 = 4$ (as in Figure 8.2). Then

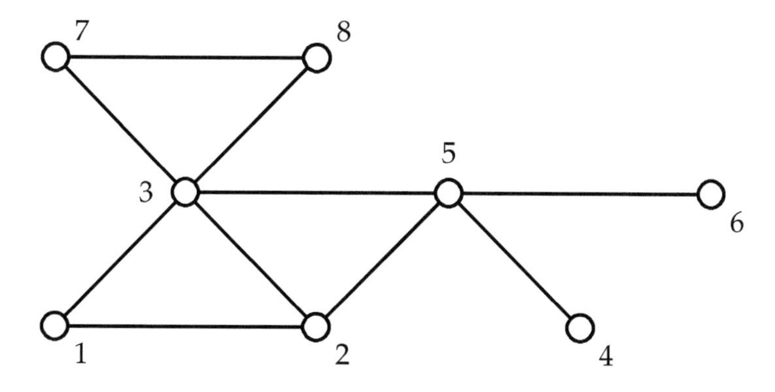

FIGURE 8.3: A graph.

there are four values that are deleted from C_3 as a consequence of (8.3), namely, $3, 4, 5$ and 8.

The backtracking algorithm to solve the n-**queens** problem is presented as Algorithm 8.3. This algorithm is identical to Algorithm 8.2, except that there are two additional modifications to $\mathcal{C}$ to account for the two additional conditions in (8.3).

8.2 Generating All Cliques

As another example of backtracking, we develop an algorithm to generate all the cliques in a given graph G. See Problem 8.2 for the problem statement.

Problem 8.2: All Cliques

Instance: An undirected graph $G = (V, E)$.

Find: All the cliques in the graph G. (A ***clique*** is a subset of vertices $W \subseteq V$ such that $uv \in E$ for all $u, v \in W$, $u \neq v$.)

When we design a backtracking algorithm for **All Cliques**, the state space tree will contain a node for every clique in the graph. For convenience, let us denote the vertices by $V = \{1, \ldots, n\}$. In order to ensure that we generate each clique only once, we stipulate that a clique $\{x_1, \ldots, x_\ell\}$ should be written in the form $X = [x_1, \ldots, x_\ell]$, where $x_1 < x_2 < \cdots < x_\ell$.

We give an example to illustrate the concepts described above. Consider the graph shown in Figure 8.3. In this graph, every vertex is a clique of size 1, the edges are cliques of size 2 and the cliques of size 3 are

$$\{1, 2, 3\}, \{2, 3, 5\}, \{3, 7, 8\}.$$

There are no cliques of size 4.

A clique is a ***maximal clique*** if it is not contained in a larger clique. Note that there can be maximal cliques of different sizes. In the graph displayed in Figure 8.3, the three cliques of size 3 are maximal, and the following edges are maximal cliques of size 2:

$$\{4,5\},\{5,6\}.$$

Let us think about how to construct the relevant choice sets in order to design a backtracking algorithm. Suppose $X = [x_1, \ldots, x_\ell]$ (this is a clique of size ℓ, where $x_1 < x_2 < \cdots < x_\ell$). The choice set $\mathcal{C}_{\ell+1}$ is defined as follows. Initially, $\ell = 0$ and $X = []$, and thus we have

$$\mathcal{C}_1 = V.$$

In general, for $\ell \geq 1$, we want to ensure that any vertex v added to the clique X is adjacent to all the vertices that are currently in X. We also require that $v > x_\ell$, as we have already observed. Therefore,

$$\mathcal{C}_{\ell+1} = \{v \in V : v > x_\ell \text{ and } \{v, x_i\} \in E \text{ for } 1 \leq i \leq \ell\}. \tag{8.4}$$

Suppose that the following sets have been precomputed, for all $v \in V$:

$$A_v = \{v \in V : \{u,v\} \in E\}.$$

We might refer to these as ***adjacency sets*** for the given graph. Then, for $\ell \geq 1$, the choice set $\mathcal{C}_{\ell+1}$ can be constructed as follows:

$$\mathcal{C}_{\ell+1} = \{x_\ell + 1, \ldots, n\} \bigcap \left(\bigcap_{j=1}^{\ell} A_{x_j} \right). \tag{8.5}$$

Algorithm 8.4 is our backtracking algorithm for **All Cliques**. The computation of the choice sets is based on (8.5). Observe that this algorithm is not able to recognize which cliques are maximal. If we want to determine the maximal cliques,

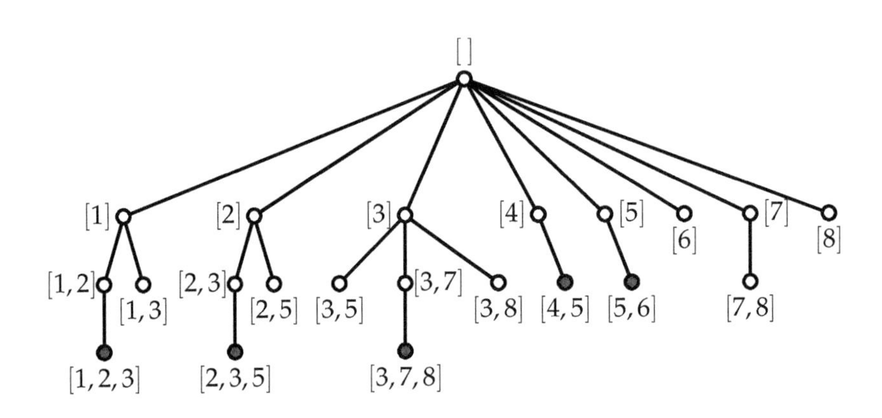

FIGURE 8.4: A state space tree for CLIQUES.

Algorithm 8.4: CLIQUES(ℓ, n)

global $A_1, \ldots, A_n, X$
if $\ell = 0$
  then *last* $\leftarrow 0$
  else *last* $\leftarrow x_\ell$
$\mathcal{C} \leftarrow \{last + 1, \ldots, n\}$
for $j \leftarrow 1$ **to** ℓ
  do $\mathcal{C} \leftarrow \mathcal{C} \cap A_{x_j}$
for all $i \in \mathcal{C}$
  do $\begin{cases} x_{\ell+1} \leftarrow i \\ \text{CLIQUES}(\ell+1, n) \end{cases}$
main
 $X = [\,]$
 CLIQUES$(0, n)$

we just need to add a couple of extra statements to Algorithm 8.4. Basically, we compute a "modified" choice set, denoted by $\mathcal{N}_{\ell+1}$, where we do not restrict the possible extensions to $\{x_\ell + 1, \ldots, n\}$. Thus, we define

$$\mathcal{N}_{\ell+1} = \{1, \ldots, n\} \bigcap \left(\bigcap_{j=1}^{\ell} A_{x_j} \right).$$

The modified choice set $\mathcal{N}_{\ell+1}$ is empty if and only if the current solution X is a maximal clique. Note that this does not affect the execution of the algorithm; the only change is that we now identify maximal cliques as the algorithm executes. Algorithm 8.5 incorporates this additional test; the new statements are highlighted in red.

Example 8.1 Suppose we want to run Algorithm 8.5 on the graph from Example 8.3. In order to do this, we would first compute the adjacency sets $A_1, \ldots, A_8$ as follows:

$$\begin{array}{ll} A_1 = \{2,3\} & A_2 = \{1,3,5\} \\ A_3 = \{1,2,5,7,8\} & A_4 = \{5\} \\ A_5 = \{2,3,4,6\} & A_6 = \{5\} \\ A_7 = \{3,8\} & A_8 = \{3,7\}. \end{array}$$

When Algorithm 8.5 is executed, we obtain the state space tree that is shown in Figure 8.4. The nodes corresponding to maximal cliques are colored red. □

Another simple modification we can consider is based on the observation that

Algorithm 8.5: CLIQUES2(ℓ, n)

```
global A_1, ..., A_n, X
if ℓ = 0
  then last ← 0
  else last ← x_ℓ
C ← {last + 1, ..., n}
N ← {1, ..., n}
for j ← 1 to ℓ
  do { C ← C ∩ A_{x_j}
     { N ← N ∩ A_{x_j}
if N = ∅
  then X is a maximal clique
  else { for all i ∈ C
       {   do { x_{ℓ+1} ← i
       {        { CLIQUES2(ℓ + 1, n)
main
 X = []
 CLIQUES2(0, n)
```

the choice sets have a simple recursive description, which allows them to be computed more efficiently. For $v \in V$, define

$$B_v = \{u \in V : u > v\}.$$

The sets B_v can also be precomputed. Then, for $\ell \geq 1$, we have

$$\mathcal{C}_{\ell+1} = \mathcal{C}_\ell \bigcap A_{x_\ell} \bigcap B_{x_\ell}$$

and

$$\mathcal{N}_{\ell+1} = \mathcal{N}_\ell \bigcap A_{x_\ell}.$$

The sets $\mathcal{C}_1, \ldots, \mathcal{C}_n$ and $\mathcal{N}_1, \ldots, \mathcal{N}_n$ are global. This allows Algorithm 8.5 to be written in a slightly simpler form; see Algorithm 8.6.

We should point out that a given backtracking algorithm may or may not be amenable to recursive computation of the choice sets. This approach is generally useful when each choice set is a subset of the choice set on the previous level. So, for example, we could conveniently compute the choice set recursively in Algorithm 8.2 (which generates all the permutations of $\{1, \ldots, n\}$). In this algorithm, each choice set could be computed from the previous choice set by removing one element. On the other hand, a recursive computation of the choice sets would not make much sense for the backtracking algorithms for the n-**queens** problem, since the choice set for one row is not really related in a direct way to the choice set for the previous row.

Algorithm 8.6: CLIQUES3(ℓ, n)

global $A_1, \ldots, A_n, B_1, \ldots, B_n, \mathcal{C}_1, \ldots, \mathcal{C}_n, \mathcal{N}_1, \ldots, \mathcal{N}_n, X$
if $\ell = 0$
  then $\begin{cases} \mathcal{C}_1 \leftarrow \{1, \ldots, n\} \\ \mathcal{N}_1 \leftarrow \{1, \ldots, n\} \end{cases}$
  else $\begin{cases} \mathcal{C}_{\ell+1} \leftarrow \mathcal{C}_\ell \cap A_{x_\ell} \cap B_{x_\ell} \\ \mathcal{N}_{\ell+1} \leftarrow \mathcal{N}_\ell \cap A_{x_\ell} \end{cases}$
if $\mathcal{N}_{\ell+1} = \varnothing$
  then X is a maximal clique
  else { **for all** $i \in \mathcal{C}_{\ell+1}$
    do $\begin{cases} x_{\ell+1} \leftarrow i \\ \text{CLIQUES3}(\ell+1, n) \end{cases}$
main
  $X = [\,]$
  CLIQUES3($0, n$)

8.3 Sudoku

Sudoku is a popular puzzle that appears in many newspapers. As described in Wikipedia:

> The objective is to fill a 9 by 9 grid with digits so that each column, each row, and each of the nine 3 by 3 subgrids that compose the grid (also called "boxes," "blocks," or "regions") contain all of the digits from 1 to 9. The puzzle setter provides a partially completed grid, which for a well-posed puzzle has a single solution.

A ***Latin square*** of ***order*** 9 is a 9 by 9 array on nine symbols such that every row and every column of the array contains every symbol. Hence, the solution to a Sudoku puzzle is a Latin square of order 9 that satisfies the additional properties that the nine blocks also contain every symbol. There is no *a priori* reason why a given partial solution cannot have multiple completions, or no completions at all. However, Sudoku puzzles are required to have *unique* completions. Figure 8.5 provides an example given in Wikipedia.

Backtracking is one possible approach that could be used to solve Sudoku puzzles. We could consider filling in all the empty cells in a 9 by 9 array A in some natural order (say left to right, top to bottom). When we examine a particular empty cell ℓ, the choice set would contain all the symbols that have not already appeared in some filled cell in the same row, column or block as ℓ. The choice sets can be computed quickly if we keep track of the symbols occurring in each row, each column and each block of the array A as the algorithm proceeds. Therefore, at any

The initial partial solution is

5	3			7				
6			1	9	5			
	9	8					6	
8				6				3
4			8		3			1
7				2				6
	6					2	8	
			4	1	9			5
				8			7	9

The complete solution is

5	3	4	6	7	8	9	1	2
6	7	2	1	9	5	3	4	8
1	9	8	3	4	2	5	6	7
8	5	9	7	6	1	4	2	3
4	2	6	8	5	3	7	9	1
7	1	3	9	2	4	8	5	6
9	6	1	5	3	7	2	8	4
2	8	7	4	1	9	6	3	5
3	4	5	2	8	6	1	7	9

FIGURE 8.5: An initial partial Sudoku solution and its unique completion.

given time during the execution of the algorithm, let us define

$$\begin{aligned} R_i &= \{\text{current entries in row } i\} \\ C_j &= \{\text{current entries in column } j\} \\ B_k &= \{\text{current entries in block } k\}, \end{aligned}$$

for $1 \leq i,j,k \leq 9$.

Suppose we number the cells from left to right, top to bottom, as $1,\ldots,81$. Then it is not difficult to compute the row, column and block containing a particular cell $\ell \in \{1,\ldots,81\}$. First, the row containing the cell ℓ is

$$\mathsf{row}(\ell) = \left\lfloor \frac{\ell - 1}{9} \right\rfloor + 1.$$

It is also not difficult to see that the column containing the cell ℓ is

$$\mathsf{column}(\ell) = (\ell - 1) \bmod 9 + 1.$$

After computing the row and column containing a cell ℓ, say

$$i = \mathsf{row}(\ell) \quad \text{and} \quad j = \mathsf{column}(\ell),$$

it is not hard to compute the block containing ℓ:

$$\mathsf{block}(\ell) = 3 \times \left\lfloor \frac{i - 1}{3} \right\rfloor + \left\lfloor \frac{j - 1}{3} \right\rfloor + 1.$$

The following function (or algorithm) ROWCOLUMNBLOCK computes these three values, where $k = \mathsf{block}(\ell)$.

Algorithm 8.7: ROWCOLUMNBLOCK(ℓ)

comment: we assume $1 \leq \ell \leq 81$
$i \leftarrow \left\lfloor \frac{\ell-1}{9} \right\rfloor + 1$
$j \leftarrow (\ell - 1) \bmod 9 + 1$
$k \leftarrow 3 \times \left\lfloor \frac{i-1}{3} \right\rfloor + \left\lfloor \frac{j-1}{3} \right\rfloor + 1$
return (i,j,k)

We also need to specify the values in the filled cells in the initial partial solution, as these are not allowed to be modified. So, for $1 \leq \ell \leq 81$, we let $a_\ell = 0$ if cell ℓ is not initially filled. If $a_\ell \neq 0$, then a_ℓ denotes the initial (fixed) value in cell ℓ.

Referring to the example given in Figure 8.5, we would have the following

Algorithm 8.8: INITIALIZE(A)

```
external RowColumnBlock
global A, R, C, B
comment: A = [a_1, ..., a_81] contains initial values
for i ← 1 to 9
  do { R[i] ← ∅
       C[i] ← ∅
       B[i] ← ∅
for ℓ ← 1 to 81
  do if a_ℓ ≠ 0
    then { (i, j, k) ← RowColumnBlock(ℓ)
           R[i] ← R[i] ∪ {a_ℓ}
           C[j] ← C[j] ∪ {a_ℓ}
           B[k] ← B[k] ∪ {a_ℓ}
```

initial values:

$a_1 = 5$	$a_2 = 3$	$a_5 = 7$	
$a_{10} = 6$	$a_{13} = 1$	$a_{14} = 9$	$a_{15} = 5$
$a_{20} = 9$	$a_{21} = 8$	$a_{26} = 6$	
$a_{28} = 8$	$a_{32} = 6$	$a_{36} = 3$	
$a_{37} = 4$	$a_{40} = 8$	$a_{42} = 3$	$a_{45} = 1$
$a_{46} = 7$	$a_{50} = 2$	$a_{54} = 6$	
$a_{56} = 6$	$a_{61} = 2$	$a_{62} = 8$	
$a_{67} = 4$	$a_{68} = 1$	$a_{69} = 9$	$a_{72} = 5$
$a_{77} = 8$	$a_{80} = 7$	$a_{81} = 9.$	

Given the initial values in A, we need to initialize the sets $R[i]$, $C[i]$ and $B[i]$. This is straightforward; see Algorithm 8.8.

The backtracking algorithm is also not difficult. We fill in the cells from left to right, top to bottom. Suppose we are looking at a cell, say cell $\ell + 1$, that has not been filled initially. Using ROWCOLUMNBLOCK, we determine the row i, the column j and the block k that contains this cell. The entries in

$$R[i] \cup C[j] \cup B[k]$$

are not available for cell $\ell + 1$, so we delete them from the choice set.

Assuming the choice set is non-empty, we consider the possible values y in the choice set one at a time. For each such value y:

- we update $R[i]$, $C[j]$ and $B[k]$,

Algorithm 8.9: SUDOKU(ℓ)

```
external RowColumnBlock, Initialize
global A, R, C, B, X
comment: X = [x_1, ..., x_81]
comment: A = [a_1, ..., a_81] contains initial values
if ℓ = 81
  then X is the solution to the Sudoku puzzle
 else if a_{ℓ+1} ≠ 0
  then { x_{ℓ+1} ← a_{ℓ+1}
       { Sudoku(ℓ + 1)
  else { (i, j, k) ← RowColumnBlock(ℓ + 1)
       { choice ← {1, ..., 9} \ (R[i] ∪ C[j] ∪ B[k])
       { for all y ∈ choice
       {    do { R[i] ← R[i] ∪ {y}
       {       { C[j] ← C[j] ∪ {y}
       {       { B[k] ← B[k] ∪ {y}
       {       { x_{ℓ+1} ← y
       {       { Sudoku(ℓ + 1)
       {       { R[i] ← R[i] \ {y}
       {       { C[j] ← C[j] \ {y}
       {       { B[k] ← B[k] \ {y}
main
 Initialize(A)
 X = []
 Sudoku(0)
```

- we assign $x_{\ell+1}$ the value y, and
- we call the backtracking algorithm recursively.

When we return from a recursive call, we restore $R[i]$, $C[j]$ and $B[k]$ back to their previous state.

When $\ell = 81$, this means that all 81 cells have been filled in and therefore we have found the complete solution. As we have mentioned, a "valid" Sudoku puzzle has exactly one valid solution, so it would be reasonable to stop as soon as we find one solution.

8.4 Pruning and Bounding Functions

In the rest of this chapter, we focus on backtracking algorithms for optimization problems. For example, suppose we want to find a clique of maximum size in a given graph. This problem is stated as follows.

Problem 8.3: Max Clique

Instance: An undirected graph $G = (V, E)$.

Find: A clique of maximum size in the graph G.

We observe that it would be trivial to modify any algorithm solving **All Cliques** so that it finds the maximum clique, simply by keeping track of the largest clique found, as the algorithm is executed. For example, Algorithm 8.6 could be modified as shown in Algorithm 8.10 (the new statements are in red). Note that we only need to check the maximal cliques, because any maximum clique is necessarily maximal.

One useful method of speeding up a backtracking algorithm is through the use of a suitable bounding function. We illustrate the basic idea with the **Max Clique** problem. At a given point in the execution of Algorithm 8.10, suppose we have a current clique X of size ℓ and we have just computed the choice set $\mathcal{C}_{\ell+1}$. The largest clique found "so far" has size *MaxSize*. Clearly, the largest possible size of any clique that contains the current clique X is

$$\ell + |\mathcal{C}_{\ell+1}|.$$

Suppose that the following condition holds:

$$\ell + |\mathcal{C}_{\ell+1}| \leq MaxSize. \tag{8.6}$$

Then it must be the case that no "extension" of X can do better than the current largest clique. Therefore, there is no point in making any recursive calls from the given ℓ-tuple X. That is, we can ***prune*** the subtree rooted at X in this situation.

Example 8.2 Let us see how this pruning technique would be applied to the graph in Figure 8.3. In Figure 8.6, we show the original state space tree (from Figure 8.4), followed by the pruned tree. The black nodes in the second tree indicate where pruning has taken place.

The first instance of pruning happens when $X = [1]$, after we have found the clique $\{1, 2, 3\}$. At this time, we have $MaxSize = 3$, $\ell = 1$ and $|\mathcal{C}_{\ell+1}| = 2$. Therefore (8.6) holds and we do not make the second recursive call (where we set $x_3 = 2$) from this node.

Next, when $X = [2]$, we have $MaxSize = 3$, $\ell = 1$ and $|\mathcal{C}_{\ell+1}| = 2$. Therefore (8.6) holds and we do not make either of the two recursive calls that we made previously from this node.

Algorithm 8.10: MAXCLIQUES(ℓ, n)

```
global A_1,...,A_n, B_1,...,B_n, C_1,...,C_n, N_1,...,N_n
global X, MaxSize, Xmax
if ℓ = 0
  then { C_1 ← {1,...,n}
       { N_1 ← {1,...,n}
  else { C_{ℓ+1} ← C_ℓ ∩ A_{x_ℓ} ∩ B_{x_ℓ}
       { N_{ℓ+1} ← N_ℓ ∩ A_{x_ℓ}
if N_{ℓ+1} = ∅
  then { X is a maximal clique
       { if ℓ > MaxSize
       {   then { MaxSize ← ℓ
       {        { Xmax ← X
  else { for all i ∈ C_{ℓ+1}
       {   do { x_{ℓ+1} ← i
       {      { MAXCLIQUES(ℓ + 1, n)
main
  X = []
  MaxSize ← 0
  MAXCLIQUES(0, n)
  Xmax is a maximum clique, having size MaxSize
```

In contrast, when $X = [3]$, we have $MaxSize = 3$, $\ell = 1$ and $|\mathcal{C}_{\ell+1}| = 3$. Therefore (8.6) does not hold, and we make all three recursive calls (setting $x_3 = 5, 7$ and 8 in turn). No pruning is done at this node.

It is easy to see that we also prune the nodes $X = [3,7], [4], [5]$ and $[7]$. □

8.4.1 Bounding Functions

Let's try to generalize the bounding technique used in Example 8.2. Suppose we have a backtracking algorithm for a maximization problem, i.e., we are trying to find a feasible solution with the maximum possible profit. If X is any node in the state space tree, let $\mathsf{P}(X)$ denote the maximum profit of any node that is a descendant of X (including X itself). We could compute $\mathsf{P}(X)$ by traversing the subtree rooted at X, making all the possible recursive calls from X.

A ***bounding function*** is any function B defined on the nodes of the state space tree that satisfies the property

$$\mathsf{B}(X) \geq \mathsf{P}(X) \tag{8.7}$$

for all nodes X in the state space tree. In the **Max Clique** example above,

$$\mathsf{B}(X) = \ell + |\mathcal{C}_{\ell+1}|, \tag{8.8}$$

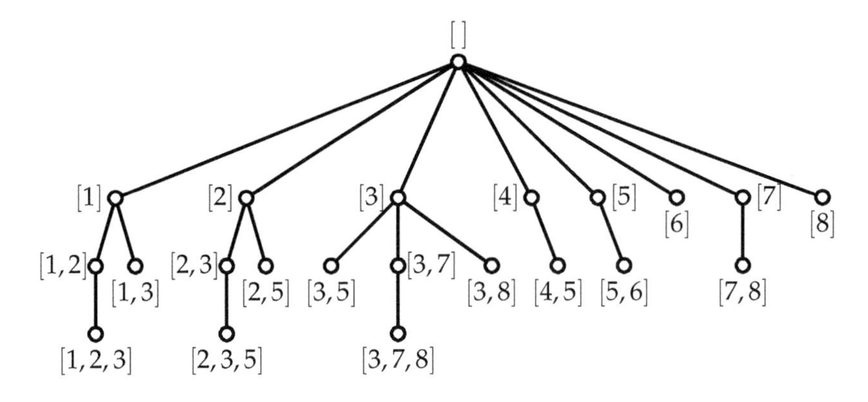

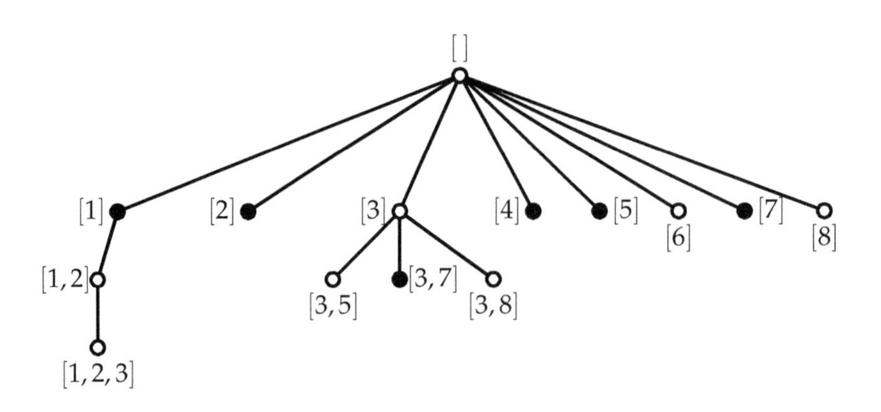

FIGURE 8.6: Non-pruned and pruned state space trees for CLIQUES.

where $X = [x_1, \ldots, x_\ell]$, is a bounding function.

In general, we incorporate a bounding function into a backtracking algorithm exactly as we did in Example 8.2: if the "current" optimal profit, say *OptP*, is at least $\mathsf{B}(X)$, then we prune the state space tree at the node X (i.e., we do not make any recursive calls from the node X). This is because the maximum achievable profit from any node that is a descendant of X in the state space tree is at most $\mathsf{B}(X)$, and we have already found a solution having profit at least $\mathsf{B}(X)$.

However, we can be a bit more clever in terms of how we use the bounding function. In general, we might make several recursive calls from a node X, as we iterate through the possible values in the current choice set. We should stop making recursive calls from X if it ever happens that $OptP \geq \mathsf{B}(X)$. We check this condition prior to *every* recursive call because the value of *OptP* might increase as the algorithm is executed.

Using a hypothetical example, let us describe in a bit more detail how this improved pruning strategy might be useful. Suppose that we have $\mathsf{B}(X) = 100$ for some node $X = [x_1, \ldots, x_\ell]$ and suppose that $OptP = 90$. Suppose also that $\mathcal{C}_{\ell+1} = \{y_1, y_2, y_3, y_4\}$.

1. We verify that $\mathsf{B}(X) > OptP$, so we proceed to make a recursive call, assigning $x_{\ell+1} = y_1$. When we explore this subtree, it might be the case that *OptP* is increased to 96.

2. We verify that it is still the case that $\mathsf{B}(X) > OptP$. Therefore we make the next recursive call, assigning $x_{\ell+1} = y_2$. Now, when we explore this second subtree, suppose that we find a feasible solution achieving profit 100, so we increase *OptP* to 100 (of course we cannot find a feasible solution extending X that has profit > 100, since $\mathsf{B}(X) = 100$).

3. Now that $OptP = 100$, we have $\mathsf{B}(X) \leq OptP$, so we can prune the last two branches at X. In other words, we do not make the last two recursive calls, which would correspond to setting $x_{\ell+1} = y_3$ or y_4.

Algorithm 8.11: BOUNDINGBACKTRACK(ℓ)

```
external B
comment: B is a bounding function
global X, OptX, OptP
if X = [x_1, ..., x_ℓ] is a complete solution
  then { compute P to be the profit of X
         if P > OptP
           then { OptP ← P
                  OptX ← X
compute C_{ℓ+1}
bound ← B(X)
for each x ∈ C_{ℓ+1}
  do { if bound ≤ OptP
         then return ()
         else { x_{ℓ+1} ← x
                BACKTRACK(ℓ + 1)
main
  X = []
  BACKTRACK(0)
```

Algorithm 8.11 is a pseudocode description of a generic backtracking algorithm that incorporates a bounding function denoted by B. We assume that we are trying to solve a maximization problem, so we are searching for a complete feasible solution having the maximum profit. This algorithm differs from Algorithm 8.1 in a few places.

First, whenever we encounter a complete solution, we compare it to the current optimal solution. We update *OptP* and *OptX* if the new solution is an improvement.

Then we compute the choice set and the bounding function B, given the current ℓ-tuple, X. We iterate through the choice set $\mathcal{C}_{\ell+1}$ as usual, but we stop making recursive calls from the given X if it ever happens that $B \leq OptP$, as we have discussed in the **Max Clique** example.

8.4.2 The Max Clique Problem

For the **Max Clique** problem, we have already given one example of a bounding function, namely, (8.8). However, there are various other bounding functions for this problem. These other bounding functions may give stronger bounds, but they also take more time to compute. Thus there is a tradeoff: better pruning may be possible, at the cost of extra computation time to compute the bounding function. In general, some experimentation might be required to see which choice of bounding function is preferred, for a given problem.

One useful bounding function for **Max Clique** is based on the **Vertex Coloring** (optimization) problem, which is defined as Problem 8.4. Note that the **Vertex Coloring** problem is a minimization problem.

Problem 8.4: Vertex Coloring

Instance: An undirected graph $G = (V, E)$.

Feasible solution: A ***vertex coloring*** (or more simply, a ***coloring***) of G, which is a function

$$f : V \rightarrow \{1, \ldots, c\}$$

(for some positive integer c) such that $f(u) \neq f(v)$ whenever $\{u, v\} \in E$. Such a coloring is termed a ***c-vertex coloring***.

Find: A vertex coloring of G that has the minimum possible number of colors.

The **Vertex Coloring** problem is not known to be solvable in polynomial time, as it is NP-hard. However, any feasible solution (i.e., any vertex coloring, optimal or not) of the **Vertex Coloring** problem can be used to obtain a bounding function for the **Max Clique** problem. So for example, we could find a vertex coloring using a simple greedy algorithm and incorporate it into a bounding function.

The main idea is as follows. Suppose that f is a vertex coloring of G that uses c colors, $\{1, \ldots, c\}$. Define the ***color classes*** as follows:

$$\mathsf{col}(j) = \{v : f(v) = j\},$$

for $1 \leq j \leq c$.

Let $X = [x_1, \ldots, x_\ell]$ and suppose that we have constructed the choice set $\mathcal{C}_{\ell+1}$. Suppose that $X' = [x_1, \ldots, x_\ell, \ldots, x_m]$ is any extension of X. Then the $m - \ell$ vertices $x_{\ell+1}, \ldots, x_m \in \mathcal{C}_{\ell+1}$ must form a clique. This means that they have to come from different color classes, since two vertices in the same color class cannot be adjacent.

We are assuming we have a particular c-coloring f. Suppose there are c' color classes that have a non-empty intersection with the set $\mathcal{C}_{\ell+1}$. That is, we compute

$$c' = |\{j : \mathsf{col}(j) \cap \mathcal{C}_{\ell+1} \neq \varnothing\}|. \tag{8.9}$$

Then

$$m - \ell \leq c'$$

and the size of the largest clique containing X is at most

$$\ell + c'.$$

Therefore, we can define a bounding function

$$\mathsf{B}(X) = \ell + c', \tag{8.10}$$

where c' is defined in (8.9). Observe that (8.10) is always at least as good as (8.8), because $c' \leq |\mathcal{C}_{\ell+1}|$. If $c' < |\mathcal{C}_{\ell+1}|$, then (8.10) improves (8.8), since it is a stronger (i.e., smaller) upper bound.

We describe a couple of ways that this kind of bounding function can be incorporated into a backtracking algorithm. Perhaps the simplest approach is to find a vertex coloring of G before we begin the execution of the backtracking algorithm. As we mentioned already, we might just use a greedy algorithm to do this. Then we could also precompute the color classes, which would expedite the computation of (8.9) whenever we want to evaluate the bounding function (8.10).

Alternatively, every time we want to compute $\mathsf{B}(X)$ for node X, we could compute $\mathcal{C}_{\ell+1}$ (as usual) and then find a vertex coloring of the ***induced subgraph*** $G[\mathcal{C}_{\ell+1}]$. This induced subgraph has vertex set $\mathcal{C}_{\ell+1}$ and the edges in this subgraph are the edges in E that have both endpoints in $\mathcal{C}_{\ell+1}$, namely,

$$E[\mathcal{C}_{\ell+1}] = \{\{u, v\} \in E : u, v \in \mathcal{C}_{\ell+1}\}.$$

Obviously this is more work than computing one "initial" vertex coloring, but the hope would be that it would result in more significant pruning, which would then lead to a smaller overall execution time.

One nice feature of bounding functions is that we can experiment with different bounding functions and the basic algorithm stays the same. Algorithm 8.12 is a version of Algorithm 8.10, modified to include an arbitrary bounding function, B. The changes are indicated in red.

8.5 0-1 Knapsack Problem

In this section, we investigate backtracking algorithms to solve the **0-1 Knapsack** problem, which was introduced as Problem 5.5. Recall that an instance of this problem consists of a list of profits, $P = [p_1, \ldots, p_n]$, a list of weights,

Algorithm 8.12: MAXCLIQUES2(ℓ, n)

```
external B
global A_1,...,A_n, B_1,...,B_n, C_1,...,C_n, N_1,...,N_n, X
if ℓ = 0
  then { C_1 ← {1,...,n}
       { N_1 ← {1,...,n}
  else { C_{ℓ+1} ← C_ℓ ∩ A_{x_ℓ} ∩ B_{x_ℓ}
       { N_{ℓ+1} ← N_ℓ ∩ A_{x_ℓ}
if N_{ℓ+1} = ∅
  then { X is a maximal clique
       { if ℓ > MaxSize
       {   then { MaxSize ← ℓ
       {        { Xmax ← X
  else { bound ← B(X)
       { for all i ∈ C_{ℓ+1}
       {   do { if bound ≤ MaxSize
       {      {   then return ()
       {      {   else { x_{ℓ+1} ← i
       {      {        { MAXCLIQUES2(ℓ + 1, n)
main
  X = []
  MaxSize ← 0
  MAXCLIQUES2(0, n)
  Xmax is a maximum clique, having size MaxSize
```

$W = [w_1, \ldots, w_n]$, and a capacity M. We are asked to find an n-tuple $[x_1, \ldots, x_n] \in \{0,1\}^n$, where

$$\sum_{i=1}^{n} w_i x_i \leq M,$$

that maximizes

$$\sum_{i=1}^{n} p_i x_i.$$

We present a basic backtracking algorithm for the **0-1 Knapsack** problem as Algorithm 8.13. Its structure is quite obvious. Suppose we have a feasible ℓ-tuple $X = [x_1, \ldots, x_\ell]$. This means that $X \in \{0,1\}^\ell$ and

$$\sum_{i=1}^{\ell} w_i x_i \leq M.$$

If $\ell = n$, then X is a complete solution. In this situation, we would compute the profit it attains (denoted by *CurP*) and compare it to the current optimal profit (denoted by *OptP*), updating *OptP* if it is appropriate to do so.

If $\ell < n$, we would compute the choice set $\mathcal{C}_{\ell+1}$. There are only two possible values for $x_{\ell+1}$, namely 0 and 1. Clearly, it is always the case that $0 \in \mathcal{C}_{\ell+1}$. It is also easy to see that $1 \in \mathcal{C}_{\ell+1}$ if and only if

$$w_{\ell+1} + \sum_{i=1}^{\ell} w_i x_i \leq M. \tag{8.11}$$

To facilitate the computation of (8.11), we keep track of the current weight, say

$$CurW = \sum_{i=1}^{\ell} w_i x_i,$$

and pass it as a parameter in Algorithm 8.13.

Algorithm 8.13: 0-1 KNAPSACK(ℓ, *CurW*)

```
global n, X, P, W, M, OptP, OptX
if ℓ = n
  then { CurP ← Σ_{i=1}^{n} x_i p_i
         if CurP > OptP
           then { OptP ← CurP
                  OptX ← X
  else { if CurW + w_{ℓ+1} ≤ M
           then C ← {0,1}
           else C ← {0}
         for all i ∈ C
           do { x_{ℓ+1} ← i
                0-1 KNAPSACK(ℓ + 1, CurW + w_{ℓ+1} x_{ℓ+1})
main
  X = []
  CurW ← 0
  0-1 KNAPSACK(0, CurW)
```

8.5.1 Bounding Function

Now we describe how we can incorporate a bounding function into Algorithm 8.13. The bounding function we use is based on the **Rational Knapsack** problem. Observe that any feasible solution to the **0-1 Knapsack** problem is also a feasible solution for the same instance of the **Rational Knapsack** problem. However, it is simple to compute the optimal solution to an instance of the **Rational Knapsack** problem by using Algorithm 5.4, which is a greedy algorithm.

Suppose that $X = [x_1, \ldots, x_\ell]$ is a partial feasible solution to the **0-1 Knapsack** problem. The current weight of this solution is

$$CurW = \sum_{i=1}^{\ell} w_i x_i,$$

which means that the remaining (as yet unused) capacity is $M - CurW$. We are free to include any subset of the last $n - \ell$ objects to use up some or all of this remaining capacity. This is just a "smaller" instance of the **0-1 Knapsack** problem. However, if we solve the corresponding instance of the **Rational Knapsack** problem, we will obtain an *upper bound* on the additional profit we can attain from the last $n - \ell$ objects.

Let

$$\mathsf{RKnap}([p_{\ell+1}, \ldots, p_n], [w_{\ell+1}, \ldots, w_n], M - CurW)$$

denote the optimal profit for this instance of the **Rational Knapsack** problem. As we mentioned above, this value is easily computed using a greedy algorithm. Then our bounding function is

$$\mathsf{B}(X) = CurP + \lfloor \mathsf{RKnap}([p_{\ell+1}, \ldots, p_n], [w_{\ell+1}, \ldots, w_n], M - CurW) \rfloor . \tag{8.12}$$

That is, we include the profit that is already achieved from the first ℓ objects, along with an upper bound on the profit we can realize from the last $n - \ell$ objects.

Note that optimal profit for an instance of the **Rational Knapsack** problem might be a rational number. Therefore, we can round this down to the nearest integer for use in our bounding function.

We can incorporate this bounding function into our backtracking algorithm, Algorithm 8.13, in the usual way. The reader can fill in the details.

Example 8.3 Suppose we have an instance of the **0-1 Knapsack** problem consisting of $n = 5$ items. The profits (i.e., the array P), the weights (i.e., the array W) and the capacity (M) are as follows:

$$\begin{aligned} P &= [23, 24, 15, 13, 16] \\ W &= [11, 12, 8, 7, 9] \\ M &= 26. \end{aligned}$$

The items in this instance are already sorted in decreasing order of profit/weight, which facilitates computation of the bounding function. We assume that we always investigate the choice $x_{\ell+1} = 1$ first (i.e., before we try $x_{\ell+1} = 0$), if this choice is permissible.

If we run the basic backtracking algorithm, Algorithm 8.13, we find that the state space tree comprises 44 nodes. The optimal solution is $[1, 0, 1, 1, 0]$, having optimal profit 51. However, if we incorporate the bounding function (8.12), then the number of nodes in the state space tree is reduced to 11. The pruned state space tree is shown in Figure 8.7. In this diagram, we have not rounded the bounding

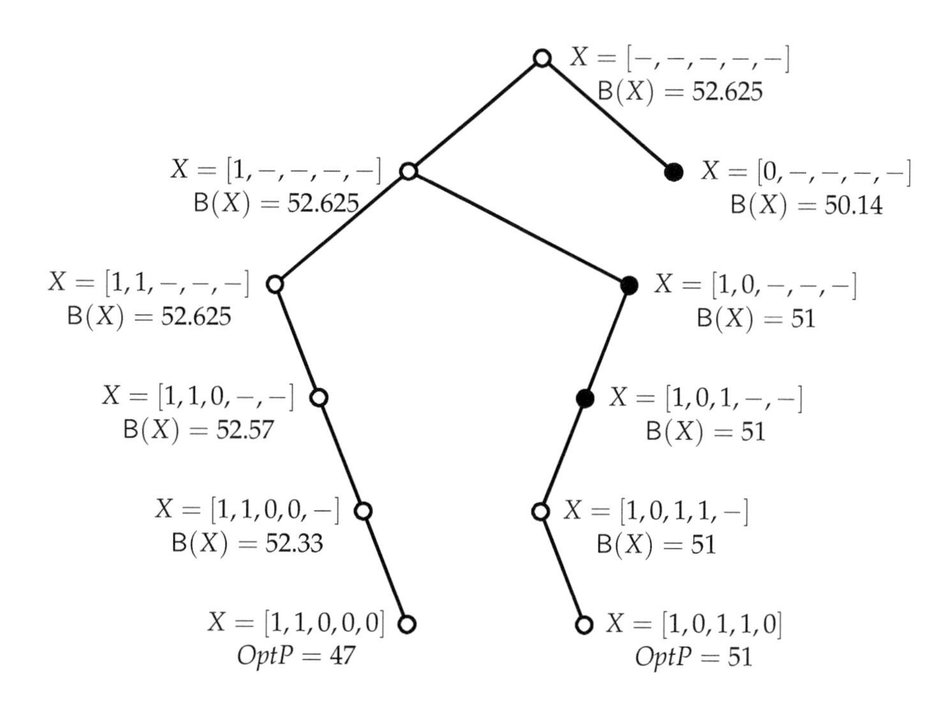

FIGURE 8.7: State space tree for a 0-1 knapsack instance with bounding function.

function down, in order to better show how the (unrounded) values of the bounding function vary from one level to the next.

Pruning takes place after the solution $X = [1, 0, 1, 1, 0]$, having profit 51, is found. This permits pruning at the blackened nodes in the diagram. It turns out that two nodes are pruned at $X = [1, 0, 1, -, -]$, six nodes are pruned at $X = [1, 0, -, -, -]$ and 25 nodes are pruned at $X = [0, -, -, -, -]$. □

8.6 Traveling Salesperson Problem

One very well-studied optimization problem is the **Traveling Salesperson** problem, which we present as Problem 8.5. We will subsequently refer to it as the **TSP-Optimization** problem. Note that it is a minimization problem.

The motivation for the name "traveling salesperson" is that a salesperson might want to visit n cities and then return home, without visiting any city more than once, in the most efficient manner.

For convenience, suppose the vertices in G are $V = \{1, \ldots, n\}$. A Hamiltonian cycle H will be represented by an n-tuple $[x_1, \ldots, x_n]$, where $\{x_1, \ldots, x_n\} = V$. The

Problem 8.5: TSP-Optimization

Instance: A graph $G = (V, E)$ and edge weights $w : E \rightarrow \mathbb{Z}^+$.

Find: A Hamiltonian cycle H in G such that

$$w(H) = \sum_{e \in H} w(e)$$

is minimized. (A ***Hamiltonian cycle*** is a cycle that passes through every vertex in V exactly once.)

edges in H are

$$x_1x_2, \quad x_2x_3, \quad \dots \quad , \quad x_{n-1}x_n, \quad x_nx_1.$$

The elements in $[x_1, \dots, x_n]$ constitute a permutation of $\{1, \dots, n\}$. Therefore, we can utilize Algorithm 8.2, which generates all the permutations of $\{1, \dots, n\}$, to construct a backtracking algorithm for **TSP-Optimization**.

However, we can use any point in the Hamiltonian cycle as the starting point, so we can assume without loss of generality that $x_1 = 1$. This reduces the size of the state space tree by a factor of n, meaning that we only need to consider $n!/n = (n-1)!$ permutations.

For convenience, we will define the choice function recursively. Suppose $X = [x_1, \dots, x_\ell]$ is a partial solution with $\ell \geq 2$. Then

$$\mathcal{C}_\ell = \{1, \dots, n\} \setminus \{x_1, \dots, x_{\ell-1}\}$$

and

$$\mathcal{C}_{\ell+1} = \{1, \dots, n\} \setminus \{x_1, \dots, x_\ell\}.$$

Therefore we have the recursive formula

$$\mathcal{C}_{\ell+1} = \mathcal{C}_\ell \setminus \{x_\ell\}$$

for $\ell \geq 2$. The first two choice sets are

$$\mathcal{C}_1 = \{1\}$$

since we stipulate that $x_1 = 1$, and

$$\mathcal{C}_2 = \{2, \dots, n\}.$$

Our basic backtracking algorithm for **TSP-Optimization** is presented as Algorithm 8.14. As usual, we keep track of the current optimal solution as the algorithm proceeds, updating it when necessary. In this algorithm, w is an n by n matrix of edge weights. We use the convention that $w[i, j] = \infty$ if $\{i, j\} \notin E$. Also, $w[i, i] = \infty$ for $1 \leq i \leq n$. Finally, $w[i, j] = w[j, i]$ for all i, j.

Algorithm 8.14: TSP(ℓ, n)

$$
\begin{array}{l}
\textbf{global } X, OptX, OptW, w, \mathcal{C}_1, \ldots, \mathcal{C}_n \\
\textbf{if } \ell = n \\
\quad \textbf{then } \begin{cases} CurW \leftarrow w[x_n, x_1] + \sum_{i=1}^{n-1} w[x_i, x_{i+1}] \\ \textbf{if } CurW < OptW \\ \quad \textbf{then } \begin{cases} OptW \leftarrow CurW \\ OptX \leftarrow X \end{cases} \end{cases} \\
\quad \textbf{else } \begin{cases} \textbf{if } \ell = 0 \\ \quad \textbf{then } \mathcal{C}_1 \leftarrow \{1\} \\ \quad \textbf{else if } \ell = 1 \\ \quad \textbf{then } \mathcal{C}_2 \leftarrow \{2, \ldots, n\} \\ \quad \textbf{else } \mathcal{C}_{\ell+1} \leftarrow \mathcal{C}_\ell \setminus \{x_\ell\} \\ \textbf{for all } i \in \mathcal{C}_{\ell+1} \\ \quad \textbf{do } \begin{cases} x_{\ell+1} \leftarrow i \\ \text{TSP}(\ell+1, n) \end{cases} \end{cases} \\
\textbf{main} \\
\quad X = [\,] \\
\quad OptW \leftarrow \infty \\
\quad \text{TSP}(0, n)
\end{array}
$$

8.6.1 Bounding Function

Now, we improve the basic backtracking algorithm for **TSP** by incorporating a useful bounding function. Suppose we have a partial solution $X = [x_1, \ldots, x_\ell]$, where $1 \leq \ell \leq n-1$. Of course, we require that the values $x_1, \ldots, x_\ell$ are all distinct. So X corresponds to the path containing edges

$$x_1x_2, \quad x_2x_3, \quad \ldots \quad , \quad x_{\ell-1}x_\ell.$$

Let

$$Y = \{1, \ldots, n\} \setminus \{x_1, \ldots, x_\ell\}.$$

We still need to choose $x_{\ell+1}, \ldots, x_n$ so that

$$\{x_{\ell+1}, \ldots, x_n\} = Y.$$

Although we have an undirected graph, it is useful to think of a Hamiltonian cycle as consisting of *directed edges* $x_1x_2, \ldots, x_{n-1}x_n, x_nx_1$. The extension of X to a complete solution will require $n - \ell + 1$ additional directed edges:

$$x_\ell x_{\ell+1}, \quad \ldots \quad , \quad x_{n-1}x_n, \quad x_nx_1. \tag{8.13}$$

The tails of these $n - \ell + 1$ directed edges are the vertices in

$$X_{\mathcal{T}} = Y \cup \{x_\ell\}$$

and the heads are the vertices in

$$X_{\mathcal{H}} = Y \cup \{x_1\}.$$

Let's look at the weight matrix w. Suppose we only retain the rows of w indexed by $X_{\mathcal{T}}$ and the columns indexed by $X_{\mathcal{H}}$; call this $n - \ell + 1$ by $n - \ell + 1$ matrix w_X. Then the costs of the as-yet-unchosen edges in (8.13) form a ***transversal*** of the matrix w_X (i.e., we have $n - \ell + 1$ entries, where there is one entry in each row and one entry in each column).

TSP-Optimization is a minimization problem. Therefore, for purposes of a bounding function, we require a *lower bound* on the sum of the entries in a transversal of the matrix w_X. We could get a lower bound by summing the smallest entry in every row, or by summing the smallest entry in every column. However, it is better to derive a lower bound that depends on the distribution of elements in both the rows and the columns. We describe a process that is sometimes referred to as ***reducing*** the matrix.

Suppose we compute the following quantities:

$$r_a = \min\{w[a, b] : b \in X_{\mathcal{T}}\}, \tag{8.14}$$

for $a \in X_{\mathcal{H}}$; and

$$c_b = \min\{w[a, b] - r_a : a \in X_{\mathcal{H}}\}, \tag{8.15}$$

for $b \in X_{\mathcal{T}}$.

LEMMA 8.1 *The sum of the entries in any transversal of w_X is at least*

$$\sum_{a \in X_{\mathcal{H}}} r_a + \sum_{b \in X_{\mathcal{T}}} c_b. \tag{8.16}$$

PROOF Suppose the transversal consists of the cells

$$(a_1, b_1), \dots, (a_{n-\ell+1}, b_{n-\ell+1}),$$

where

$$\{a_1, \dots, a_{n-\ell+1}\} = X_{\mathcal{T}}$$

and

$$\{b_1, \dots, b_{n-\ell+1}\} = X_{\mathcal{H}}.$$

For $1 \leq i \leq n - \ell + 1$, we have

$$c_{b_i} \leq w[a_i, b_i] - r_{a_i}$$

from (8.15), so

$$r_{a_i} + c_{b_i} \leq w[a_i, b_i] \tag{8.17}$$

Summing (8.17) over i, we have

$$\sum_{i=1}^{n-\ell+1} \left(r_{a_i} + c_{b_i}\right) \leq \sum_{i=1}^{n-\ell+1} w[a_i, b_i].$$

Since the $n - \ell + 1$ cells form a transversal, we have

$$\sum_{a \in X_{\mathcal{H}}} r_a + \sum_{b \in X_{\mathcal{T}}} c_b \leq \sum_{i=1}^{n-\ell+1} w[a_i, b_i],$$

as desired. ∎

We have one additional observation before defining our bounding function, namely, that we cannot use the edge $x_\ell x_1$ in the completion of the Hamiltonian cycle, even though $x_\ell \in X_{\mathcal{T}}$ and $x_1 \in X_{\mathcal{H}}$. This is because inclusion of this edge would close the cycle prematurely, before all the vertices have been visited (since $\ell \leq n-1$). Therefore, we modify the weight matrix by defining $w[x_\ell, x_1] = \infty$ before computing (8.16). (After computing the bounding function, we should restore $w[x_\ell, x_1]$ to its previous value.)

We therefore define our bounding function, computed on the modified weight matrix, to be

$$\mathsf{B}(X) = \sum_{i=1}^{\ell-1} w[x_i, x_{i+1}] + \sum_{a \in X_{\mathcal{H}}} r_a + \sum_{b \in X_{\mathcal{T}}} c_b. \tag{8.18}$$

The first sum accounts for the edges that have already been chosen. The latter two sums are computed as in (8.16); they provide a lower bound on the sum of the weights of the remaining edges that have not yet been defined.

Example 8.4 Suppose we have the complete graph K_5 with weight matrix

$$w = \begin{pmatrix} \infty & 3 & 4 & 6 & 1 \\ 3 & \infty & 5 & 7 & 2 \\ 4 & 5 & \infty & 8 & 4 \\ 6 & 7 & 8 & \infty & 6 \\ 1 & 2 & 4 & 6 & \infty \end{pmatrix}.$$

Suppose $\ell = 3$ and $X = [1, 2, 5]$. First we set $w[5, 1] = \infty$. Then we compute

$$X_{\mathcal{T}} = \{3, 4, 5\}$$

and

$$X_{\mathcal{H}} = \{1, 3, 4\}.$$

Retaining the relevant rows and columns of the modified weight matrix w, we have

$$w_X = \begin{pmatrix} 4 & \infty & 8 \\ 6 & 8 & \infty \\ \infty & 4 & 6 \end{pmatrix}.$$

Then $r_3 = 4$, $r_4 = 6$ and $r_5 = 4$, $c_1 = 0$, $c_3 = 0$ and $c_4 = 2$, so the bound computed by (8.16) is equal to 16. (The minimum sum of the entries in a transversal of w_X is equal to 18, so the bound of Lemma 8.1 is not tight in this instance.)

Algorithm 8.15: TSPBOUND(ℓ, n, X)

external w
comment: we assume $1 \leq \ell \leq n$
if $\ell < n$
  then $\begin{cases} v \leftarrow w[x_\ell, x_1] \\ w[x_\ell, x_1] \leftarrow \infty \end{cases}$
$Y \leftarrow \{1, \ldots, n\} \setminus \{x_1, \ldots, x_\ell\}$
$X_{\mathcal{T}} \leftarrow Y \cup \{x_\ell\}$
$X_{\mathcal{H}} \leftarrow Y \cup \{x_1\}$
bound $\leftarrow 0$
for $i \leftarrow 1$ **to** $\ell - 1$
  do *bound* $\leftarrow$ *bound* $+ w[x_i, x_{i+1}]$
for $i \in X_{\mathcal{T}}$
  do {
    min $\leftarrow \infty$
    for $j \in X_{\mathcal{H}}$
      do { **if** $w[i,j] <$ *min*
        then *min* $\leftarrow w[i,j]$ }
    bound $\leftarrow$ *bound* $+$ *min*
    $R_i \leftarrow$ *min* }
for $j \in X_{\mathcal{H}}$
  do {
    min $\leftarrow \infty$
    for $i \in X_{\mathcal{T}}$
      do { **if** $w[i,j] - R_i <$ *min*
        then *min* $\leftarrow w[i,j] - R_i$ }
    bound $\leftarrow$ *bound* $+$ *min* }
if $\ell < n$
  then $w[x_\ell, x_1] \leftarrow v$

Using (8.18), the bounding function is computed as

$$\begin{aligned} \mathsf{B}(X) &= w[1,2] + w[2,5] + 16 \\ &= 3 + 2 + 16 \\ &= 21. \end{aligned}$$

□

As a point of interest, let us examine what happens if we extend the definition of the bounding function to include the case where $\ell = n$. In this case, we would have $X_{\mathcal{T}} = \{x_\ell\}$ and $X_{\mathcal{H}} = \{x_1\}$. Here we must include the edge $x_\ell x_1$ to complete the Hamiltonian cycle. The value of the bounding function will then be equal to the weight of the Hamiltonian cycle $x_1, \ldots, x_n, x_1$. Thus, $\mathsf{B}(X) = \mathsf{P}(X)$ when $\ell = n$.

Algorithm 8.15 computes the bounding function (8.18); the computations are

Algorithm 8.16: TSP2(ℓ, n)

```
external TSPBOUND
global X, OptX, OptW, w, C_1, ..., C_n
if ℓ = n
  then { CurW ← w[x_n, x_1] + Σ_{i=1}^{n-1} w[x_i, x_{i+1}]
       { if CurW < OptW
       {   then { OptW ← CurW
       {        { OptX ← X
else if ℓ = 0
  then { C_1 ← {1}
       { TSP2(ℓ + 1, n)
  else { if ℓ = 1
       {   then C_2 ← {2, ..., n}
       {   else C_{ℓ+1} ← C_ℓ \ {x_ℓ}
       { bound ← TSPBOUND(ℓ, n, X)
       { for all i ∈ C_{ℓ+1}
       {   do { if bound ≥ OptW
       {      {   then return ()
       {      {   else { x_{ℓ+1} ← i
       {      {        { TSP2(ℓ + 1, n)
main
 X = []
 OptW ← ∞
 TSP2(0, n)
```

straightforward. This algorithm assumes that $\ell \geq 1$. Note that when $\ell = 1$, we have not yet chosen any edges in the Hamiltonian cycle. We have just chosen a starting point (which we have assumed is $x_1 = 1$). In this case, we have $X_{\mathcal{T}} = X_{\mathcal{H}} = \{1, \ldots, n\}$ and the value of the bounding function is obtained by reducing the entire weight matrix w. Thus, computing the bounding function when $\ell = 1$ results in a lower bound on the weight of *any* Hamiltonian cycle.

We incorporate the bounding function TSPBOUND into Algorithm 8.14 in the usual way, yielding Algorithm 8.16. Note that we do not evaluate the bounding function when $\ell = 0$. We just proceed to make a single recursive call with $x_1 = 1$. It is also important to observe that we continue to iterate through $\mathcal{C}_{\ell+1}$ so long as $bound < OptW$. This is because **TSP-Optimization** is a minimization problem and therefore we stop examining elements in $\mathcal{C}_{\ell+1}$ if $bound \geq OptW$.

Example 8.5 We execute Algorithms 8.14 and 8.16 on the graph from Example 8.4. An optimal Hamiltonian cycle is $1, 3, 4, 2, 5, 1$, which has weight 22. Even for a small graph such as this one, the use of a bounding function significantly reduces the size of the state space tree, from 66 to 17 nodes.

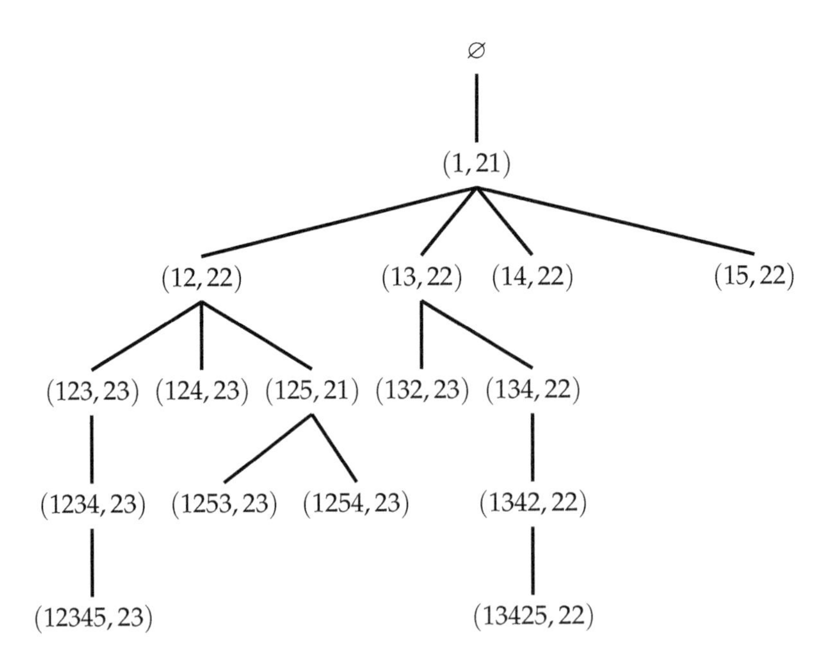

FIGURE 8.8: State space tree for a TSP instance with a bounding function.

The state space tree for Algorithm 8.14 generates a total of $4! = 24$ permutations (namely, all permutations of $\{1, \dots, 5\}$ that start with 1). There is one node at level 0; one node at level 1; four nodes at level 2; 12 nodes at level 3; 24 nodes at level 4; and 24 nodes at level 5.

The state space tree for Algorithm 8.16 is shown in Figure 8.8. In this example, we examine the vertices in each choice set in increasing order. To save space, we just indicate the assigned co-ordinates of X at each node of the tree. We also indicate the value of the bounding function at each node. (When $\ell = 5$, the value of the bounding function is the cost attained at that node.) So, for example, the bounding function at the node corresponding to the partial solution $X = [1, 2, 3, -, -]$ has the value 23. This node is therefore labeled by $(123, 23)$.

It is not difficult to see how the algorithm executes step-by-step by looking at this tree. The first feasible solution that is generated is $X = [1, 2, 3, 4, 5]$, which has weight 23. This is sufficient to prune the node $[1, 2, 4]$ since the bounding function has the value 23 at that node. However, at node $[1, 2, 5]$, the bounding function takes on the value 21, so further exploration is necessary.

A bit later in the search, the solution $X = [1, 3, 4, 2, 5]$, which has weight 22, is generated. This enables all subsequent nodes to be pruned. $X = [1, 3, 4, 2, 5]$ is the optimal solution produced by this algorithm. □

8.7 Branch-and-bound

We continue using the same notation as before. As usual, the current (partial or complete) solution X to a given problem is an ℓ-tuple, denoted as $X = [x_1, \ldots, x_\ell]$. So far, we have evaluated the bounding function, $\mathsf{B}(X)$, and compared its value to the current optimal solution before making any recursive calls for the possible values $x_{\ell+1} \in \mathcal{C}_{\ell+1}$. The recursive calls are made as long as the value of the bounding function is greater than (in the case of a maximization problem) or less than (in the case of a minimization problem) the current optimal solution.

We have made these recursive calls in some arbitrary order (e.g., in increasing order of $x_{\ell+1}$ in the backtracking algorithm for **TSP-Optimization**). However, it might be a better strategy to make the relevant recursive calls in a more intelligent order. Intuitively, it seems that we would be more likely to find the optimal solution in a branch of the state space tree where there is a larger "gap" between the value of the bounding function and the current optimal solution (of course, there is no guarantee that this will in fact happen).

Branch-and-bound incorporates the following three steps, which are carried out for any $X = [x_1, \ldots, x_\ell]$:

1. Evaluate the bounding function

$$\mathsf{B}([x_1, \ldots, x_\ell, x_{\ell+1}])$$

for all possible choices $x_{\ell+1} \in \mathcal{C}_{\ell+1}$.

2. Sort the list of elements $x_{\ell+1} \in \mathcal{C}_{\ell+1}$ according to the values of the bounding functions just computed in step 1. For a maximization problem, these values should be sorted in decreasing order; for a minimization problem, these values should be sorted in increasing order.

3. Make the recursive calls according to the sorted order as computed in step 2.

Algorithm 8.17 incorporates branch-and-bound into our backtracking algorithm for **TSP-Optimization**. The main changes from Algorithm 8.16 are indicated in red. Specifically, after we compute the choice set, we form a list $\mathcal{N}$ of ordered pairs. Each ordered pair has the form $(i, bound)$, where $i \in \mathcal{C}_{\ell+1}$ and *bound* is the value of the bounding function obtained by setting $x_{\ell+1} = i$. Then we sort the list $\mathcal{N}$ in increasing order of the second co-ordinates of the ordered pairs in $\mathcal{N}$. Finally, we process the choice set in this particular order. That is, we use the values of the bounding function to determine the order in which the recursive calls will be made.

Example 8.6 Branch-and-bound (Algorithm 8.17) can result in a considerably smaller state space tree as compared to Algorithm 8.16. For the following weight matrix for the complete graph K_{10}, Algorithm 8.17 resulted in a state space tree

Algorithm 8.17: TSP3(ℓ, n)

```
external TSPBOUND
global X, OptX, OptW, w, C_1, ..., C_n
if ℓ = n
  then { CurW ← w[x_n, x_1] + Σ_{i=1}^{n-1} w[x_i, x_{i+1}]
       { if CurW < OptW
       {   then { OptW ← CurW
       {        { OptX ← X
else if ℓ = 0
  then { C_1 ← {1}
       { TSP3(ℓ + 1, n)
  else { if ℓ = 1
       {   then C_2 ← {2, ..., n}
       {   else C_{ℓ+1} ← C_ℓ \ {x_ℓ}
       { N ← [(i, TSPBOUND(ℓ + 1, n, [x_1, ..., x_ℓ, i])) : i ∈ C_{ℓ+1}]
       { sort N in increasing order by second co-ordinates
       { for all (i, bound) ∈ N
       {   do { if bound ≥ OptW
       {      {   then return ()
       {      {   else { x_{ℓ+1} ← i
       {      {        { TSP3(ℓ + 1, n)
main
  X = []
  OptW ← ∞
  TSP3(0, n)
```

having 128 nodes, as compared to 1874 nodes in the state space tree when Algorithm 8.16 is used.

$$w = \begin{pmatrix} \infty & 864 & 524 & 338 & 839 & 576 & 600 & 76 & 394 & 974 \\ 864 & \infty & 553 & 645 & 576 & 937 & 758 & 317 & 733 & 705 \\ 524 & 553 & \infty & 111 & 6 & 909 & 478 & 41 & 50 & 852 \\ 338 & 645 & 111 & \infty & 624 & 507 & 137 & 372 & 926 & 884 \\ 839 & 576 & 6 & 624 & \infty & 60 & 419 & 557 & 841 & 250 \\ 576 & 937 & 909 & 507 & 60 & \infty & 276 & 639 & 907 & 77 \\ 600 & 758 & 478 & 137 & 419 & 276 & \infty & 736 & 620 & 344 \\ 76 & 317 & 41 & 372 & 557 & 639 & 736 & \infty & 897 & 106 \\ 394 & 733 & 50 & 926 & 841 & 907 & 620 & 897 & \infty & 504 \\ 974 & 705 & 852 & 884 & 250 & 77 & 344 & 106 & 504 & \infty \end{pmatrix}.$$

To compare, Algorithm 8.14 is an exhaustive search that would give rise to a state space tree having 986411 nodes. □

We ran Algorithms 8.16 and 8.17 on complete graphs of various sizes and some data is presented in Table 8.2. Ten complete graphs K_n of each size n were tested, where each edge had a random weight between 1 and 1000. There is certainly a very wide variation among the different samples of each size. The branch-and-bound strategy generally results in a significantly smaller state space tree. However, when comparing Algorithms 8.16 and 8.17, we should also recognize that Algorithm 8.17 incurs more overhead in the sorting that is constantly being done.

Both strategies improve greatly on an exhaustive search (Algorithm 8.14). The number of nodes in a state space tree using an exhaustive search for the complete graph K_{15} is 236975164806, and the number of nodes for K_{20} is 330665665962404001.

TABLE 8.2: Size of state space trees for Algorithms 8.16 and 8.17.

n	Algorithm 8.16			Algorithm 8.17		
	min.	max.	avg.	min.	max.	avg.
10	666	2672	1314	37	305	144
15	25930	282715	78452	858	18937	5124
20	243650	5160637	1574524	3975	250046	57327

8.8 Notes and References

In this chapter, we have followed the basic treatment of backtrack algorithms given by Kreher and Stinson [44].

Backtracking algorithms have existed since the 19th century (if not earlier), but the term "backtracking" is generally attributed to D. Lehmer in the 1950s. Two relatively early surveys on backtracking algorithms are Bitner and Reingold [8] and by Golomb and Baumert [26]. The 1965 paper by Golomb and Baumert [26] discusses the n-**queens** problem, among others.

The branch-and-bound technique seems to have been developed in the early 1960s. The 1960 paper by Land and Doig [47] is one of the earliest to describe this technique. The first paper to explicitly use the term "branch-and-bound" is the 1963 paper by Little, Murty, Sweeney and Karel [48]. This paper describes branch-and-bound in the context of the **TSP** in detail. It also presents the bounding function that we described in Section 8.6.1, which is based on reducing a matrix.

We discussed generating permutations using backtracking. There are, in fact, a large number of efficient algorithms to generate combinatorial structures such as permutations, combinations, etc. See Kreher and Stinson [44], for example, for more information.

Exercises

8.1 (a) Suppose $1 \le k \le n$. Design a basic backtrack algorithm to compute all the k-subsets of $\{1,\ldots,n\}$. You can assume that a k-subset is represented as $X = [x_1,\ldots,x_k]$, where $1 \le x_1 < \cdots < x_k \le n$. You can make use of the fact that

$$\mathcal{C}_{\ell+1} = \{x_\ell + 1,\ldots,n\},$$

for $1 \le \ell \le n-1$.

(b) Prove that a partial solution $X = [x_1,\ldots,x_\ell]$, where $\ell < k$ and $1 \le x_1 < \cdots < x_\ell \le n$, can be completed to a k-subset of $\{1,\ldots,n\}$ only if $x_\ell \le n-k+\ell$. Then use this observation to incorporate a simple form of pruning into your algorithm.

8.2 On a chessboard, a knight is allowed to move one cell horizontally and two cells vertically, or two cells horizontally and one cell vertically. An ***open knight's tour*** on an n by n chessboard will start at any cell and make a sequence of n^2-1 moves, so that every cell on the chessboard is visited. Devise a backtracking algorithm to find all the open knight's tours on an n by n chessboard starting at a given cell. Write a program to implement your algorithm.

(a) Use your algorithm to determine the number of tours on a 5 by 5 chessboard.

(b) Modify your algorithm and use it to determine the number of tours on a 6 by 6 chessboard that start at the top left corner cell.

HINT When the knight occupies a given cell, there are at most eight cells to which it can move. However, some of these cells might have been visited already. The remaining cells would comprise the choice set.

8.3 Let n be a positive integer. An ***addition chain*** with target n is a sequence of increasing integers $x_0,\ldots,x_\ell$ that satisfies the following properties:

- $x_0 = 1$
- $x_\ell = n$
- for $1 \le k \le \ell$, we can express x_k in the form $x_k = x_i + x_j$ for indices i and j, where $0 \le i \le j \le k-1$.

The ***length*** of the addition chain is ℓ (but note that there are $\ell+1$ integers in the sequence).

For example, the following is an addition chain of length 8 with target 47:

$$1, 2, 3, 5, 10, 20, 23, 46, 47.$$

Devise a backtracking algorithm to find all the addition chains with a given target n and thus determine the addition chain with target n that has the minimum length. Write a program to implement your algorithm and run it on the values

$$n \in \{71, 127, 191, 379, 607\}.$$

HINT Your algorithm will choose integers $x_1, x_2, \ldots$ in turn. Given that $x_1, \ldots, x_k$ have been chosen, the choice set for x_{k+1} will contain all the integers of the form $x_i + x_j$ (where $0 \leq i \leq j \leq k$) that are at least $x_k + 1$ and at most n. It is useful to keep track of the shortest addition chain found as the algorithm is executed. This allows a simple form of pruning. Finally, it is also advantageous to consider the elements in the choice set in deceasing order (i.e., from largest to smallest).

8.4 Let n be a positive integer. Suppose the integers $1, \ldots, n$ are placed in a circle, with one integer in each position. We can think of this arrangement as a ***cyclic permutation*** of the set $\{1, \ldots, n\}$. Among all the n possible sums of three cyclically consecutive values, find the maximum sum, S. We wish to minimize the value of S.

For example, when $n = 6$, if we place the numbers $1, \ldots, 6$ in the cyclic order $5, 4, 1, 6, 3, 2$, then the sums are $10, 11, 10, 11, 10, 11$ and hence the maximum sum is $S = 11$.

(a) Show that the average sum is $3(n+1)/2$ and hence

$$S \geq \left\lceil \frac{3(n+1)}{2} \right\rceil.$$

(b) Given n and a target value $T \geq \lceil \frac{3(n+1)}{2} \rceil$, devise a backtracking algorithm to find all examples of cyclic permutations of $\{1, \ldots, n\}$ in which the maximum sum of three cyclically consecutive elements is at most T.

HINT Use Algorithm 8.2 as a starting point. The choice sets can be further reduced by the following observations:

- without loss of generality, $x_1 = 1$ (because the permutation is cyclic)
- for $2 \leq \ell \leq n-2$, it must be the case that

$$x_{\ell+1} \leq T - x_\ell - x_{\ell-1}.$$

- The last element, x_n, must satisfy the following inequality:

$$x_n \leq \min\{T - x_{n-1} - x_{n-2}, T - x_1 - x_{n-1}, T - x_2 - x_1\}.$$

(c) Write a program to implement your algorithm and run it with the following (n, T)-pairs: $n = 12$ and $T = 21$; $n = 13$ and $T = 23$; $n = 14$ and $T = 24$; and $n = 15$ and $T = 25$.

8.5 Let k and v be positive integers. A (v,k)-***difference packing*** is a set of k distinct elements of $\mathbb{Z}_v$, say $D = \{x_1, \ldots, x_k\}$, such that the differences $(x_i - x_j) \bmod v$ (for $i \neq j$) are all distinct. For example, $\{0,1,3,9\}$ is a $(13,4)$-difference packing. A necessary condition for the existence of (v,k)-difference packing is that $v \geq k^2 - k + 1$. A difference packing is also known as a ***modular Golomb ruler***.

(a) Design a backtracking algorithm to enumerate all (v,k)-difference packings for specified values of v and k.

HINT It is perhaps not so convenient to construct choice sets here, so we might just consider all the possible extensions one at a time. We can suppose without loss of generality that

$$0 = x_1 < \cdots < x_k \leq v - 1.$$

It will also be useful to keep track of all the differences that occur in a given partial solution $[x_1, \ldots, x_\ell]$. Say this set is denoted by $\mathcal{D}$. Then, when we choose a value for $x_{\ell+1}$, we need to check that

- $x_\ell < x_{\ell+1} \leq v$, and
- the 2ℓ "new differences," namely, $\pm(x_{\ell+1} - x_i) \bmod v$ (for $1 \leq i \leq \ell$), are all distinct, and none of them appear in $\mathcal{D}$.

(b) Use your algorithm to find $(21,5)$-, $(31,6)$- and $(48,7)$-difference packings.

8.6 Find the optimal solution to the instance of **TSP** given by the following weight matrix that is defined on the complete graph K_{10}.

$$\begin{pmatrix}
\infty & 864 & 524 & 338 & 839 & 576 & 600 & 76 & 394 & 974 \\
864 & \infty & 553 & 645 & 576 & 937 & 758 & 317 & 733 & 705 \\
524 & 553 & \infty & 111 & 6 & 909 & 478 & 41 & 50 & 852 \\
338 & 645 & 111 & \infty & 624 & 507 & 137 & 372 & 926 & 884 \\
839 & 576 & 6 & 624 & \infty & 60 & 419 & 557 & 841 & 250 \\
576 & 937 & 909 & 507 & 60 & \infty & 276 & 639 & 907 & 77 \\
600 & 758 & 478 & 137 & 419 & 276 & \infty & 736 & 620 & 344 \\
76 & 317 & 41 & 372 & 557 & 639 & 736 & \infty & 897 & 106 \\
394 & 733 & 50 & 926 & 841 & 907 & 620 & 897 & \infty & 504 \\
974 & 705 & 852 & 884 & 250 & 77 & 344 & 106 & 504 & \infty
\end{pmatrix}$$

Chapter 9

Intractability and Undecidability

In this chapter, we present the theory of NP-completeness, which is an attempt to categorize decision problems as tractable (i.e., solvable in polynomial time) or intractable. We define complexity classes, including **P**, **NP** and **NPC**, which are abbreviations (respectively) for polynomial, non-deterministic polynomial, and NP-complete. We give proofs that a number of familiar problems are NP-complete. We also extend the analysis of intractability to include NP-hard problems, which are not required to be decision problems. Approximation algorithms are considered as a way to deal with NP-hard problems in practice. Finally, we briefly address the concept of undecidable problems, which are problems that cannot be solved by any algorithm.

9.1 Decision Problems and the Complexity Class P

One of the main goals of computer science is to determine which problems can be solved in polynomial time. It turns out that there is a large class of problems (the NP-complete problems), which are equivalent in the sense that either they all can be solved in polynomial time, or none of them can be solved in polynomial time. It is widely believed, though not proven, that none of the NP-complete problems can be solved in polynomial time. The NP-complete problems include many familiar problems of practical interest.

In this chapter, we develop the theory of NP-completeness, which is based on decision problems. A ***decision problem*** is any problem in which the solution is an answer of the form *yes* or *no*. That is, given a problem instance I, there is a question that is to be answered *yes* or *no*. Thus we can classify every problem instance I of a decision problem as being a ***yes-instance*** (if the correct answer is *yes*) or a ***no-instance*** (if the correct answer is *no*.).

An algorithm for a decision problem is not required to specify a "solution" to a problem. It just has to reply *yes* or *no* correctly. Therefore, an algorithm A is said to ***solve*** a decision problem Π provided that A finds the correct answer (*yes* or *no*) for every instance I of Π in finite time.

As usual, an algorithm A for a decision problem Π is said to be a ***polynomial-time algorithm*** provided that the complexity of A is $O(n^k)$, where k is a positive integer and $n = \mathsf{size}(I)$. The complexity class **P** denotes the set of all decision

DOI: 10.1201/9780429277412-9

problems that have polynomial-time algorithms solving them. We write $\Pi \in \mathbf{P}$ if the decision problem Π is in the complexity class **P**.

Note that, if a problem is not a decision problem, it cannot be in the class **P**. Thus, **Sorting** is *not* in **P**, etc.

To further expand on these concepts, we examine Problems 9.1, 9.2 and 9.3, which are three graph-theoretic decision problems involving cycles and circuits in graphs.

Problem 9.1: Cycle

Instance: An undirected graph $G = (V, E)$.

Question: Does G contain a cycle?

Problem 9.2: Hamiltonian Cycle

Instance: An undirected graph $G = (V, E)$.

Question: Does G contain a Hamiltonian cycle? (Recall that a ***Hamiltonian cycle*** is a cycle that passes through every vertex in V exactly once.)

Problem 9.3: Eulerian Circuit

Instance: An undirected graph $G = (V, E)$.

Question: Does G contain an Eulerian circuit? (Recall that an ***Eulerian circuit*** is a circuit that passes through every edge in E exactly once.)

Since these three problems are all phrased as decision problems, we are not required to actually output a cycle (or circuit) in the case where the desired structure exists.

First, we observe that **Cycle** $\in$ **P**. For example, we can use a breadth-first or depth-first search to determine if a graph contains a cycle. If we encounter a non-tree edge during the search, we can immediately answer *yes* and quit. If we complete the search without encountering a non-tree edge, then the correct answer is *no*. Note, that using this approach, we do not actually identify the edges in a cycle in the graph, but we know that a cycle must exist.

It is also clear that **Eulerian Circuit** $\in$ **P**. Theorem 7.17 (Euler's Theorem) asserts that a connected graph has an Eulerian circuit if and only if every vertex has even degree. Thus, it suffices to first check if the graph is connected, and then, if it is, compute the degrees of all the vertices. If any vertex in a connected graph has odd degree, then we give the answer *no*, and if all vertices in a connected graph have even degree, we output *yes*. Note that this approach determines whether or not a graph has an Eulerian circuit without actually finding one.

On the other hand, **Hamiltonian Cycle** is one of the so-called NP-complete problems, which will be introduced in Section 9.5. For now, we just mention that there is, at the present time, no known polynomial-time algorithm to solve this problem (or any NP-complete problem, for that matter).

We next state two knapsack problems, Problem 9.4 and 9.5, which are phrased as decision problems.

Problem 9.4: 0-1 Knapsack-Dec

Instance: A list of ***profits***, $P = [p_1, \ldots, p_n]$; a list of ***weights***, $W = [w_1, \ldots, w_n]$; a ***capacity***, M; and a ***target profit***, T.

Question: Is there an n-tuple $[x_1, x_2, \ldots, x_n] \in \{0,1\}^n$ such that

$$\sum w_i x_i \leq M \quad \text{and} \quad \sum p_i x_i \geq T?$$

Problem 9.5: Rational Knapsack-Dec

Instance: A list of ***profits***, $P = [p_1, \ldots, p_n]$; a list of ***weights***, $W = [w_1, \ldots, w_n]$; a ***capacity***, M; and a ***target profit***, T.

Question: Is there an n-tuple $[x_1, x_2, \ldots, x_n] \in [0,1]^n$ such that

$$\sum w_i x_i \leq M \quad \text{and} \quad \sum p_i x_i \geq T?$$

Recall that we introduced optimization versions of knapsack problems in Section 5.7. We have "converted" the optimization problems to decision problems by specifying a target profit as part of the problem instance and then asking if there is a feasible solution that attains the specified target profit. This is a standard method of converting an optimization problem to a decision problem.

It is not hard to see that **Rational Knapsack-Dec** $\in$ **P**. We have already described an efficient greedy algorithm, namely, Algorithm 5.4, to find the optimal n-tuple, X, for an instance of (the optimization version of) **Rational Knapsack**. Suppose we use this algorithm to solve the instance consisting of P, W and M. The output is the n-tuple X that yields the optimal profit. Then, we can easily compute the profit T^* attained by X. Finally, we compare T^* to T. This leads to a polynomial-time algorithm solving Problem 9.4.

On the other hand, we will show in Section 9.6 that **0-1 Knapsack-Dec** is NP-complete.

9.2 Polynomial-time Turing Reductions

The theory of NP-completeness involves reductions in an essential way. Recall that reductions were introduced in Section 2.3. The idea of a Turing reduction is to use an algorithm designed to solve one problem in order to solve a different problem. We review the basics of Turing reductions now.

Suppose Π_1 and Π_2 are problems (not necessarily decision problems). A (hypothetical) algorithm O_2 to solve Π_2 is called an ***oracle*** for Π_2. Suppose that A_1 is an algorithm that solves Π_1, assuming the existence of an oracle O_2 for Π_2. (O_2 is used as a subroutine within the algorithm A_1.) Then we say that A_1 is a ***Turing reduction*** from Π_1 to Π_2, which we denote using the notation $\Pi_1 \leq^T \Pi_2$.

In this chapter, we are mainly concerned with polynomial-time algorithms and reductions. Thus we define polynomial-time Turing reductions. A Turing reduction A_1 is a ***polynomial-time Turing reduction*** if the running time of A_1 is a polynomial-time algorithm, under the assumption that the oracle O_2 has *unit cost* running time. The "unit cost" assumption means that we are just accounting for

1. the amount of work done in the reduction "outside" of the oracle, and

2. the number of calls to the oracle.

If there is a polynomial-time Turing reduction from Π_1 to Π_2, we write $\Pi_1 \leq_P^T \Pi_2$.

If there is a polynomial-time reduction A_1 from Π_1 to Π_2 and there is a polynomial-time algorithm A_2 to solve Π_2, then there is a polynomial-time algorithm to solve Π_1. We just replace the oracle O_2 by the "real" algorithm A_2. Thus, the existence of a polynomial-time Turing reduction means that, if we can solve Π_2 in polynomial time, then we can also solve Π_1 in polynomial time. Another way to think of a polynomial-time Turing reduction is that it is an *implication* establishing the relative difficulty of solving two given problems.

We defined optimization problems in Section 5.1. We have also defined the notion of a decision problem in Section 9.1. In addition, it is sometimes useful to consider ***optimal value problems*** that are associated with optimization problems. We illustrate by defining three variants of the **Traveling Salesperson** problem as Problems 9.6, 9.7 and 9.8. Note that the optimization version of this problem was previously defined as Problem 8.5; we restate it here for convenience.

Problem 9.6: TSP-Optimization

Instance: A graph G and edge weights $w : E \rightarrow \mathbb{Z}^+$.

Find: A Hamiltonian cycle H in G such that

$$w(H) = \sum_{e \in H} w(e)$$

is minimized.

Problem 9.7: TSP-Optimal Value

Instance: A graph G and edge weights $w : E \rightarrow \mathbb{Z}^+$.

Find: The minimum integer T such that there exists a Hamiltonian cycle H in G with $w(H) = T$.

Problem 9.8: TSP-Dec

Instance: A graph G, edge weights $w : E \rightarrow \mathbb{Z}^+$, and a target T.
Question: Does there exist a Hamiltonian cycle H in G with $w(H) \leq T$?

Note that the three problems defined above are asking for different types of outputs:

- The output of an algorithm that solves **TSP-Optimization** is a Hamiltonian cycle.
- The output of an algorithm that solves **TSP-Optimal Value** is a positive integer.
- The output of an algorithm that solves **TSP-Dec** is *yes* or *no*.

We will use polynomial-time Turing reductions to show that the three different versions of the **Traveling Salesperson** problem are ***polynomially equivalent***: if one of them can be solved in polynomial time, then all of them can be solved in polynomial time. (However, we should emphasize that it is generally believed that *none* of them can be solved in polynomial time.)

There are six possible reductions that can be considered between these three problems. Three of them are rather trivial and the other three are proven using some perhaps not quite so obvious techniques.

The three "trivial" reductions are

- **TSP-Dec** $\leq_P^T$ **TSP-Optimal Value**,
- **TSP-Optimal Value** $\leq_P^T$ **TSP-Optimization**, and
- **TSP-Dec** $\leq_P^T$ **TSP-Optimization**.

We leave these as exercises for the reader and proceed to the more interesting reductions.

First, we show that **TSP-Optimal Value** $\leq_P^T$ **TSP-Dec**. We assume the existence of an oracle TSP-DEC-SOLVER that solves **TSP-Dec**. Given an instance of **TSP-Dec**, this oracle simple returns a yes/no answer.

The reduction uses a binary search technique. To get the process started, we need to determine an initial finite interval in which the optimal value must occur. The trivial lower bound is 0 (because all edge weights are positive) and the trivial upper bound is the sum of all the edge weights. These upper and lower bounds on the optimal value are denoted by *hi* and *lo*, respectively. Then we perform a binary search of the interval $[lo, hi]$ to find the exact optimal value. In each iteration of the binary search, we call the oracle TSP-DEC-SOLVER once. The response of the oracle allows us to restrict our attention to either the upper half or the lower half of the interval.

Algorithm 9.1: TSP-OPTIMALVALUE-SOLVER(G, w)

```
external TSP-DEC-SOLVER
hi ← Σ_{e∈E} w(e)
lo ← 0
if not TSP-DEC-SOLVER(G, w, hi)
  then return (∞)
while hi > lo
  do { mid ← ⌊(hi+lo)/2⌋
       if TSP-DEC-SOLVER(G, w, mid)
         then hi ← mid
         else lo ← mid + 1
return (hi)
```

The complexity of the binary search is logarithmic in the length of the interval being searched, i.e., a polynomial function of

$$O\left(\log_2 \sum_{e \in E} w(e)\right).$$

The size of the problem instance is

$$\mathsf{size}(I) = n^2 + \sum_{e \in E} \log_2(w(e)+1).$$

The first term accounts for an n by n adjacency matrix, and the second term accounts for the binary representations of all the edge weights. We observe that

$$\begin{aligned} \sum_{e \in E} \log_2(w(e)+1) &= \log_2 \left(\prod_{e \in E}(w(e)+1)\right) \\ &\geq \log_2 \sum_{e \in E} w(e). \end{aligned}$$

It follows that the complexity of the binary search is $O(\mathsf{size}(I))$. On the other hand, note that a linear search would *not* yield a polynomial-time reduction.

We next show that **TSP-Optimization** $\leq_P^T$ **TSP-Dec**. The reduction is given in Algorithm 9.2.

We begin with an intuitive explanation of this reduction. In Algorithm 9.2, setting $w_0[e] = \infty$ has the effect of deleting e from G (in practice, we could use some large number instead of ∞). We are deleting each edge e in turn, one after the other, checking to see if a minimum-weight Hamiltonian cycle still exists when an edge E is deleted. If the answer is "no," then we restore the edge e and include e in the minimum-weight Hamiltonian cycle H.

We give a more detailed proof of correctness for Algorithm 9.2. At the termination of the algorithm, it is clear that H contains a Hamiltonian cycle of minimum

Algorithm 9.2: TSP-OPTIMIZATION-SOLVER($G = (V, E), w$)

external TSP-OPTIMALVALUE-SOLVER, TSP-DEC-SOLVER
$T^* \leftarrow$ TSP-OPTIMALVALUE-SOLVER(G, w)
if $T^* = \infty$
  then return ("no Hamiltonian cycle exists")
$w_0 \leftarrow w$
$H \leftarrow \varnothing$
for all $e \in E$
  do $\begin{cases} w_0[e] \leftarrow \infty \\ \textbf{if not } \text{TSP-DEC-SOLVER}(G, w_0, T^*) \\ \quad \textbf{then } \begin{cases} w_0[e] \leftarrow w[e] \\ H \leftarrow H \cup \{e\} \end{cases} \end{cases}$
return (H)

weight T^*, where H consists of the edges that are not deleted from G. We claim that H is precisely a Hamiltonian cycle. Suppose not; then $C \cup \{e\} \subseteq H$, where C is a Hamiltonian cycle of weight T^* and $e \in G \setminus C$. Consider the particular iteration when e was added to H. Let G' denote the edges in the graph at this point in time. G' contains a Hamiltonian cycle of weight T^*, but $G' \setminus \{e\}$ does not, so e is included in H. We are assuming that

$$C \cup \{e\} \subseteq H,$$

which implies that

$$C \subseteq H \backslash \{e\}.$$

Since $H \subseteq G'$, we have

$$C \subseteq H \backslash \{e\} \subseteq G' \backslash \{e\}.$$

Therefore e would *not* have been added to H, which is a contradiction.

It is interesting to note that Algorithm 9.2 finds the "lexicographically last" minimum-weight Hamiltonian cycle (if there is more than one).

The complexity of Algorithm 9.2 is not difficult to analyze. The complexity of the **for** loop is $\Theta(m)$, which is $O(\mathsf{size}(I))$. We can avoid the use of the subroutine TSP-OPTIMALVALUE-SOLVER, by simply plugging the code for TSP-OPTIMALVALUE-SOLVER into TSP-OPTIMIZATION-SOLVER.

9.3 The Complexity Class NP

We now begin to develop the basic theory of NP-completeness. We will focus on decision problems for the time being. First, we define several important concepts.

certificate

Informally, a certificate for a yes-instance I is some "extra information" denoted by *Cert*, which makes it easy to *verify* that the instance I is a yes-instance. For many of the problems we will consider, it is straightforward to specify the structure of a certificate for a yes-instance that can be verified in polynomial time.

certificate verification algorithm

Suppose that VER is an algorithm that verifies certificates for yes-instances. Then $\text{VER}(I, \mathit{Cert})$ outputs *yes* if I is a yes-instance and *Cert* is a valid certificate for I. If $\text{VER}(I, \mathit{Cert})$ outputs *no*, then either I is a no-instance, or I is a yes-instance and *Cert* is an invalid certificate.

polynomial-time certificate verification algorithm

A certificate verification algorithm VER is a polynomial-time certificate verification algorithm if the complexity of VER is $O(n^k)$, where k is a positive integer and $n = \mathsf{size}(I)$.

solving a decision problem

A certificate verification algorithm VER is said to ***solve*** a decision problem Π provided that

- *for every* yes-instance I, *there exists* a certificate *Cert* such that $\text{VER}(I, \mathit{Cert})$ outputs *yes*.
- *for every* no-instance I and *for every* certificate *Cert*, $\text{VER}(I, \mathit{Cert})$ outputs *no*.

Beware the use of the term *solve* here! We do not claim that there is any efficient method to *find* a valid certificate for a yes-instance.

the complexity class NP

NP denotes the set of all decision problems that have polynomial-time certificate verification algorithms solving them. We write $\Pi \in \mathbf{NP}$ if the decision problem Π is in the complexity class **NP**. We note that **NP** is an abbreviation for ***nondeterministic polynomial time***.[1]

[1]There is an alternative, but equivalent, definition for **NP** that is formally phrased in terms of nondeterministic Turing machines. The certificate-based definition that we are using is more useful for our purposes, however, due to its more intuitive nature.

finding certificates vs. verifying certificates

It is *not required* to be able to *find* a certificate *Cert* for a yes-instance in polynomial time in order to say that a decision problem $\Pi \in \mathbf{NP}$. When we say that a problem $\Pi \in \mathbf{NP}$, there may or may not be a way to define certificates for no-instances that can be verified in polynomial time.

As we proceed, we should always keep the following in mind: *finding* a certificate can be much more difficult than *verifying* the correctness of a given certificate. As a rough analogy, *finding a proof* for a theorem can be much harder than *verifying the correctness of someone else's proof.*

As an example, we give a certificate verification algorithm for **Hamiltonian Cycle**, which we introduced as Problem 9.2. Let $G = (V, E)$ be a graph, where $n = |V|$. Suppose we define a certificate to consist of an n-tuple of vertices, say $X = [x_1, \ldots, x_n]$. We can think of X as a possible Hamiltonian cycle for the graph G, and the verification algorithm will check to see if X really is a Hamiltonian cycle.

Note that we have to check that

1. the certificate X contains exactly n vertices, where $n = |V|$,
2. no vertex occurs more than once in X, and
3. every pair of (cyclically) consecutive vertices in X is an edge in E.

Algorithm 9.3: HAMILTONIAN CYCLE VERIFICATION(G, X)

```
comment: X is the certificate that is being verified
flag ← true
Used ← {x_1}
j ← 2
while (j ≤ n) and flag
     ⎧ flag ← (x_j ∉ Used) and ({x_{j-1}, x_j} ∈ E)
     ⎪ if (j = n)
  do ⎨   then flag ← flag and ({x_n, x_1} ∈ E)
     ⎪ Used ← Used ∪ {x_j}
     ⎩ j ← j + 1
return (flag)
```

These verifications are not difficult to carry out; see Algorithm 9.3. This algorithm examines the vertices $x_1, x_2, \ldots, x_n$ in order. We verify that every consecutive pair of vertices in X (including $x_n x_1$) is an edge in E. Also, the set *Used* allows us to ensure that no vertex occurs more than once in X.

Suppose we represent the graph G using an adjacency matrix. Then Algorithm

9.3 takes time $O(n)$ since we can verify if any pair of vertices is an edge in $O(1)$ time. Because $\mathsf{size}(I) = n^2$, the verification algorithm is polynomial-time.

It is intuitively obvious that verifying a certificate for a yes-instance is not harder than solving a given decision problem. We have the following important result.

THEOREM 9.1 *$P \subseteq NP$.*

PROOF Suppose that A is a polynomial-time algorithm solving the decision problem Π. In our certificate verification algorithm, certificates are defined to be empty (i.e., $Cert = \varnothing$). The certificate verification algorithm simply runs A on a given instance I, ignoring any certificate that may be present. ∎

9.4 Polynomial Transformations

We now define polynomial transformations for decision problems. (Recall that we already introduced these types of reductions in Section 2.3.3.) For a decision problem Π, let $\mathcal{I}(\Pi)$ denote the set of all instances of Π. Let $\mathcal{I}_{yes}(\Pi)$ and $\mathcal{I}_{no}(\Pi)$ denote the set of all yes-instances and no-instances (respectively) of Π.

Suppose that Π_1 and Π_2 are decision problems. We say that there is a ***polynomial transformation*** from Π_1 to Π_2 (denoted $\Pi_1 \leq_P \Pi_2$) if there exists a function $f : \mathcal{I}(\Pi_1) \rightarrow \mathcal{I}(\Pi_2)$ such that the following properties are satisfied:

1. $f(I)$ is computable in polynomial time (as a function of $\mathsf{size}(I)$, for any $I \in \mathcal{I}(\Pi_1)$),
2. if $I \in \mathcal{I}_{yes}(\Pi_1)$, then $f(I) \in \mathcal{I}_{yes}(\Pi_2)$, and
3. if $I \in \mathcal{I}_{no}(\Pi_1)$, then $f(I) \in \mathcal{I}_{no}(\Pi_2)$.

Properties 2 and 3 are illustrated in Figure 9.1.

In practice, when we want to prove the implication stated in property 3, we usually prove the contrapositive statement

"If $f(I) \in \mathcal{I}_{yes}(\Pi_2)$, then $I \in \mathcal{I}_{yes}(\Pi_1)$."

Polynomial transformations are also known as ***Karp reductions*** or ***many-one reductions***. A polynomial transformation can be thought of as a (simple) special case of a polynomial-time Turing reduction, i.e., if $\Pi_1 \leq_P \Pi_2$, then $\Pi_1 \leq_P^T \Pi_2$. Given a polynomial transformation f from Π_1 to Π_2, the corresponding Turing reduction is as follows:

1. Given $I \in \mathcal{I}(\Pi_1)$, construct $f(I) \in \mathcal{I}(\Pi_2)$.
2. Run $O_2(f(I))$, where O_2 is an oracle for Π_2.

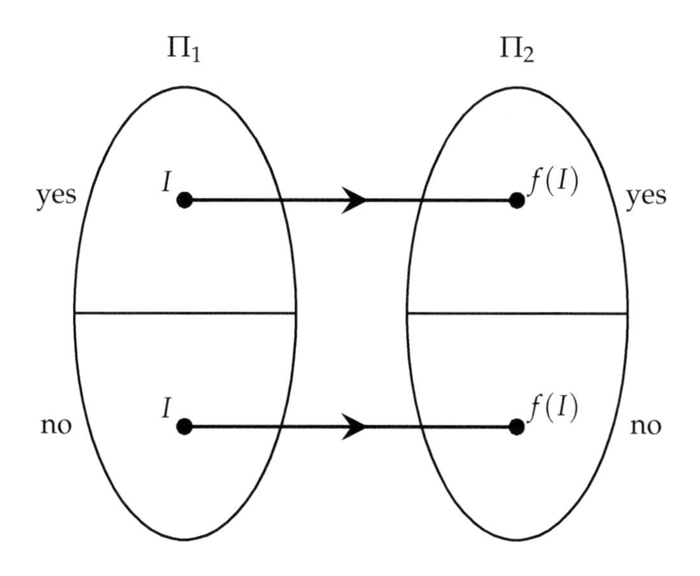

FIGURE 9.1: A polynomial transformation.

Thus, we transform the instance I to $f(I)$, and then we make a single call to the oracle O_2 with input $f(I)$.

We will see numerous examples of polynomial transformations in subsequent sections. One very important point to keep in mind is that, in general, we do not need to know whether I is a yes-instance or a no-instance of Π_1 when we transform it to an instance $f(I)$ of Π_2. The transformation is required to work "correctly," regardless of whether I is a yes-instance or a no-instance.

9.4.1 Clique $\leq_P$ Vertex-Cover

We begin by describing a simple polynomial transformation from one graph theoretic decision problem (**Clique**, which is Problem 9.9) to another graph theoretic decision problem (**Vertex Cover**, which is Problem 9.10). We will see later that both of these problems are NP-complete.

Problem 9.9: Clique

Instance: An undirected graph $G = (V, E)$ and an integer k, where $1 \leq k \leq |V|$.

Question: Does G contain a clique of size $\geq k$? (Recall that a ***clique*** is a subset of vertices $W \subseteq V$ such that $uv \in E$ for all $u, v \in W$, $u \neq v$.)

Problem 9.10: Vertex Cover

Instance: An undirected graph $G = (V, E)$ and an integer k, where $1 \leq k \leq |V|$.

Question: Does G contain a vertex cover of size $\leq k$? (A ***vertex cover*** is a subset of vertices $W \subseteq V$ such that $\{u, v\} \cap W \neq \varnothing$ for all edges $uv \in E$.)

We now show that **Clique** $\leq_P$ **Vertex Cover**. Suppose that $I = (G, k)$ is an instance of **Clique**, where

- $G = (V, E)$,
- $V = \{v_1, \ldots, v_n\}$, and
- $1 \leq k \leq n$.

Then construct an instance $f(I) = (H, \ell)$ of **Vertex Cover**, where

- $H = (V, F)$,
- $\ell = n - k$, and
- $v_i v_j \in F \Leftrightarrow v_i v_j \notin E$.

The graph H is called the ***complement*** of G, because every edge of G is a non-edge of H and every non-edge of G is an edge of H.

We will assume that G and H are represented using adjacency matrices. Then we have

$$\mathsf{size}(I) = n^2 + \log_2 k \in \Theta(n^2).$$

Computing H takes time $\Theta(n^2)$ and computing ℓ takes time $\Theta(\log n)$, so $f(I)$ can be computed in time $\Theta(\mathsf{size}(I))$, which is polynomial time.

We still need to prove that properties 2 and 3 hold. Suppose I is a yes-instance of **Clique**. Therefore there exists a set of k vertices W such that $uv \in E$ for all $u, v \in W$. Define $W' = V \setminus W$. Clearly $|W'| = n - k = \ell$. We claim that W' is a vertex cover of H. Suppose $uv \in F$ (so $uv \notin E$). If $\{u, v\} \cap W' \neq \varnothing$, we're done, so assume $u, v \notin W'$. Therefore $u, v \in W$. But $uv \notin E$, so W is not a clique. This is a contradiction and hence $f(I)$ is a yes-instance of **Vertex Cover**. So we have proven that property 2 holds.

We leave the proof that property 3 holds as an exercise for the reader.

Here is a small example illustrating how this polynomial transformation would be computed.

Example 9.1 Suppose that $I = (G, k)$, where G is the following graph and $k = 3$. The transformed instance $f(I) = (H, \ell)$, where H is depicted below and $\ell = 3$.

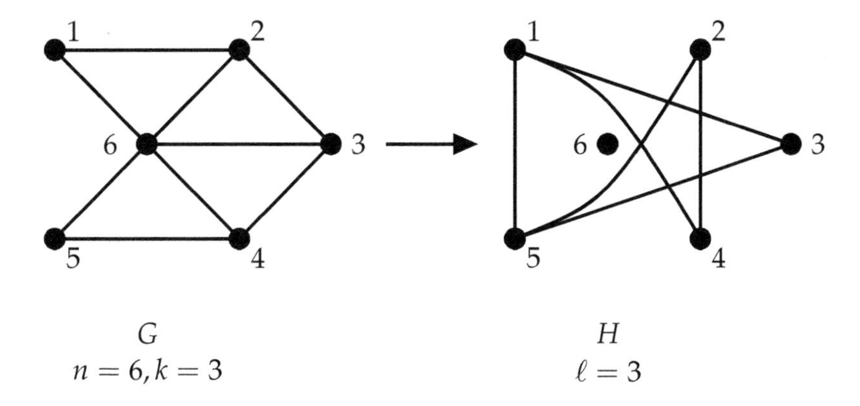

In the graph G, $W = \{2,3,6\}$ is a clique of size three. The set of vertices $W' = \{1,4,5\}$ is a vertex cover in H having size three.

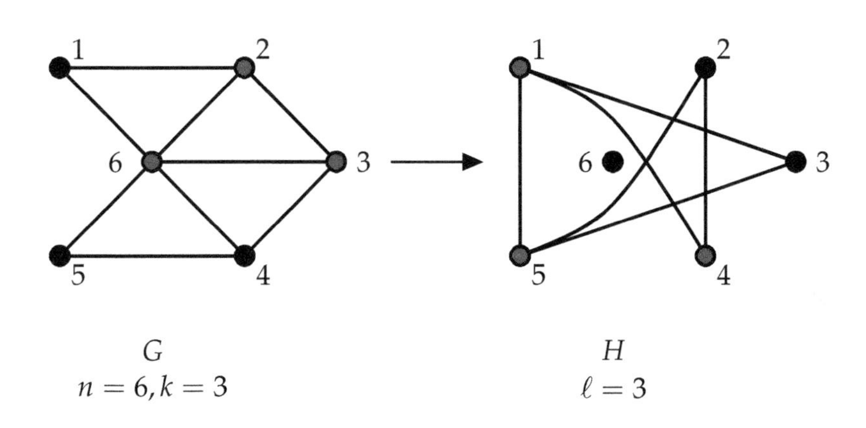

□

9.4.2 Properties of Polynomial Transformations

We now prove some fundamental properties of polynomial transformations.

THEOREM 9.2 *Suppose that the following three conditions hold:*

1. Π_1 *and* Π_2 *are decision problems,*
2. $\Pi_1 \leq_P \Pi_2$*, and*
3. $\Pi_2 \in \boldsymbol{P}$.

Then $\Pi_1 \in \boldsymbol{P}$.

PROOF Suppose A_2 is a poly-time algorithm for Π_2, having complexity $O(m^\ell)$ on an instance of size m. Suppose f is a transformation from Π_1 to Π_2 having complexity $O(n^k)$ on an instance of size n. We solve Π_1 as follows:

1. Given $I \in \mathcal{I}(\Pi_1)$, construct $f(I) \in \mathcal{I}(\Pi_2)$.
2. Run $A_2(f(I))$.

It is clear that this yields the correct answer. We need to show that these two steps can be carried out in polynomial time as a function of $n = \mathsf{size}(I)$. Step (1) can be executed in time $O(n^k)$ and it yields an instance $f(I)$ having size $m \in O(n^k)$. Step (2) takes time $O(m^\ell)$. Since $m \in O(n^k)$, the time for step (2) is $O(n^{k\ell})$, as is the total time to execute both steps. ∎

In other words, Theorem 9.2 is saying that we can solve Π_1 in polynomial time if we have a polynomial transformation from Π_1 to Π_2 and Π_2 can be solved in polynomial time.

Theorem 9.3 establishes a "transitivity" property for polynomial transformations.

THEOREM 9.3 *Suppose that Π_1, Π_2 and Π_3 are decision problems. If $\Pi_1 \leq_P \Pi_2$ and $\Pi_2 \leq_P \Pi_3$, then $\Pi_1 \leq_P \Pi_3$.*

PROOF We have a polynomial transformation f from Π_1 to Π_2, and another polynomial transformation g from Π_2 to Π_3. We define $h = f \circ g$, i.e., $h(I) = g(f(I))$ for all instances I of Π_1. (As an exercise, the reader can fill in the details of the rest of the proof.) ∎

9.5 NP-completeness

A decision problem Π is ***NP-complete*** if it satisfies the following two properties:

- $\Pi \in \mathbf{NP}$
- For all $\Pi' \in \mathbf{NP}$, $\Pi' \leq_P \Pi$.

That is, a particular problem $\Pi \in \mathbf{NP}$ is NP-complete if *any* problem in **NP** can be polynomially transformed to Π. Note that the definition does not immediately imply that NP-complete problems exist![2] Fortunately, for our purposes, there are in fact many NP-complete problems, as we will show a bit later.

[2]In the article [21], Fryer illustrates this point amusingly by defining a ***cowraffe*** to be a cow having a neck that is five feet long. But of course, as he says, "no such animal exists."

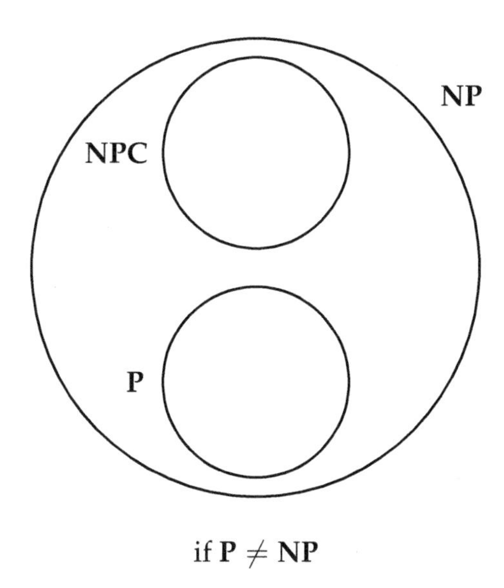

FIGURE 9.2: **P**, **NP** and **NPC**, if **P** $\neq$ **NP**.

The complexity class **NPC** denotes the set of all NP-complete problems. **NPC** is just an abbreviation for "NP-complete."

The following important result states that, if any NP-complete problem can be solved in polynomial time, then all problems in **NP** can be solved in polynomial time. This theorem follows from basic properties of polynomial transformations.

THEOREM 9.4 *If* $\mathbf{P} \cap \mathbf{NPC} \neq \varnothing$*, then* $\mathbf{P} = \mathbf{NP}$*.*

PROOF From Theorem 9.1, we know that $\mathbf{P} \subseteq \mathbf{NP}$, so it suffices to show that $\mathbf{NP} \subseteq \mathbf{P}$. Suppose $\Pi \in \mathbf{P} \cap \mathbf{NPC}$ and let $\Pi' \in \mathbf{NP}$. We will show that $\Pi' \in \mathbf{P}$.

1. Since $\Pi' \in \mathbf{NP}$ and $\Pi \in \mathbf{NPC}$, it follows that $\Pi' \leq_P \Pi$ (this is just the definition of NP-completeness).

2. Since $\Pi' \leq_P \Pi$ and $\Pi \in \mathbf{P}$, it follows that $\Pi' \in \mathbf{P}$ (see Theorem 9.2).

∎

Most computer scientists believe that $\mathbf{P} \neq \mathbf{NP}$. In fact, the most famous open conjecture in theoretical computer science is that $\mathbf{P} \neq \mathbf{NP}$.

Theorem 9.4 states that if *any* NP-complete problem can be solved in polynomial time, then $\mathbf{P} = \mathbf{NP}$, which would imply that *all* NP-complete problems can be solved in polynomial time. Equivalently, if $\mathbf{P} \neq \mathbf{NP}$, then $\mathbf{P} \cap \mathbf{NPC} = \varnothing$ (this is just the contrapositive statement). See Figure 9.2.

We note that ***Ladner's Theorem*** states that, if $\mathbf{P} \neq \mathbf{NP}$, then there are problems in **NP** that are not in $\mathbf{P} \cup \mathbf{NPC}$. Such problems are termed ***NP-intermediate***.

9.5.1 Satisfiability and the Cook-Levin Theorem

The "satisfiability" problems are of fundamental importance in the theory of NP-completeness. We introduce a particular satisfiability problem called **CNF-Satisfiability** as Problem 9.11.

> **Problem 9.11: CNF-Satisfiability**
>
> **Instance:** A ***boolean formula*** F in n ***boolean variables*** $x_1, \ldots, x_n$ in ***conjunctive normal form***. Thus, F is the ***conjunction*** (logical "and") of m ***clauses***, where each clause is the ***disjunction*** (logical "or") of literals. (A ***literal*** is a boolean variable or its negation.)
>
> **Question:** Is there a truth assignment such that F evaluates to *true*?

We note that "CNF" is an abbreviation for "conjunctive normal form." Also, **Satisfiability** is often abbreviated to **SAT**, so the above problem is sometimes referred to as **CNF-SAT**.

Example 9.2 We give an example of a small instance of **CNF-Satisfiability**. Suppose we have boolean variables x, y, z and a CNF formula

$$(x \vee y \vee z) \wedge (\overline{x} \vee \overline{y} \vee z) \wedge (\overline{x} \vee y \vee \overline{z}).$$

This formula contains three clauses:

$$x \vee y \vee z, \quad \overline{x} \vee \overline{y} \vee z \quad \text{and} \quad \overline{x} \vee y \vee \overline{z}.$$

In order to satisfy the first clause, at least one of x, y or z must have the value *true*. In order to satisfy the second clause, we require

$$x = \mathit{false},\ y = \mathit{false} \text{ or } z = \mathit{true}.$$

In order to satisfy the third clause, we require

$$x = \mathit{false},\ y = \mathit{true} \text{ or } z = \mathit{false}.$$

It turns out that this instance is a yes-instance, as can be seen by verifying the truth assignment

$$x = \mathit{true}, \quad y = \mathit{false} \quad \text{and} \quad z = \mathit{false}.$$

In the first clause, $x = \mathit{true}$; in the second clause, $\overline{y} = \mathit{true}$; and in the third clause, $\overline{z} = \mathit{true}$. □

REMARK Here and elsewhere, we use the notation "$\vee$" to denote a logical "or" and we use "$\wedge$" to denote a logical "and." A bar will indicate the negation of a boolean variable, e.g., $\overline{x}$ is the negation of the boolean variable x. It is read as "not x." ∎

It is easy to see that **CNF-Satisfiability** $\in$ **NP**, because we can take any satisfying truth assignment to be a certificate for a yes-instance and then verify it in polynomial time.

The following fundamental result is attributed independently to Cook and Levin. It establishes the existence of at least one NP-complete problem, namely, **CNF-SAT**.

THEOREM 9.5 (Cook-Levin Theorem) *CNF-Satisfiability* $\in$ *NPC.*

To prove Theorem 9.5, it is necessary to give a "generic" polynomial transformation from an arbitrary problem in **NP** to **CNF-SAT**. We give a very informal proof sketch that a certain type of satisfiability problem is NP-complete. The problem is named **Circuit-SAT** and it is defined as follows. We consider ***boolean circuits*** that contain an arbitrary number of ***boolean inputs*** and one ***boolean output***. The ***gates*** in a boolean circuit are AND, OR and NOT gates. An instance of **Circuit-SAT** consists of a boolean circuit C having n boolean inputs, say $x_1, \ldots, x_n$. The question to be answered is if there is a truth assignment that assigns a boolean value (*true* or *false*) to each input, such that the output of the circuit C is *true*.

Note that a conjunctive normal form boolean formula gives rise to a boolean circuit in a natural way. Such a circuit would have three levels of gates:

1. The first level consists of NOT gates. There is potentially one NOT gate for each boolean input.

2. The second level consists of OR gates, one such gate for each clause.

3. The third level consists of a single AND gate having m inputs (namely the m outputs from the second level).

To prove from first principles that **Circuit-SAT** is NP-complete, we need to show two things:

1. **Circuit-SAT** is in **NP**, and

2. for any $\Pi \in$ **NP**, there is a polynomial transformation from Π to **Circuit-SAT**.

It is fairly obvious that **Circuit-SAT** is in **NP** (a certificate for a yes-instance just consists of a satisfying truth assignment), so we concentrate on proving 2.

Suppose Π is in **NP**. Then there is a polynomial-time certificate verification algorithm for Π, say VER$(I, Cert)$, where I is an instance of Π and *Cert* is a certificate. The running time of VER is a polynomial function of size(I).

In a rigorous proof, the main work involved is to show that the algorithm VER can be converted in polynomial time to a boolean circuit C that performs the same computation. Here we will just make the (plausible?) assumption that this conversion is possible. There is one technical detail that we need to address. A circuit has a fixed number of inputs, whereas an algorithm allows inputs of any size. So we are really assuming the existence of a *family* of circuits, one for each possible value of $n = \text{size}(I)$.

Therefore, at this point, we assume we can construct (in polynomial time) a circuit C that evaluates $\text{VER}(I, Cert)$ for any instance I (of the decision problem Π) having any (fixed) size n. We need to describe the polynomial transformation from Π to **Circuit-SAT**. That is, given an instance I of Π, we need to show how to construct (in polynomial time) an instance $f(I)$ of **Circuit-SAT**, such that $f(I)$ is a yes-instance if and only if I is a yes-instance.

Given I with $\text{size}(I) = n$, we construct $f(I)$ in two steps:

1. Construct the circuit C that evaluates VER for instances of Π having size n.

2. Fix (i.e., hardwire) the boolean inputs to C that correspond to the given instance I. Call the resulting circuit C' and define $f(I) = C'$. Note that the values of the boolean inputs to C' that correspond to a certificate *Cert* are still unspecified.

The two steps are illustrated in Figure 9.3.

We claim that the mapping $I \rightarrow f(I)$ is a polynomial transformation. First, it is clear that $f(I)$ can be computed in polynomial time. So we need to show that $f(I)$ is a yes-instance if and only if I is a yes-instance.

Suppose that I is a yes-instance of Π. Then there must exist a certificate *Cert* such that $\text{VER}(I, Cert) = true$. If we now use *Cert* to set the inputs to C', then C' will evaluate to *true*.

Finally, suppose that $C' = f(I)$ is a yes-instance of **Circuit-SAT**. This means that there is a truth assignment for the inputs to C' that causes C' to output *true*. This truth assignment corresponds to a certificate *Cert* such that $\text{VER}(I, Cert) = true$. Therefore I is a yes-instance of Π.

9.6 Proving Problems NP-complete

Given any NP-complete problem, say Π_1, other problems in **NP** can be proven to be NP-complete via polynomial transformations *starting from* Π_1, as stated in Theorem 9.6.

THEOREM 9.6 *Suppose that the following conditions are satisfied:*

- $\Pi_1 \in \textbf{\textit{NPC}}$,
- $\Pi_1 \leq_P \Pi_2$, *and*
- $\Pi_2 \in \textbf{\textit{NP}}$.

Then $\Pi_2 \in \textbf{\textit{NPC}}$.

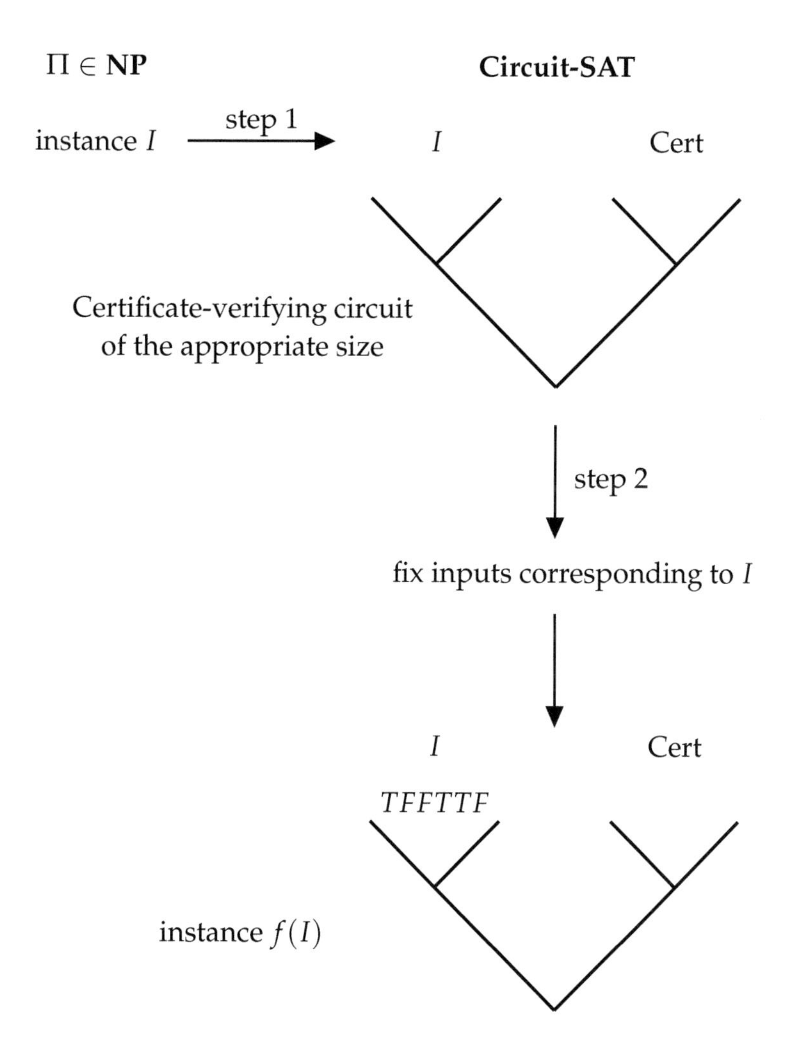

FIGURE 9.3: Transforming an instance of $\Pi \in$ **NP** to an instance of **Circuit-SAT**.

PROOF By hypothesis, we have $\Pi_2 \in \mathbf{NP}$. So it suffices to prove that $\Pi \leq_P \Pi_2$ for all $\Pi \in \mathbf{NP}$.

1. For any $\Pi \in \mathbf{NP}$, we have that $\Pi \leq_P \Pi_1$ since $\Pi_1 \in \mathbf{NPC}$.
2. By hypothesis, $\Pi_1 \leq_P \Pi_2$.

Applying Theorem 9.3, we have that $\Pi \leq_P \Pi_2$ and the proof is complete. ∎

Thus, using one NP-complete problem as a starting point, it is possible to prove a second problem NP-complete by means of a *single* polynomial transformation. It suffices to start with a "known" NP-complete problem and polynomially transform it to another problem in **NP**.

9.6.1 More Satisfiability Problems

Problems 9.12 and 9.13 are two additional satisfiability problems that are special cases of **CNF-SAT**.

Problem 9.12: 3-CNF-Satisfiability

Instance: A boolean formula F in n boolean variables, such that F is the conjunction of m clauses, where each clause is the disjunction of exactly *three* literals.

Question: Is there a truth assignment such that F evaluates to *true*?

Problem 9.13: 2-CNF-Satisfiability

Instance: A boolean formula F in n boolean variables, such that F is the conjunction of m clauses, where each clause is the disjunction of exactly *two* literals.

Question: Is there a truth assignment such that F evaluates to *true*?

There is a significant difference between **3-CNF-Satisfiability** (i.e., **3-CNF-SAT**) and **2-CNF-Satisfiability** (i.e., **2-CNF-SAT**). To be specific, **3-CNF-SAT** $\in$ **NPC**, while **2-CNF-SAT** $\in$ **P**.

First, we describe briefly (without proof) how **2-CNF-SAT** can be solved in polynomial time. This algorithm is due to Aspvall, Plass and Tarjan. Suppose we are given an instance F of **2-CNF-SAT** on a set of boolean variables X. Then Algorithm 9.4 can be used to solve the instance F.

9.6.2 CNF-Satisfiability $\leq_P$ 3-CNF-Satisfiability

As we mentioned above, even though **2-CNF-SAT** $\in$ **P**, it turns out that **3-CNF-SAT** is NP-complete. We will prove the NP-completeness of **3-CNF-SAT** by giving a reduction from **CNF-SAT**. Suppose that $(X, \mathcal{C})$ is an instance of **CNF-SAT**, where $X = \{x_1, \ldots, x_n\}$ and $\mathcal{C} = \{C_1, \ldots, C_m\}$. For each clause C_j, do the following:

Algorithm 9.4: APT(F)

1. For every clause $x \vee y$ in F (where x and y are literals), construct two directed edges (arcs) $\overline{x}y$ and $\overline{y}x$. We get a directed graph on vertex set $X \cup \overline{X}$.
2. Determine the strongly connected components of this directed graph, e.g., using Algorithm 7.11.
3. F is a yes-instance if and only if there is no strongly connected component containing x and $\overline{x}$, for any $x \in X$.

case 1

If $|C_j| = 1$, say $C_j = \{z\}$, then construct four clauses

$$\{z,a,b\},\{z,a,\overline{b}\},\{z,\overline{a},b\},\{z,\overline{a},\overline{b}\},$$

where a and b are new boolean variables.

case 2

If $|C_j| = 2$, say $C_j = \{z_1, z_2\}$, then construct two clauses

$$\{z_1,z_2,c\},\{z_1,z_2,\overline{c}\},$$

where c is a new boolean variable.

case 3

If $|C_j| = 3$, then leave C_j unchanged.

case 4

If $|C_j| \geq 4$, say $C_j = \{z_1, z_2, \dots, z_k\}$, then construct $k-2$ new clauses

$$\{z_1,z_2,d_1\},\{\overline{d_1},z_3,d_2\},\{\overline{d_2},z_4,d_3\},\dots,\{\overline{d_{k-4}},z_{k-2},d_{k-3}\},$$
$$\{\overline{d_{k-3}},z_{k-1},z_k\},$$

where $d_1, \dots, d_{k-3}$ are new boolean variables.

This case can be depicted in the following way:

$$\begin{array}{ccc} z_1 & z_2 & d_1 \\ \overline{d_1} & z_3 & d_2 \\ \overline{d_2} & z_4 & d_3 \\ \vdots & \vdots & \vdots \\ \overline{d_{k-4}} & z_{k-2} & d_{k-3} \\ \overline{d_{k-3}} & z_{k-1} & z_k \end{array}$$

Note that the literals in C_j are shown in red.

Clearly the instance $f(I)$ can be constructed in polynomial time. Thus, we proceed to prove that I is a yes-instance of **CNF-SAT** if and only if $f(I)$ is a yes-instance of **3-CNF-SAT**.

Suppose I is a yes-instance of **CNF-SAT**. We show that $f(I)$ is a yes-instance of **3-CNF-SAT**. Fix a truth assignment for X in which every clause contains a true literal. We consider each clause C_j of the instance I.

case 1

If $C_j = \{z\}$, then z must be true. The corresponding four clauses in $f(I)$ each contain z, so they are all satisfied.

case 2

If $C_j = \{z_1, z_2\}$, then at least one of the z_1 or z_2 is true. The corresponding two clauses in $f(I)$ each contain z_1, z_2, so they are both satisfied.

case 3

If $C_j = \{z_1, z_2, z_3\}$, then C_j occurs unchanged in $f(I)$.

case 4

Suppose $C_j = \{z_1, z_2, z_3, \ldots, z_k\}$ where $k > 3$ and suppose $z_t \in C_j$ is a true literal. Define $d_i = \mathit{true}$ for $1 \leq i \leq t-2$ and define $d_i = \mathit{false}$ for $t-1 \leq i \leq k$. It is straightforward to verify that the $k-2$ corresponding clauses in $f(I)$ each contain a true literal.

Conversely, suppose $f(I)$ is a yes-instance of **3-CNF-SAT**. We show that I is a yes-instance of **CNF-SAT**.

case 1

Four clauses in $f(I)$ having the form $\{z, a, b\}$, $\{z, a, \overline{b}\}$, $\{z, \overline{a}, \overline{b}\}$ $\{z, \overline{a}, \overline{b}\}$ are all satisfied if and only if $z = \mathit{true}$. Then the corresponding clause $\{z\}$ in I is satisfied.

case 2

Two clauses in $f(I)$ having the form $\{z_1, z_2, c\}$, $\{z_1, z_2, \overline{c}\}$ are both satisfied if and only if at least one of $z_1, z_2 = \mathit{true}$. Then the corresponding clause $\{z_1, z_2\}$ in I is satisfied.

case 3

If $C_j = \{z_1, z_2, z_3\}$ is a clause in $f(I)$, then C_j occurs unchanged in I.

case 4

Finally, consider the $k-2$ clauses in $f(I)$ arising from a clause $C_j = \{z_1, z_2, z_3, \ldots, z_k\}$ in I, where $k > 3$. We show that at least one of $z_1, z_2, \ldots, z_k = \mathit{true}$ if all $k-2$ of these clauses contain a true literal.

Assume all of $z_1, z_2, \ldots, z_k = \mathit{false}$. In order for the first clause to contain a true literal, $d_1 = true$. Then, in order for the second clause to contain a true literal, $d_2 = true$. This pattern continues, and in order for the second to last clause to contain a true literal, $d_{k-3} = true$. But then the last clause contains no true literal, which is a contradiction.

The following diagram might be useful in understanding this argument:

$$\begin{array}{ccc} z_1 & z_2 & d_1 \\ \overline{d_1} & z_3 & d_2 \\ \overline{d_2} & z_4 & d_3 \\ \vdots & \vdots & \vdots \\ \overline{d_{k-4}} & z_{k-2} & d_{k-3} \\ \overline{d_{k-3}} & z_{k-1} & z_k \end{array} \quad \rightarrow \quad \begin{array}{ccc} \mathit{false} & \mathit{false} & d_1 \\ \overline{d_1} & \mathit{false} & d_2 \\ \overline{d_2} & \mathit{false} & d_3 \\ \vdots & \vdots & \vdots \\ \overline{d_{k-4}} & \mathit{false} & d_{k-3} \\ \overline{d_{k-3}} & \mathit{false} & \mathit{false} \end{array}$$

We have shown that at least one of $z_1, z_2, \ldots, z_k = true$, which says that the clause $\{z_1, z_2, z_3, \ldots, z_k\}$ contains a true literal, as required.

It is important to note that I is an *arbitrary* instance of **CNF-SAT**, but $f(I)$ is the instance of **3-CNF-SAT** that is *constructed from* I. Thus $f(I)$ is *not* an arbitrary instance of **3-CNF-SAT**. When we prove the correctness of the transformation, the instances I and $f(I)$ are specified. Then we prove that I is a yes-instance if and only if $f(I)$ is a yes-instance.

9.6.3 3-CNF-Satisfiability $\leq_P$ Clique

We describe a reduction from **3-CNF-SAT** to **Clique** (which is Problem 9.9). Let I be the instance of **3-CNF-SAT** consisting of n variables, $x_1, \ldots, x_n$, and m clauses, $C_1, \ldots, C_m$. Let

$$C_i = \{z_1^i, z_2^i, z_3^i\},$$

for $1 \leq i \leq m$.

Define $f(I) = (G, k)$, where $G = (V, E)$ according to the following rules:

1. $V = \{v_j^i : 1 \leq i \leq m, 1 \leq j \leq 3\}$,
2. $v_j^i v_{j'}^{i'} \in E$ if and only if $i \neq i'$ and $z_j^i \neq \overline{z_{j'}^{i'}}$, and
3. $k = m$.

Observe that *non-edges* of the constructed graph correspond to

1. "inconsistent" truth assignments of literals from two different clauses; or
2. any two literals in the same clause.

The m "rows" of vertices correspond to the m clauses. The three "columns" of vertices correspond to the three literals in each clause.

Suppose I is a yes-instance of **3-CNF-SAT**. We show that $f(I)$ is a yes-instance of **Clique**. Start with a satisfying truth assignment and choose a true literal from each clause. We claim that the corresponding vertices in V form a clique of size m in G. Suppose not; then we have two vertices on different "levels" that are not joined by an edge. This happens only if the two corresponding literals are negations of each other. But this would mean that the two literals cannot both be true, which is a contradiction.

Conversely, suppose $f(I)$ is a yes-instance of **Clique**. We show that I is a yes-instance of **3-CNF-SAT**. A clique of size m in $f(I)$ must contain one v_j^i, for $1 \leq i \leq m$ (i.e., one vertex from each level). We claim that it is possible to set the corresponding literals to *true*. (The only way that this could *not* be done would be if two of the literals are negations of each other. But such a pair of literals would correspond to a non-edge in $f(I)$.) The resulting truth assignment yields a true literal in each clause, so it is a satisfying truth assignment.

Example 9.3 Suppose we have the following instance I of **3-CNF-SAT**:

$$C_1 = \{x_1, \overline{x_2}, \overline{x_3}\}$$
$$C_2 = \{\overline{x_1}, x_2, x_3\}$$
$$C_3 = \{x_1, x_2, x_3\}$$

This instance is easily seen to be a yes-instance;

$$x_1 = true, \quad x_2 = true \quad \text{and} \quad x_3 = false$$

is one satisfying truth assignment. We have indicated a true literal in each clause above in red: $\overline{x_3}$ from C_1, x_2 from C_2 and x_1 from C_3.

The constructed instance $f(I)$ is shown in Figure 9.4. The vertices in this graph that "correspond" to the selected true literals, namely, vertices v_3^1, v_2^2 and v_1^3, form a clique.

□

9.6.4 Vertex Cover $\leq_P$ Subset Sum

We now define a new problem known as **Subset Sum** in Problem 9.14. Then we give a reduction from **Vertex Cover** to **Subset Sum**.

Problem 9.14: Subset Sum

Instance: A list of ***sizes*** $S = [s_1, \ldots, s_n]$; and a ***target sum***, T. These are all positive integers.

Question: Does there exist a subset $J \subseteq \{1, \ldots, n\}$ such that

$$\sum_{i \in J} s_i = T?$$

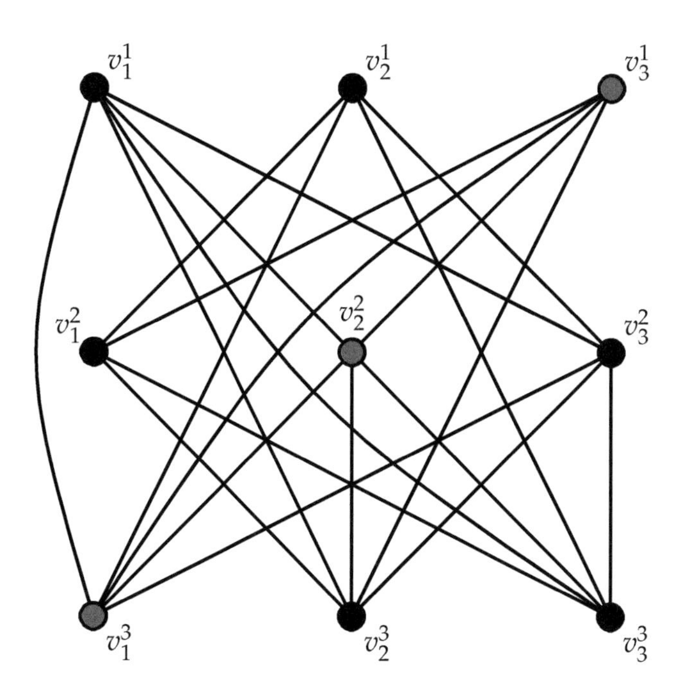

FIGURE 9.4: An instance of **Clique** constructed from an instance of **3-CNF-SAT**.

Suppose $I = (G, k)$ is an instance of **Vertex Cover**, where

$$G = (V, E), |V| = n, |E| = m \text{ and } 1 \leq k \leq n.$$

It will be convenient to denote

$$V = \{v_1, \ldots, v_n\} \text{ and } E = \{e_0, \ldots, e_{m-1}\}.$$

For $1 \leq i \leq n, 0 \leq j \leq m-1$, let $C = (c_{ij})$, where

$$c_{ij} = \begin{cases} 1 & \text{if } e_j \text{ is incident with } v_i \\ 0 & \text{otherwise.} \end{cases}$$

Define $n + m$ sizes and a target sum W as follows:

$$a_i = 10^m + \sum_{j=0}^{m-1} c_{ij} 10^j \quad (1 \leq i \leq n)$$

$$b_j = 10^j \quad (0 \leq j \leq m-1)$$

$$W = k \cdot 10^m + \sum_{j=0}^{m-1} 2 \cdot 10^j.$$

Then define $f(I) = (a_1, \ldots, a_n, b_0, \ldots, b_{m-1}, W)$.

Suppose I is a yes-instance of **Vertex Cover**. Then there is a vertex cover $V' \subseteq V$ such that $|V'| = k$. For $i = 1, 2$, let E^i denote the edges having exactly i vertices in V'. Then $E = E^1 \cup E^2$ because V' is a vertex cover. Let

$$A' = \{a_i : v_i \in V'\} \quad \text{and} \quad B' = \{b_j : e_j \in E^1\}.$$

The sum of the sizes in A' is

$$k \cdot 10^m + \sum_{\{j: e_j \in E^1\}} 10^j + \sum_{\{j: e_j \in E^2\}} 2 \times 10^j.$$

The sum of the sizes in B' is

$$\sum_{\{j: e_j \in E^1\}} 10^j.$$

Therefore, the sum of all the chosen sizes is

$$\begin{aligned} k \cdot 10^m + \sum_{\{j: e_j \in E\}} 2 \cdot 10^j &= k \cdot 10^m + \sum_{j=1}^{m} 2 \cdot 10^j \\ &= W. \end{aligned}$$

Conversely, suppose $f(I)$ is a yes-instance of **Subset Sum**. We show that I is a yes-instance of **Vertex Cover**.

Let $A' \cup B'$ be the subset of chosen sizes. Define

$$V' = \{v_i : a_i \in A'\}.$$

We claim that V' is a vertex cover of size k. In order for the coefficient of 10^m to be equal to k, we must have $|V'| = k$ (because there can't be any carries occurring). The coefficient of any other term 10^j $(0 \leq j \leq m-1)$ must be equal to 2. Suppose that $e_j = v_i v_{i'}$. There are two possible situations that can occur:

1. a_i and $a_{i'}$ are both in A'. Then V' contains both vertices incident with e_j.
2. Exactly one of a_i or $a_{i'}$ is in A' and $b_j \in B'$. In this case, V' contains exactly one vertex incident with e_j.

In both cases, e_j is incident with at least one vertex in V'.

Example 9.4 We provide an example of the above-described transformation in Figure 9.5. We start with a graph, followed by the matrix C, the sizes and the target sum.

In Figure 9.5, $\{v_2, v_3\}$ is a vertex cover (these vertices are colored red). We have

$$E^2 = \{e_2\} \text{ and } E^1 = \{e_1, e_3, e_4, e_5\}.$$

Thus

$$A' = \{a_2, a_3\} \text{ and } B' = \{b_1, b_3, b_4, b_5\}.$$

These sizes, which are indicated in red, sum to W. □

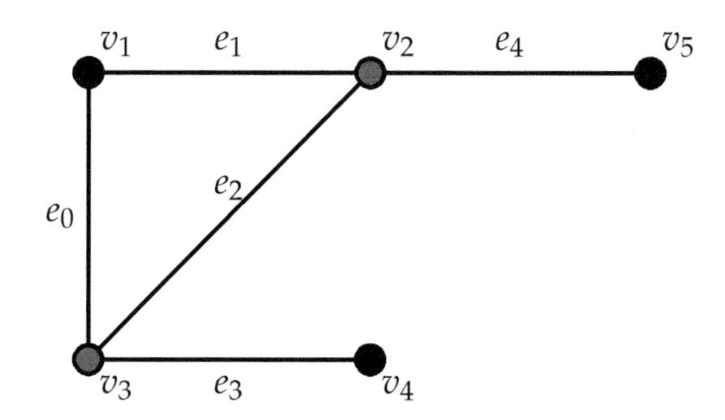

$$C = \begin{pmatrix} 0 & 0 & 0 & 1 & 1 \\ 1 & 0 & 1 & 1 & 0 \\ 0 & 1 & 1 & 0 & 1 \\ 0 & 1 & 0 & 0 & 0 \\ 1 & 0 & 0 & 0 & 0 \end{pmatrix}$$

Note that the columns of C are indexed by e_4, e_3, e_2, e_1, e_0 going from left to right.

$$\begin{aligned} a_1 &= 100011 & b_0 &= 1 \\ a_2 &= 110110 & b_1 &= 10 \\ a_3 &= 101101 & b_2 &= 100 \\ a_4 &= 101000 & b_3 &= 1000 \\ a_5 &= 110000 & b_4 &= 10000 \end{aligned}$$

$$W = 222222 = a_2 + a_3 + b_0 + b_1 + b_3 + b_4$$

FIGURE 9.5: A reduction from **Vertex Cover** to **Subset Sum**.

9.6.5 Subset Sum $\leq_P$ 0-1 Knapsack-Dec

Next, we describe a reduction from **Subset Sum** to **0-1 Knapsack-Dec**. This is a quite simple reduction.

Let I be an instance of **Subset Sum** consisting of sizes $[s_1, \ldots, s_n]$ and target sum T. Define

$$\begin{aligned} p_i &= s_i, \quad 1 \leq i \leq n \\ w_i &= s_i, \quad 1 \leq i \leq n \\ M &= T. \end{aligned}$$

Then define $f(I)$ to be the instance of **0-1 Knapsack-Dec** consisting of profits $[p_1, \ldots, p_n]$, weights $[w_1, \ldots, w_n]$, capacity M, and target profit T.

Suppose that I is a yes-instance of **Subset Sum**, so there is a subset $J \subseteq I$ such that $\sum_{i \in J} s_i = T$. Define

$$x_i = \begin{cases} 1 & \text{if } i \in J \\ 0 & \text{if } i \notin J. \end{cases}$$

Then

$$\sum_{i=1}^{n} x_i p_i = \sum_{i=1}^{n} x_i w_i = T,$$

so $f(I)$ is a yes-instance of **0-1 Knapsack-Dec**.

Conversely, suppose $f(I)$ is a yes-instance of **0-1 Knapsack-Dec**. We have

$$\begin{aligned} \sum_{i=1}^{n} x_i p_i &= \sum_{i=1}^{n} x_i s_i \\ &\geq T \end{aligned}$$

and

$$\begin{aligned} \sum_{i=1}^{n} x_i w_i &= \sum_{i=1}^{n} x_i s_i \\ &\leq M \\ &= T, \end{aligned}$$

so it follows that

$$\sum_{i=1}^{n} x_i s_i = T.$$

Define $J = \{i : x_i = 1\}$; then

$$\sum_{i \in J} s_i = T.$$

Hence, I is a yes-instance of **Subset Sum**.

9.6.6 Subset Sum $\leq_P$ Partition

We define a new problem, **Partition** (Problem 9.15), and show that there is a reduction from **Subset Sum** to **Partition**.

Let $[s_1, \ldots, s_n]$ and T define an instance I of **Subset Sum**. We define an instance $f(I)$ of partition to consist of the following $n + 2$ sizes:

$$s_1, \ldots, s_n, T, \quad \text{and} \quad Sum + 1 - T.$$

We leave a proof of correctness of this transformation as an exercise for the reader. It is a fairly straightforward proof.

Problem 9.15: Partition

Instance: A list of ***sizes*** $[s_1, \ldots, s_n]$ such that

$$Sum = \sum_{i=1}^{n} s_i$$

is even. These values are all positive integers.

Question: Does there exist a subset $J \subseteq \{1, \ldots, n\}$ such that

$$\sum_{i \in J} s_i = \sum_{i \in \{1,\ldots,n\} \setminus J} s_i = \frac{Sum}{2}?$$

9.6.7 Partition $\leq_P$ Multiprocessor Scheduling-Dec

The **Multiprocessor Scheduling** optimization problem was defined as Problem 5.7. Here is the decision version of this problem.

Problem 9.16: Multiprocessor Scheduling-Dec

Instance: A list of n positive integers,

$$T = [t_1, \ldots, t_n]$$

and positive integers m and F. For $1 \leq i \leq n$, t_i is the length of the ith ***task***, and m is the number of ***processors***. F is a desired ***finishing time***.

Question: Does there exist a schedule for the n tasks on m processors, such that each processor executes only one task at a time and the finishing time of the schedule is $\leq F$?

It is very easy to give a reduction from **Partition** to **Multiprocessor Scheduling-Dec**. Let $[s_1, \ldots, s_n]$ be an instance of **Partition**. For $1 \leq i. \leq n$, define $t_i = s_i$. Let $m = 2$ and define

$$F = \frac{\sum_{i=1}^{n} s_i}{2}.$$

The proof details are left as an exercise for the reader.

9.6.8 Vertex Cover $\leq_P$ Hamiltonian Cycle

In this section, we describe a reduction from **Vertex Cover** to **Hamiltonian Cycle**. This a very complicated reduction. The transformation is based on a small graph that is informally called a ***widget***. Let $u \neq v$. A widget $W_{u,v}$ has 12 vertices, denoted $(u, v, 1), \ldots, (u, v, 6)$ and $(v, u, 1), \ldots, (v, u, 6)$. The widget contains 14 edges:

- A path of five edges passing through the following six vertices:

$$(u, v, 1), \ldots, (u, v, 6).$$

- A path of five edges passing through the following six vertices:

$$(v, u, 1), \ldots, (v, u, 6).$$

- Four additional edges:

$$\begin{array}{ll} (u,v,1)(v,u,3) & (u,v,3)(v,u,1) \\ (u,v,4)(v,u,6) & (u,v,6)(v,u,4). \end{array}$$

Observe that $W_{u,v}$ is isomorphic to $W_{v,u}$.

Here is a depiction of the widget $W_{u,v}$:

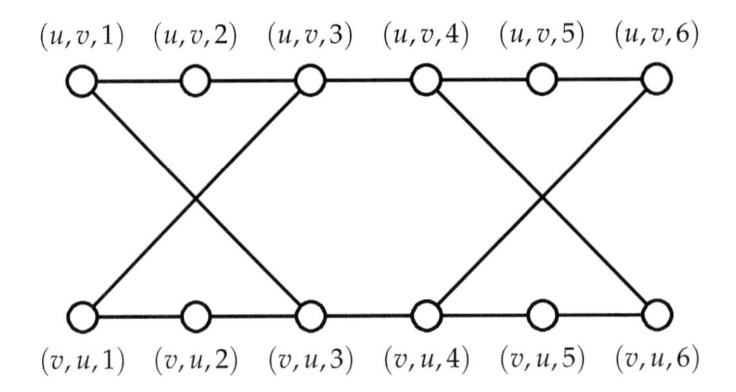

There are precisely three ways to traverse all the vertices in a widget $W_{u,v}$:

1. Traverse the two paths of length five separately:

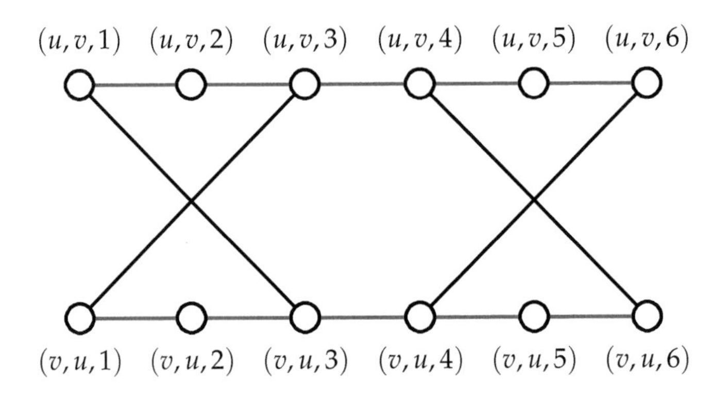

2. Traverse the following path of red edges:

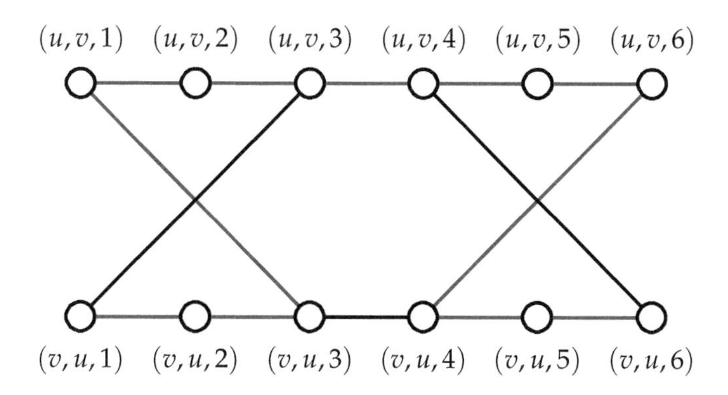

3. Traverse the following path of red edges:

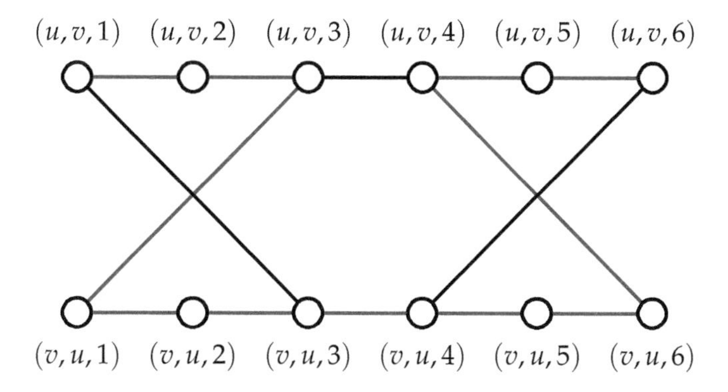

Observe that, in the three cases considered, we "enter" and "leave" on the top (case 3), on the bottom (case 2), or both (case 1).

A widget $W_{u,v}$ will be associated with an edge uv of a given graph. We interpret the three types of paths enumerated above in reference to a given vertex cover as follows:

1. two separate paths correspond to a situation where both u and v are in the cover,
2. a single path that has endpoints $(u, v, 1)$ and $(u, v, 6)$ corresponds to a situation where u is in the cover and v is not in the cover, and
3. a single path that has endpoints $(v, u, 1)$ and $(v, u, 6)$ corresponds to a situation where v is in the cover and u is not in the cover.

Now we can describe the transformation. Suppose $I = (G, k)$ is an instance of **Vertex Cover**, where $G = (V, E)$.

1. For each edge $uv \in E$, construct a widget $W_{u,v}$.
2. For each vertex $u \in V$, let the vertices adjacent to u be denoted $u_1, \ldots, u_\ell$ (the ordering of these ℓ vertices is arbitrary). Link the ℓ corresponding widgets $W_{u,u_1}, \ldots, W_{u,u_\ell}$ (which we call a ***chain***) in the same order by introducing edges
$$
\begin{array}{c}
(u, u_1, 6)(u, u_2, 1), \\
(u, u_2, 6)(u, u_3, 1), \\
\ldots \\
(u, u_{\ell-1}, 6)(u, u_\ell, 1).
\end{array}
$$
3. Create k new ***selector vertices***, denoted $s_1, \ldots, s_k$.
4. Connect each selector vertex to the beginning and end of each chain. (Given the chain above, we would connect each s_i to $(u, u_1, 1)$ and $(u, u_\ell, 6)$.)

Carrying out this process for every vertex u, a graph G' is constructed. This graph G' defines the instance $f(I)$. Note that every widget will be contained in two chains in the graph G'.

Suppose I is a yes-instance of **Vertex Cover**. Given a vertex cover consisting of k vertices, we describe how to construct a Hamiltonian cycle in the graph G' defined in the instance $f(I)$ of **Hamiltonian Cycle**.

1. For each edge uv in the graph G, traverse the corresponding widget $W_{u,v}$ according to whether one or two endpoints are in the cover, as in the three possible cases described above.
2. Link together the paths for the widgets corresponding to edges incident with each vertex. This creates a ***chain*** corresponding to each vertex.
3. Link these k chains together using the k selector vertices.

Example 9.5 Suppose we have the following graph G and suppose we seek a vertex cover of size $k = 2$.

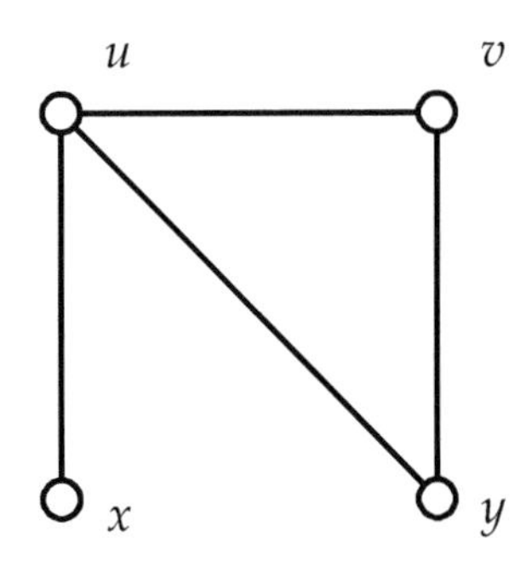

There will be four widgets that are connected into chains as follows:

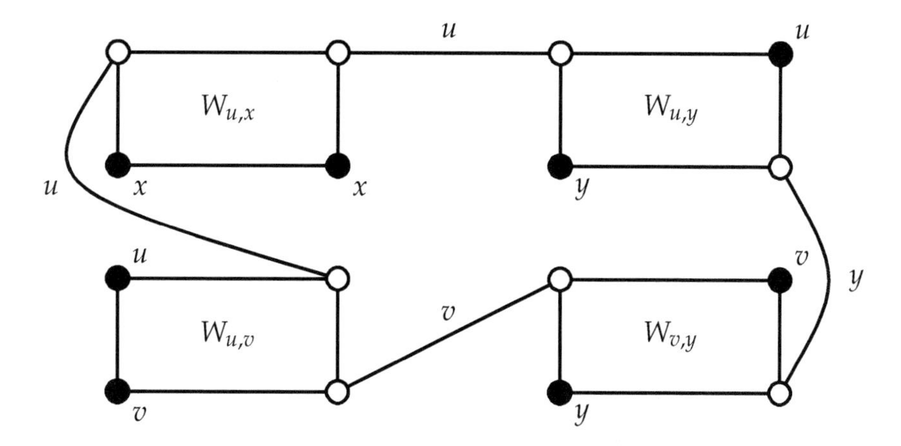

Each of the black vertices are connected to two selector vertices (these vertices and edges are not shown in the diagram, for clarity). There are four chains, corresponding to the four vertices in the original graph:

vertex	chain
u	$W_{u,v}, W_{u,x}, W_{u,y}$
v	$W_{u,v}, W_{v,y}$
x	$W_{u,x}$
y	$W_{u,y}, W_{v,y}$

We have labeled the edges and the black vertices in each chain with their corresponding vertices from G to make it clear which vertices and edges in G' correspond to which chain. There are two connecting edges in the u chain, one connecting edge in each of the v and y chains, and no connecting edge in the x chain.

It is easy to verify that $\{u, y\}$ is a vertex cover of size two in G. We now show how this vertex cover gives rise to a Hamiltonian cycle in the graph G'. In G', we will incorporate the chains corresponding to u and y. Note that edge uy contains two vertices of the cover, while the other three edges of G each contain one vertex of the cover: edge uv and ux contain vertex u and edge vy contains vertex y.

These two chains give rise to the following edges that will be included in the Hamiltonian cycle:

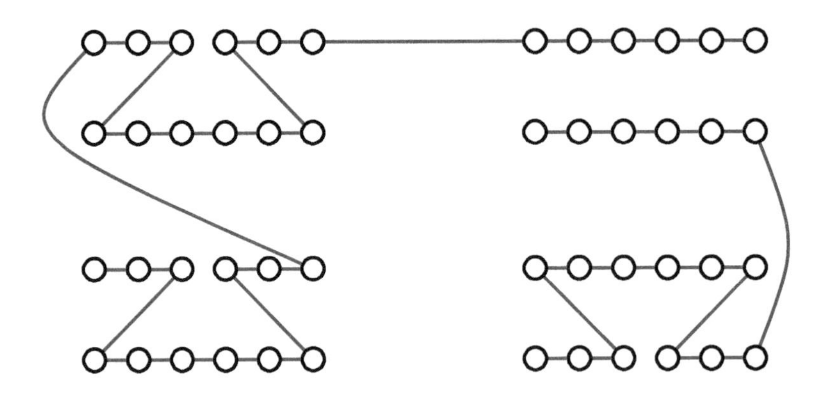

The final step is to link the ends of the two chains to two selector vertices to form a Hamiltonian cycle:

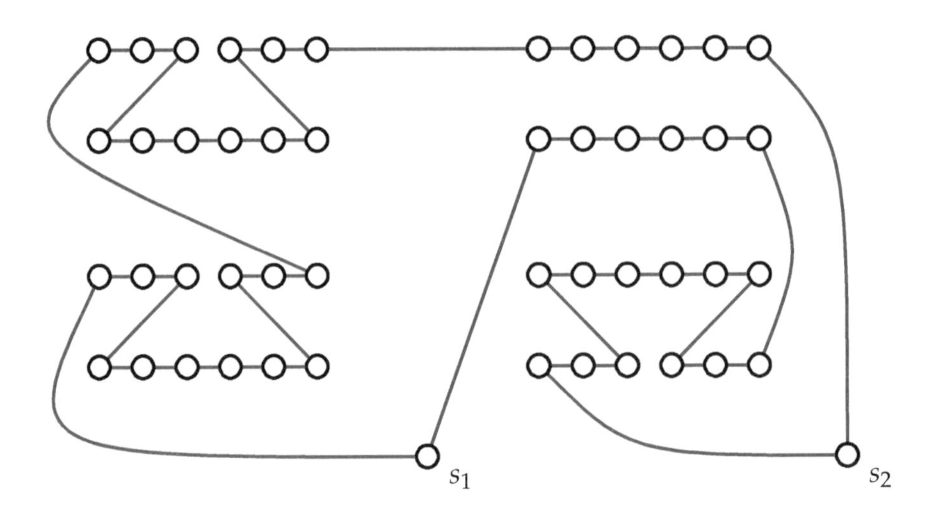

□

Conversely, suppose $f(I)$ is a yes-instance of **Hamiltonian Cycle**. Let H be a Hamiltonian cycle in the graph G'. We describe how to construct a vertex cover of size k in the graph G.

1. A portion of H that connects any two consecutive selector vertices is a chain of paths corresponding to a particular vertex of the graph in I. Include this vertex in the cover.

2. The k selected vertices will form the vertices in the vertex cover.

We leave the details for the reader to verify as an exercise.

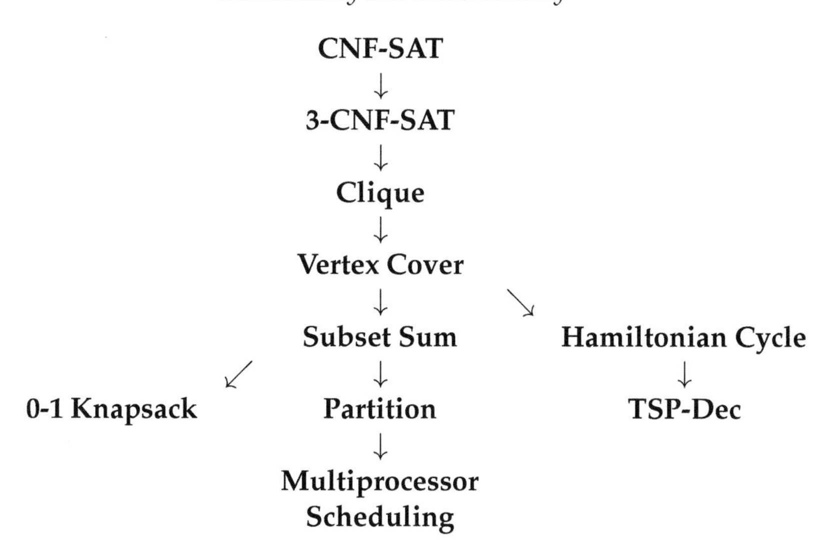

FIGURE 9.6: Some polynomial transformations between NP-complete problems.

9.6.9 Hamiltonian Cycle $\leq_P$ TSP-Dec

Our final reduction is from **Hamiltonian Cycle** to **TSP-Dec**. Let I be an instance of **Hamiltonian Cycle** consisting of a graph $G = (V, E)$.

For the complete graph K_n, where $n = |V|$, define edge weights as follows:

$$w(uv) = \begin{cases} 1 & \text{if } uv \in E \\ 2 & \text{if } uv \notin E. \end{cases}$$

Then define $f(I)$ to be the instance of **TSP-Dec** consisting of the graph K_n, edge weights w and target $T = n$.

Let $I = G = (V, E)$ be an instance of **Hamiltonian Cycle**. Suppose that I is a yes-instance, so there is a Hamiltonian cycle C in the graph G. Then, in the instance $f(I)$ of **TDP-Dec**, the Hamiltonian cycle C has weight n, and hence $f(I)$ is a yes-instance.

Conversely, suppose $f(I)$ is a yes-instance of **TDP-Decision**. Let C be a Hamiltonian cycle of weight at most n. Then C must consist of n edges of weight 1. Therefore C is a Hamiltonian cycle in G and I is a yes-instance of **Hamiltonian Cycle**.

9.6.10 Summary of Polynomial Transformations

In Figure 9.6, we summarize the polynomial transformations between NP-complete problems that we have proven in this chapter. In this figure, arrows denote polynomial transformations.

Since we started from an NP-complete problem (**CNF-SAT**), it follows that all of these problems are NP-complete.

9.7 NP-hard Problems

By definition, only decision problems can be NP-complete. However, there are many non-decision problems, such as optimization and optimal value problems, that are closely related to NP-complete problems. We would like to have a way to specify the difficulty of these problems in relation to the NP-complete problems. This is captured by the notion of NP-hardness: a problem Π is defined to be ***NP-hard*** if there exists a problem $\Pi' \in$ **NPC** such that $\Pi' \leq_P^T \Pi$. It is easy to see that any NP-complete problem is NP-hard.

In general, when we show that a problem is ***NP-hard***, we know that it is *at least as difficult* as the NP-complete problems. This provides a way to show that certain non-decision problems are probably intractable.

To prove that a problem is NP-hard, we find a polynomial-time Turing reduction *from* a known NP-complete problem (which is necessarily a decision problem) *to* a problem that may or may not be a decision problem and which may or may not be in **NP**.

For example, it is trivial to show that **TSP-Dec** $\leq_P^T$ **TSP-Optimization**, as we observed in Section 9.2. We know that **TSP-Dec** $\in$ **NPC**, so it is immediate that **TSP-Optimization** is NP-hard. This trivial Turing reduction is all that is required to prove that **TSP-Optimization** is NP-hard. In general, any optimization (or optimal value) problem "corresponding to" an NP-complete decision problem will be NP-hard.

Recall that, in Section 9.2 we showed that there is also a Turing reduction in the reverse direction, namely, **TSP-Optimization** $\leq_P^T$ **TSP-Dec**. This is a much more complicated reduction. Along with the NP-hardness of **TSP-Optimization**, this reduction proves that **TSP-Optimization** and **TSP-Dec** are of equivalent difficulty, in the sense that one problem can be solved in polynomial time if and only if the other one can. These kinds of problems are sometimes called ***NP-equivalent***. More precisely, an NP-hard problem Π is NP-equivalent if there is a polynomial-time Turing reduction from $\Pi \leq_P^T \Pi'$ for some NP-complete problem Π'.

9.8 Approximation Algorithms

NP-hard problems cannot be solved in polynomial time if **P** $\neq$ **NP**. However, for some NP-hard optimization problems, it is possible to find feasible solutions that are "close to optimal" in polynomial time. We have already seen an example of such an algorithm in Section 5.9. For the **Multiprocessor Scheduling** algorithm, we proved in Theorem 5.8 that a certain greedy algorithm could be used to find a schedule whose relative error is at most $1/3 - 1/(3m)$, where m is the number of processors in the given instance. Consequently, we can find in polynomial time a feasible solution whose finishing time is at most $4/3$ times the optimal finishing time. Such an algorithm is termed a 4/3-approximation algorithm. In general, an algorithm for a minimization problem that has a relative error that is at most ϵ will be termed a $(1+\epsilon)$-***approximation algorithm***.[3]

9.8.1 Vertex Cover

We consider several polynomial-time approximation algorithms for NP-hard optimization problems in this section. As a simple example to start with, we look at the optimization version of the **Vertex Cover** problem, which is defined as follows.

Problem 9.17: Vertex Cover-Optimization

Instance: An undirected graph $G = (V, E)$.

Find: A vertex cover of minimum possible size.

The decision version of **Vertex Cover**, which was defined as Problem 9.10, is NP-complete. It is trivial to show that

$$\textbf{Vertex Cover} \leq_P^T \textbf{Vertex Cover-Optimization},$$

so **Vertex Cover-Optimization** is indeed NP-hard.

A simple polynomial-time approximation algorithm is based on a result concerning maximal matchings. Two edges in a graph are ***independent edges*** provided that there is no vertex incident with both of them. A ***matching*** in a graph G is any subset M of independent edges. A matching M is a ***maximal matching*** if there does not exist a matching M' such that $M \subseteq M'$ and $|M| < |M'|$.

LEMMA 9.7 *Suppose that $M = \{e_1, \dots, e_k\}$ is any maximal matching in a graph $G = (V, E)$. Let $e_i = \{u_i, v_i\}$ for $1 \leq i \leq k$. Then*

$$W = \{u_1, \dots, u_k\} \cup \{v_1, \dots, v_k\}$$

is a vertex cover of G that has relative error at most 1.

PROOF First we show that W is a vertex cover. Suppose not; then there is an edge $\{u, v\}$ such that neither u nor v is in W. But then $M \cup \{uv\}$ would be a matching, which contradicts the assumption that M is a maximal matching. Therefore W is a vertex cover of size $2k$.

Now suppose that W' is any vertex cover. For $1 \leq i \leq k$, W' must contain at least one of u_i and v_i. It follows that the size of the optimal cover, say W^*, is at least k. Hence, the relative error of W is at most

$$\frac{|W|}{|W^*|} - 1 = \frac{2k}{k} - 1 = 1.$$

∎

[3]We defined "relative error" in Section 5.9.

Thus, we just need to find any maximal matching in a given graph $G = (V, E)$. But this is easy—we just keep choosing independent edges from E until we can't continue any further. In more detail, we can use the following greedy algorithm to find a maximal matching.

Algorithm 9.5: MAXIMAL MATCHING($G = (V, E)$)

$M \leftarrow \varnothing$
while $|E| > 0$
do $\begin{cases} \text{choose any edge } e = \{u, v\} \in E \\ M \leftarrow M \cup \{e\} \\ \text{delete all edges from } E \text{ that are incident with } u \text{ or } v \end{cases}$
return (M)

9.8.2 Metric TSP

Problem 9.18 describes a special case of **TSP-Optimization** known as **Metric TSP-Optimization**. The decision version of **Metric TSP** is NP-complete and therefore the optimization version (as defined in Problem 9.18) is NP-hard.

Problem 9.18: Metric TSP-Optimization

Instance: A complete graph K_n and edge weights $w : E \to \mathbb{Z}^+$ that satisfy the ***triangle inequality***:

$$w(u_1, u_2) \leq w(u_1, v) + w(v, u_2)$$

for any three distinct vertices u_1, u_2, v.

Find: A Hamiltonian cycle H in G such that

$$w(H) = \sum_{e \in H} w(e)$$

is minimized.

It turns out to be fairly straightforward to obtain a polynomial-time 2-approximation algorithm for **Metric TSP-Optimization**. Here is a high-level description of the process that is used.

1. Find a minimum spanning tree T of the given (complete) graph.
2. Construct a new "graph" T' consisting of two copies of each edge in T. This yields a ***multigraph*** (so-called because a graph is not allowed to have multiple edges connecting the same pair of vertices).

3. Every vertex in T' has even degree, so T' has an Eulerian circuit C that can be found in polynomial time. (This can be done using a suitable adaptation of Algorithm 7.14.)

4. Traverse the vertices and edges in C, starting at an arbitrary vertex v. Every time an already visited vertex is encountered, proceed directly to the next unvisited vertex in C. This process of using "shortcuts" converts C into a Hamiltonian cycle H of the original graph K_n.

We claim that the Hamiltonian cycle created in step 4 of the above algorithm has relative error at most 1 (that is, its weight is at most twice the weight of an optimal Hamiltonian cycle). First, we observe that

$$w(H) \leq 2w(T). \tag{9.1}$$

This property holds because the weights satisfy the triangle inequality, which means that the shortcuts do not increase the total weight. Let H^* be an optimal Hamiltonian cycle; If we delete any edge e from H^*, we obtain a spanning tree. Since T is a minimum spanning tree, we have

$$w(H^*) > w(H^* \setminus \{e\}) \geq w(T). \tag{9.2}$$

Combining (9.1) and (9.2), we obtain

$$w(H) < 2w(H^*).$$

Example 9.6 Suppose a graph on vertex set $\{a,b,c,d,e,f,g,h,i\}$ has a minimum spanning tree consisting of the following edges:

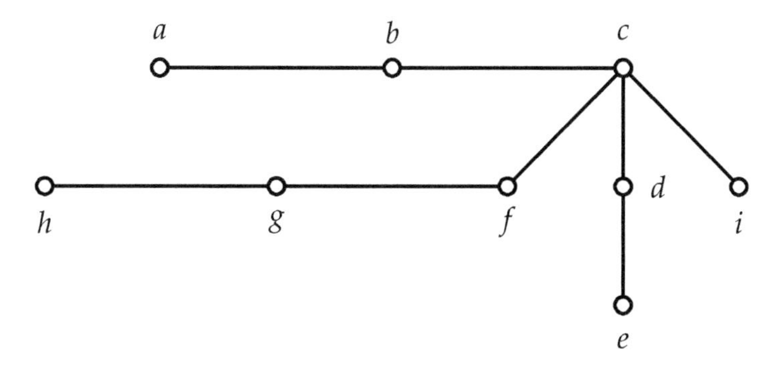

After taking two copies of every edge, an Eulerian circuit is given by

$$C = a,b,c,d,e,d,c,f,g,h,g,f,c,i,c,b,a.$$

Now, we start at a and follow the vertices in C, marking all vertices that are visited more than once:

$$C = a,b,c,d,e,d,c,f,g,h,g,f,c,i,c,b,a.$$

Then we obtain the Hamiltonian cycle

$$H = a,b,c,d,e,f,g,h,i,a.$$

□

9.8.3 Maximum Satisfiability

Maximum Satisfiability (or **MAX-SAT** for short) is a variation of **CNF-SAT** in which we seek to maximize the number of clauses that are satisfied by a truth assignment.

> **Problem 9.19: Maximum Satisfiability**
>
> **Instance:** A boolean formula F in conjunctive normal form, having n boolean variables, denoted $x_1, \ldots, x_n$, and m clauses, denoted $C_1, \ldots, C_m$.
>
> **Find:** A truth assignment the maximizes the number of clauses that contain at least one true literal.

The original version of this problem, **CNF-SAT**, which we first considered in Section 9.5.1, asked if all m clauses could be satisfied. It should be clear that **MAX-SAT** is NP-hard, since there is a trivial reduction

$$\textbf{CNF-SAT} \leq_P^T \textbf{MAX-SAT}.$$

We are going to analyze random truth assignments. By a random truth assignment, we mean that $\mathbf{x_1}, \ldots, \mathbf{x_n}$ are n independent random variables, each taking on the values *true* and *false* with probability $1/2$. We will focus on **MAX-3-SAT** (i.e., instances of **MAX-SAT** where every clause contains exactly three literals) in detail. Throughout our analysis, we assume that there are no clauses containing a variable and its negation (such clauses are always satisfied, so they are not of interest anyway).

For a given clause $C_j = \{z_1, z_2, z_3\}$, the probability that C_j contains a true literal is $7/8$. This is clear because $z_1 = z_2 = z_3 = \mathit{false}$ (which is the only "bad" case) occurs with probability $1/2 \times 1/2 \times 1/2 = 1/8$. For $1 \leq j \leq m$, let $\mathbf{Y_j}$ denote a random variable that takes on the values 0 or 1 according to whether or not clause C_j is satisfied under a random truth assignment. Thus, we have

$$\begin{aligned} \mathbf{Pr}[\mathbf{Y_j} = 1] &= \frac{7}{8} \\ \mathbf{Pr}[\mathbf{Y_j} = 0] &= \frac{1}{8}. \end{aligned}$$

We can compute the expected value of $\mathbf{Y_j}$ as follows:

$$\begin{aligned} E[\mathbf{Y_j}] &= 1 \times \mathbf{Pr}[\mathbf{Y_j} = 1] + 0 \times \mathbf{Pr}[\mathbf{Y_j} = 0] \\ &= 1 \times \frac{7}{8} + 0 \times \frac{1}{8} \\ &= \frac{7}{8}. \end{aligned}$$

Define

$$\mathbf{Y} = \sum_{j=1}^{m} \mathbf{Y_j}.$$

Y is a random variable that records the number of clauses that are satisfied under a random truth assignment. From linearity of expectation (Theorem 1.17), we see that

$$\begin{aligned} E[\mathbf{Y}] &= \sum_{j=1}^{m} E[\mathbf{Y_j}] \\ &= \sum_{j=1}^{m} \frac{7}{8} \\ &= \frac{7m}{8}. \end{aligned}$$

Thus, on average, 7/8 of the m clauses will be satisfied under a random truth assignment. In particular, there exists at least one truth assignment such that at least 7/8 of the m clauses are satisfied.

The above argument is an example of the so-called ***probabilistic method,*** which was originally due to Erdös. It gives a nice existence result, but it does not immediately allow us to identify a specific *good truth assignment*, i.e., one that results in at least 7/8 of the m clauses being satisfied. However, it turns out to be not so difficult to actually find a good truth assignment.

We originally defined "relative error" for minimization problems. Here we have a maximization problem, so we define the relative error of a given feasible solution to be

$$1 - \frac{P}{P**},$$

where P^* is the optimal "profit" and P is the "profit" of the given solution. Here, the profit is just the number of clauses that are satisfied.

We clearly have $P^* \leq m$ (because m is the total number of clauses) and $P \geq 7m/8$ (because $7m/8$ is a lower bound on the number of clauses satisfied by a good truth assignment, which exists by Theorem 9.8). Therefore, if we can *find* a good truth assignment, then we have an approximate solution to **MAX-3-SAT** whose relative error is at most

$$1 - \frac{\frac{7m}{8}}{m} = \frac{1}{8}.$$

In order to obtain a practical algorithm, we establish a useful lower bound on the probability that a random truth assignment is good.

LEMMA 9.8 *Suppose that an instance of **Maximum 3-Satisfiability** has n boolean variables, denoted* $x_1, \ldots, x_n$*, and m clauses, denoted* $C_1, \ldots, C_m$*. Then the probability that a random truth assignment is good is at least*

$$\frac{1}{8m}.$$

PROOF From the definition of expectation, we have

$$E[\mathbf{Y}] = \sum_{j=0}^{m} (j \times \mathbf{Pr}[\mathbf{Y} = j]).$$

Denote

$$L = \left\lceil \frac{7m}{8} \right\rceil$$

and split the above sum into two parts:

$$A = \sum_{j=0}^{L-1} (j \times \mathbf{Pr}[\mathbf{Y} = j])$$
$$B = \sum_{j=L}^{m} (j \times \mathbf{Pr}[\mathbf{Y} = j]).$$

Also, let

$$p = \mathbf{Pr}\left[\mathbf{Y} \geq \frac{7m}{8}\right].$$

Of course,

$$E[\mathbf{Y}] = A + B. \tag{9.3}$$

It is clear that

$$\begin{aligned} B &\leq \sum_{j=L}^{m} (m \times \mathbf{Pr}[\mathbf{Y} = j]) \\ &= mp \end{aligned}$$

and

$$\begin{aligned} A &\leq \sum_{j=0}^{L-1} ((L-1) \times \mathbf{Pr}[\mathbf{Y} = j]) \\ &= (L-1) \times (1-p) \\ &\leq L-1. \end{aligned}$$

From these upper bounds on A and B, it follows from (9.3) that

$$E[\mathbf{Y}] \leq mp + L - 1.$$

However, we have already proven that

$$E[\mathbf{Y}] = \frac{7m}{8}.$$

Therefore,

$$\frac{7m}{8} \leq mp + L - 1.$$

Using the fact that $L = \lceil \frac{7m}{8} \rceil$, it is not hard to see that

$$\frac{7m}{8} - (L-1) \geq \frac{1}{8},$$

so we have

$$mp \geq \frac{1}{8}$$

and hence

$$\mathbf{Pr}\left[\mathbf{Y} \geq \frac{7m}{8}\right] \geq \frac{1}{8m},$$

as desired. ∎

Theorem 9.8 proves that a random truth assignment will be good with probability at least $1/(8m)$. Suppose we generate a series of independent random truth assignments until we find one that is good. Theorem 1.12 tells us that the expected number of trials is at most $8m$. Therefore we can find a good truth assignment in ***expected polynomial time***.

9.9 Undecidability

A decision problem Π is ***undecidable*** if there *does not exist* an algorithm that solves Π. If Π is undecidable, then for every potential "algorithm" A, there exists at least one instance $I \in \mathcal{I}(\Pi)$ such that $\text{A}(I)$ does not find the correct answer (*yes* or *no*) in finite time.

The most famous undecidable problem is the **Halting** problem. This problem, which is presented as Problem 9.20, was proved to be undecidable by Alan Turing in 1936.

Problem 9.20: Halting	
Instance:	A program P and input x for the program P.
Question:	When program P is executed with input x, will it halt in finite time?

Note that we refer to P as a *program* rather than an *algorithm*, because P is not required to terminate in a finite amount of time for every input.

We give a somewhat informal proof that Problem 9.20 is undecidable. Suppose that HALT is any algorithm that solves the **Halting Problem**. We will derive a contradiction, which will therefore prove that the algorithm HALT cannot exist.

Suppose that P is any program and x is any input to P. The statement "HALT solves the **Halting** problem" means that

$$\text{HALT}(\text{P}, x) = \begin{cases} \textit{yes} & \text{if } \text{P}(x) \text{ halts} \\ \textit{no} & \text{if } \text{P}(x) \text{ doesn't halt.} \end{cases}$$

The answer *yes* or *no* must be found in finite time by HALT. Note that P (the "program") and I (the "input" to P) are both strings over some finite alphabet.

Consider the following program STRANGE, which uses the algorithm HALT as a subroutine. The input to STRANGE is a program P.

Program 9.6: STRANGE(P)

external HALT
if not HALT(P, P)
  then return (0)
  else $\begin{cases} i \leftarrow 1 \\ \textbf{while } i \neq 0 \textbf{ do } i \leftarrow i+1 \end{cases}$

Note that the **while** loop in STRANGE is an infinite loop. Thus, if STRANGE enters this loop, it runs forever.

Consider what happens when we run STRANGE(STRANGE). We have

$$\begin{aligned} \text{STRANGE(STRANGE) halts} &\Leftrightarrow \text{HALT(STRANGE, STRANGE)} = no \\ &\Leftrightarrow \text{STRANGE(STRANGE) does not halt.} \end{aligned}$$

This is a contradiction. Hence, the only conclusion we can make is that the algorithm HALT does not exist!

Here is another proof that **Halting** is undecidable, based on a "diagonalization" argument. Consider an enumeration (i.e., a listing) of all the possible programs, represented as finite strings over a finite alphabet, say $A_1, A_2, \ldots$.[4] Every possible program will occur somewhere in this infinite list.

Also, enumerate all the possible inputs as $x_1, x_2, \ldots$. Suppose the algorithm HALT exists and define STRANGE as above. Define a matrix M by the rule

$$M[i,j] = \begin{cases} 1 & \text{if HALT}(A_i, x_j) = yes \\ 0 & \text{if HALT}(A_i, x_j) = no. \end{cases}$$

Suppose we encode *yes* as "1" and *no* as "0." Then

$$\text{STRANGE}(j) = 1 - M[j,j],$$

and hence $\text{STRANGE}(j) \neq M[j,j]$, for all integers $j \geq 0$. It follows that STRANGE cannot correspond to any row of M, and hence STRANGE does not exist.

We give another example of an undecidable problem. The problem **Halt-All** takes a program P as input and asks if P halts on *all inputs x*. We describe a Turing reduction **Halting** $\leq^T$ **Halt-All**, which proves that **Halt-All** is undecidable.

Assume we have an algorithm HALTALLSOLVER that solves **Halt-All**. For a fixed program P and input x, let $P_x()$ be the program that executes $P(x)$ (so P_x has no input). Here is the reduction:

[4] There are a countably infinite number of such strings.

1. Given P and x (which together define an instance of **Halting**), construct the program P_x.

2. Run HALTALLSOLVER(P_x).

We have

$$\text{HALTALLSOLVER}(P_x) = yes \Leftrightarrow \text{P}(x) \text{ halts},$$

so we can solve the halting problem. If **Halt-All** were decidable, then the reduction would prove that **Halting** is decidable. But **Halting** is undecidable, so we conclude that **Halt-All** is undecidable.

Problem 9.21 is known as the **Post Correspondence** problem. It was first posed by Emil Post in 1946. This problem is also undecidable; however, we do not give a proof of its undecidability.

Problem 9.21: Post Correspondence

Instance: Two finite lists $\alpha_1, \ldots, \alpha_N$ and $\beta_1, \ldots, \beta_N$ of words over some alphabet A of size ≥ 2.

Question: Does there exist a finite list of indices, say $i_1, \ldots, i_K$, where $i_j \in \{1, \ldots, N\}$ for $1 \leq j \leq K$, such that

$$\alpha_{i_1} \cdots \alpha_{i_K} = \beta_{i_1} \cdots \beta_{i_K},$$

where a "product" of words denotes their concatenation?

Here is an example of **Post Correspondence** from Wikipedia. Suppose the instance I consists of the following two lists of three words:

$$\alpha_1 = a, \qquad \alpha_2 = ab, \qquad \alpha_3 = bba$$

and

$$\beta_1 = baa, \qquad \beta_2 = aa, \qquad \beta_3 = bb.$$

Consider the list of indices $(3, 2, 3, 1)$. It is straightforward to verify that

$$\alpha_3\alpha_2\alpha_3\alpha_1 = bbaabbbaa = \beta_3\beta_2\beta_3\beta_1.$$

Therefore I is a yes-instance.

Of course, **Post Correspondence** can be solved by exhaustive search if K is specified as part of the problem instance.

9.10 The Complexity Class EXPTIME

Define **EXPTIME** to be the set of all decision problems that can be solved by (deterministic) exponential-time algorithms, i.e., in time $O(2^{p(n)})$ where $p(n)$ is a polynomial in n and $n = \mathsf{size}(I)$.

We observe that **NP** $\subseteq$ **EXPTIME**. The idea is to generate all possible certificates of an appropriate length and check them for correctness using the given certificate verification algorithm. As an example, for **Hamiltonian Cycle**, we could generate all $n!$ certificates and check each one in turn.

It is not known if there are problems in **NP** that *cannot* be solved in polynomial time (because the **P** $=$ **NP**? conjecture is not yet resolved). However, it is possible to prove that there exist problems in **EXPTIME** $\setminus$ **P**. One such problem is the **Bounded Halting** problem. An instance of **Bounded Halting** has the form (P, x, t), where P is a program, x is an input to P, and t is a positive integer (given in *binary notation*). The question to be answered is if $\mathrm{P}(x)$ halts after at most t computation steps.

The **Bounded Halting** problem can be solved in time $O(t)$ simply by running the program P for t steps. However, this is *not* a polynomial-time algorithm, because

$$\mathsf{size}(I) = |\mathrm{P}| + |x| + \log_2 t,$$

and t is exponentially large compared to $\log_2 t$.

It is actually fairly easy to prove that **Bounded Halting** is an ***EXPTIME-complete*** problem. That is, for any decision problem Π that can be solved in exponential time, there is a polynomial transformation from Π to **Bounded Halting**.

We prove this as follows. Suppose that $\Pi \in$ **EXPTIME**, so there is an algorithm Π-SOLVER solving Π, whose running time is $2^{p(n)}$ on inputs of size n, where $p(n)$ is a polynomial in n. That is, after $2^{p(n)}$ computation steps, Π-SOLVER returns the correct answer (*yes* or *no*) for any instance x such that $\mathsf{size}(x) = n$.

Let $I = x$ be an instance of Π of size n. We construct a program P_x as follows:

Program 9.7: $\mathrm{P}_x(x)$

```
external Π-SOLVER
if Π-SOLVER(x) = no
  then { i ← 1
       { while i ≠ 0 do i ← i + 1
```

Now, we create an instance $f(I)$ of **Bounded Halting** consisting of

1. the program P_x
2. the input x, and
3. the bound $t = 2^{p(n)}$, where $n = \mathsf{size}(x)$.

An algorithm that solves instance $f(I)$ of **Bounded Halting** will return *yes* if and only if $P_x(x)$ halts after at most t steps. From the structure of P_x, this happens if and only if Π-SOLVER$(x) = yes$ (note that, if Π-SOLVER$(x) = no$, then P_x enters an infinite loop, so it does not halt at all). Therefore this is a valid reduction.

We also observe that the instance I can be computed in polynomial time as a function of $n = \mathsf{size}(x)$. The most costly step is computing $t = 2^{p(n)}$.

It can be proven that **EXPTIME** $\neq$ **P**, so there are problems that are in **EXPTIME** $\setminus$ **P**. Since **Bounded Halting** is EXPTIME-complete, it follows that **Bounded Halting** $\in$ **EXPTIME** $\setminus$ **P**.

9.11 Notes and References

The classic reference for much of the material discussed in this chapter is

- *Computers and Intractability: A Guide to the Theory of NP-Completeness*, by Garey and Johnson [25].

Another book that covers various topics in this chapter in considerable depth is

- *Computational Complexity*, by Papadimitriou [50].

For an advanced treatment of approximation algorithms, see

- *Approximation Algorithms*, by Vazirani [61].

Algorithm 9.4 (which gives a polynomial-time algorithm to solve **2-CNF-SAT**) is from [2].

Cook's 1971 paper introducing the concept and theory of NP-Completeness is [14]. The summary of the paper states:

> It is shown that any recognition problem solved by a polynomial time-bounded nondeterministic Turing machine can be "reduced" to the problem of determining whether a given propositional formula is a tautology.

In an influential 1972 paper, Karp [38] used polynomial transformations to provide a list of 21 NP-complete problems. In the abstract of this paper, he stated:

> We show that a large number of classic unsolved problems of covering, matching, packing, routing, assignment and sequencing are equivalent, in the sense that either each of them possesses a polynomial-bounded algorithm or none of them does.

We described a simple 2-approximation algorithm for **Metric TSP** in Section

9.8.2. There is a slightly more complicated 1.5-approximation algorithm, which was discovered independently by Christofides and Serdyukov in 1976.

In Section 9.8.3, we presented a randomized algorithm to find a truth assignment for an instance of **MAX-3-SAT** such at least $7/8$ of the clauses are satisfied. The algorithm runs in expected polynomial time. The so-called *method of conditional expectation* can be used to create a modified algorithm, with the same guarantees, that has worst-case polynomial-time complexity. This method is described in [61, Chapter 16].

Turing proved that **Halting** is undecidable in a 1937 paper [60]. One year earlier, Church [13] had proven a similar result in a slightly different setting.

Exercises

9.1 Complete the proof that **Clique** $\leq_P$ **Vertex Cover** (based on the transformation that was described in Section 9.4.1) by showing that I is a yes-instance of **Clique** whenever $f(I)$ is a yes-instance of **Vertex Cover**.

9.2 Draw the boolean circuit that will evaluate the CNF formula given in Example 9.2.

9.3 Prove that Algorithm 9.4 is correct.

9.4 Prove that **Subset Sum** $\leq_P$ **Partition**, based on the transformation that was described in Section 9.6.6.

9.5 A ***Hamiltonian path*** in a graph G is a simple path of length $n - 1$, where n is the number of vertices in G. Problem 9.22 asks if a graph G contains a Hamiltonian path.

> **Problem 9.22: Hamiltonian Path**
>
> **Instance:** An undirected graph $G = (V, E)$.
>
> **Question:** Does G contain a Hamiltonian path?

We describe a polynomial transformation **Hamiltonian Cycle** $\leq_P$ **Hamiltonian Path**. Suppose $G = (V, E)$ is a graph (an instance of **Hamiltonian Cycle**). Construct a graph $H = f(G)$ (an instance of **Hamiltonian Path**) as follows:

(1) Choose a vertex $x \in V$ and create a new vertex x'.

(2) For every edge $xy \in E$, create an edge $x'y$. Call this set of new edges E'.

(3) Create two new vertices z and z' and two new edges xz and $x'z'$.

(4) The graph $H = f(G)$ consists of vertices $V \cup \{x', z, z'\}$ and edges $E \cup E' \cup \{xz, x'z'\}$.

(a) Assume that G does not consist of a single edge. Then prove that G contains a Hamiltonian cycle if and only if H contains a Hamiltonian path.

(b) Prove that **Hamiltonian Path** is NP-complete.

9.6 We define a decision problem known as **Dominating Set**.

> **Problem 9.23: Dominating Set**
>
> **Instance:** An undirected graph $G = (V, E)$ and a positive integer $k \leq |V|$.
>
> **Question:** Does there exist a set of k vertices $U \subseteq V$ such that every vertex $v \in V \setminus U$ is adjacent to at least one vertex $u \in U$? (Such a set U is called a ***dominating set*** of size k.)

(a) Prove that **Dominating Set** $\in$ **NP**, by defining a suitable certificate and describing a polynomial-time algorithm to verify a certificate.

(b) We want to prove that **Dominating Set** is NP-complete, using a polynomial transformation from **Vertex Cover** (this problem was defined as Problem 9.10). Here is the transformation: given an instance $I = (G, k)$ of **Vertex Cover** where $G = (V, E)$, we define $f(I) = (H, L)$ (an instance of **Dominating Set**) as follows:

- H is a graph having vertex set V_H and edge set E_H.
- $V_H = X \cup Y$, where

$$X = \{x_v : v \in V\} \quad \text{and} \quad Y = \{y_e : e \in E\}.$$

- $E_H = \{x_u x_v : u, v \in V\} \cup \{x_u y_e : u \in V, u \in e \in E\}$.
- $L = k$.

Your task is to prove that the mapping f is indeed a polynomial transformation.

HINT When you prove that I is a yes-instance of **Vertex Cover** whenever $f(I)$ is a yes-instance of **Dominating Set**, you will begin by assuming that there is a dominating set of a certain size in the constructed graph H. You may find it useful to prove the following lemma: If $U \subseteq V_H$ is a dominating set in H, then there exists a dominating set $U' \subseteq X$ for the graph H such that $|U'| \leq |U|$.

9.7 This question explores the complexity of **Meeting Scheduling**. Specifically, we want to schedule n meetings (denoted $M_1, \ldots, M_n$) for ℓ employees (denoted $E_1, \ldots, E_\ell$) within a period of K possible days, denoted $1, \ldots, K$. Each meeting is scheduled on a particular day, but we assume that none of the meetings scheduled on a given day interfere with each other. For each employee E_i, let $S_i \subseteq \{M_1, \dots, M_n\}$ denote the meetings that E_i is required to attend. We require that no employee has all of their meetings on the same day.

This problem is stated formally as follows.

Problem 9.24: Meeting Scheduling

Instance: A set $\mathcal{M}$ of n elements, ℓ subsets $S_1, \dots, S_\ell \subseteq \mathcal{M}$, and a positive integer $K \geq 2$

Question: Does there exist a mapping $f : \mathcal{M} \rightarrow \{1, \dots, K\}$ such that

$$\{c \in S_i : f(c) = j\} \neq S_i$$

for every $i \in \{1, \dots, \ell\}$ and $j \in \{1, \dots, K\}$?

For example, suppose we have

$$\begin{aligned} S_1 &= \{M_1, M_2, M_3\} \\ S_2 &= \{M_1, M_2, M_4\} \\ S_3 &= \{M_2, M_3, M_5\} \\ S_4 &= \{M_3, M_4, M_5\} \end{aligned}$$

and $K = 2$. Then we can schedule the meetings as follows:

$$\begin{aligned} &f(M_1) = f(M_2) = f(M_5) = 1 \\ &f(M_3) = f(M_4) = 2. \end{aligned}$$

Prove that **Meeting Scheduling** is NP-complete via a polynomial transformation from **3-CNF-SAT**.

HINT It suffices to take $K = 2$. For each clause in a given instance of **3-CNF-SAT**, construct a subset involving the three literals in the clause and one additional new element. You may want to construct additional subsets of size two, as well.

9.8 We define the following set covering problem.

Problem 9.25: Set Cover

Instance: A finite set X, a set $\mathcal{B}$ of subsets of X, and a positive integer L.

Question: Do there exist L sets, in the collection $\mathcal{B}$, whose union is equal to X?

As an example, consider the instance consisting of $X = \{1,\ldots,6\}$, $\mathcal{B} = \{B_1, B_2, B_3, B_4\}$, where

$$B_1 = \{1,2,3,4\}, B_2 = \{1,3,5\}, B_3 = \{3,4,6\}, B_4 = \{2,5,6\},$$

and $L = 2$. This is a yes-instance, as can be seen by choosing B_1 and B_4.

(a) Prove that **Set Cover** $\in$ **NP**.

(b) Prove that **Set Cover** is NP-complete by finding a polynomial transformation from the known NP-complete problem **Vertex Cover** (Problem 9.10).

HINT Let $(G = (V, E), K)$ be an instance of **Vertex Cover**. Construct an appropriate instance $f(I) = (X, \mathcal{B}, L)$ of **Set Cover**, where $X = E$. You should also construct a certain set $B_v \subseteq X$ for every $v \in V$.

9.9 Consider the following decision version of Problem 9.19.

Problem 9.26: MAX-SAT-Dec

Instance: A boolean formula F in conjunctive normal form, having n boolean variables, denoted $x_1, \ldots, x_n$, and m clauses, denoted $C_1, \ldots, C_m$, and a positive integer $k \leq m$.

Question: Does there exist a truth assignment such that at least k clauses contain at least one true literal?

(a) Prove that **MAX-SAT-Dec** is NP-complete via a polynomial transformation from **CNF-SAT**.

(b) Prove that **MAX-SAT** $\leq_P^T$ **MAX-SAT-Dec**.

HINT First determine the maximum number of clauses that can be satisfied. Then, find a truth assignment that satisfies the required number of clauses by considering each boolean variable x_i in turn, determining if $x_i = \textit{true}$ or $x_i = \textit{false}$.

Bibliography

[1] G.M. Adel'son-Vel'skii and E.M. Landis. An algorithm for organization of information, *Dokl. Akad. Nauk SSSR* **146** (1962), 263–266.

[2] B. Aspvall, M.P. Plass and R.E. Tarjan. A linear-time algorithm for testing the truth of certain quantified boolean formulas. *Information Processing Letters* **8** (1979) 121–123.

[3] R. Bayer. Symmetric binary B-Trees: Data structure and maintenance algorithms. *Acta Informatica* **1** (1972) 290–306.

[4] R. Bellman. The theory of dynamic programming. *Bulletin of the American Mathematical Society* **60** (1954), 503–515.

[5] J.L. Bentley, D. Haken and J.B. Saxe. A general method for solving divide-and-conquer recurrences. *SIGACT News*, Fall 1980, 36–44.

[6] J.L. Bentley and M.I. Shamos. Divide-and-conquer in multidimensional space. In *Proceedings of the Eighth Annual ACM Symposium on Theory of Computing*, 1976, pp. 220–230.

[7] L. Bergroth, H. Hakonen and T. Raita. A survey of longest common subsequence algorithms. In *Proceedings of the Seventh International Symposium on String Processing and Information Retrieval*, 2000, pp 39–48.

[8] J.R. Bitner and E.M. Reingold. Backtrack programming techniques. *Communications of the ACM*, **18** (1975), 651–656.

[9] M. Blum, R.W. Floyd, V.R. Pratt, R.L. Rivest and R.E. Tarjan. Time bounds for selection. *Journal of Computer and System Sciences* **7** (1973), 448–461.

[10] A. Bondy and U.S.R. Murty. *Graph Theory*. Springer, 2007.

[11] G. Brassard and P. Bratley. *Fundamental of Algorithmics*. Prentice-Hall, 1996.

[12] T.M. Chan. More logarithmic-factor speedups for 3SUM, (median,+)-convolution, and some geometric 3SUM-hard problems. In *Proceedings of the 2018 Annual ACM-SIAM Symposium on Discrete Algorithms*, 2018, pp. 881–897.

[13] A. Church. An unsolvable problem of elementary number theory. *American Journal of Mathematics* **58** (1936), 345–363.

[14] S.A. COOK. The complexity of theorem-proving procedures. In *Proceedings of the Third Annual ACM Symposium on Theory of Computing*, 1971, pp. 151–158.

[15] T.H. CORMEN, C.E. LEISERSON AND R.L. RIVEST. *Introduction to Algorithms.* MIT Press, 1990.

[16] T.H. CORMEN, C.E. LEISERSON, R.L. RIVEST AND C. STEIN. *Introduction to Algorithms. Third Edition.* MIT Press, 2009.

[17] G.B. DANTZIG. Discrete-variable extremum problems. *Operations Research* **5** (1957), 266–277.

[18] S. DASGUPTA, C.H. PAPADIMITRIOU AND U. VAZIRANI. *Algorithms.* McGraw-Hill, 2006.

[19] S. DREYFUS. Richard Bellman on the birth of dynamic programming. *Operations Research* **50** (2002), 48–51.

[20] R.W. FLOYD. Algorithm 245. Treesort 3. *Communications of the ACM*, **7** (1964), 701.

[21] K.D. FRYER. Vector spaces and matrices. In *Topics in Modern Mathematics*, Prentice Hall, 1964, pp. 65–83.

[22] D. GALE AND L.S. SHAPLEY. College admissions and the stability of marriage. *American Mathematical Monthly* **69** (1962), 9–14.

[23] Z. GALIL AND G.F. ITALIANO. Data structures and algorithms for disjoint set union problems. *ACM Computing Surveys* **23** (1991), 319–344.

[24] B.A. GALLER AND M.J. FISHER. An improved equivalence algorithm. *Communications of the ACM* **7** (1964), 301–303.

[25] M. GAREY AND D.S. JOHNSON. *Computers and Intractability: A Guide to the Theory of NP-Completeness.* W.H. Freeman and Company, 1979.

[26] S.W. GOLOMB AND L.D. BAUMERT. Backtrack programming. *Journal of the ACM* **12** (1965), 516–524.

[27] M.T. GOODRICH AND R. TAMASSIA. *Algorithm Design and Applications*, Wiley, 2014.

[28] R.L. GRAHAM. Bounds on multiprocessing timing anomalies. *SIAM Journal on Applied Mathematics* **17** (1969) 416–429.

[29] A. GRØNLUND AND S. PETTIE. Threesomes, degenerates, and love triangles. *Journal of the ACM*, April 2018, Article No. 22.

[30] L.J. GUIBAS AND R. SEDGEWICK. A dichromatic framework for balanced trees. In *19th Annual Symposium on Foundations of Computer Science*, 1978, pp. 8–21.

[31] D. GUSFIELD AND R.W. IRVING. *The Stable Marriage Problem: Structure and Algorithms*. MIT Press, 2003.

[32] D. HARVEY AND J. VAN DER HOEVEN. Integer multiplication in time $O(n \log n)$, HAL archives, HAL Id: hal-02070778, April 2019.

[33] C.A.R. HOARE. Algorithm 64. Quicksort. *Communications of the ACM*, **4** (1961), 321.

[34] C.A.R. HOARE. Algorithm 65. Find. *Communications of the ACM*, **4** (1961), 321–322.

[35] R.R. HOWELL. On asymptotic notation with multiple variables. Technical Report 2007-4, January 18, 2008, Dept. of Computing and Information Sciences, Kansas State University.

[36] A.B. KAHN. Topological sorting of large networks. *Communications of the ACM* **5** (1962), 558–562.

[37] A. KARATSUBA AND YU. OFMAN. Multiplication of many-digital numbers by automatic computers. *Proceedings of the USSR Academy of Sciences* **145** (1962), 293–294.

[38] R.M. KARP. Reducibility among combinatorial problems. In *Complexity of Computer Computations*, pp. 85–103. Plenum Press, 1972.

[39] J. KLEINBERG AND É. TARDOS. *Algorithm Design*. Pearson, 2005.

[40] G.T. KLINCSEK. Minimal triangulations of polygonal domains. *Annals of Discrete Mathematics* **9** (1980), 121–123.

[41] D.E. KNUTH. Notes on open addressing, 1963.

[42] D.E. KNUTH. *The Art of Computer Programming. Volume 1: Fundamental Algorithms, Third Edition*. Addison-Wesley, 1997.

[43] W. KOCAY AND D.L. KREHER. *Graphs, Algorithms, and Optimization. Second Edition*. Chapman & Hall/CRC, 2016.

[44] D.L. KREHER AND D.R. STINSON. *Combinatorial Algorithms: Generation, Enumeration and Search*. CRC Press, 1999.

[45] H.T. KUNG, F. LUCCIO AND F.P. PREPARATA. On finding the maxima of a set of vectors. *Journal of the ACM* **22** (1975), 469–476.

[46] F. LE GALL. Powers of tensors and fast matrix multiplication. In *Proceedings of the 39th International Symposium on Symbolic and Algebraic Computation*, 2014, pp. 296–303.

[47] A.H. LAND AND A.G. DOIG. An automatic method of solving discrete programming problems. *Econometrica* **28** (1960), 497–520.

[48] J.D.C. LITTLE, K.G. MURTY, D.W. SWEENEY AND C. KAREL. An algorithm for the traveling salesman problem. *Operations Research* **11** (1963), 972–989.

[49] D. MICHIE. Memo Functions and Machine Learning. *Nature* **218** (1968), 19–22.

[50] C.H. PAPADIMITRIOU. *Computational Complexity*. Addison-Wesley, 1994.

[51] I. PAPARRIZOS. A tight bound on the worst-case number of comparisons for Floyd's heap construction algorithm. In *Optimization Theory, Decision Making, and Operations Research Applications*, 2013, pp. 153–162. (Springer Proceedings in Mathematics & Statistics, vol. 31, Springer.)

[52] W.W. PETERSON. Addressing for random-access storage. *IBM Journal of Research and Development* **1** (1957), 130–146.

[53] A. SCHRIJVER. On the history of the shortest path problem. *Documenta Math.* Extra Volume: Optimization Stories (2012), 155–167.

[54] R. SEDGEWICK AND K. WAYNE. *Algorithms. Fourth Edition.* Addison-Wesley, 2011.

[55] M. SHARIR. A strong-connectivity algorithm and its applications in data flow analysis. *Computers and Mathematics with Applications* **7** (1981), 67–72.

[56] V. STRASSEN. Gaussian elimination is not optimal. *Numer. Math.* **13** (1969), 354–356.

[57] M.A. SUCHENEK. Elementary yet precise worst-case analysis of Floyd's heap-construction program. *Fundamenta Informaticae* **120** (2012) 75–92.

[58] R.E. TARJAN. Depth-first search and linear graph algorithms. *SIAM J. Computing* **1** (1972), 146–160.

[59] R.E. TARJAN. Edge-disjoint spanning trees and depth-first search. *Acta Informatica* **6** (1976), 171–185.

[60] A.M. TURING. On computable numbers, with an application to the Entscheidungsproblem. *Proceedings of the London Mathematical Society* **52** (1937), 230–265

[61] V.V. VAZIRANI. *Approximation Algorithms*. Springer, 2003.

[62] J.W.J. WILLIAMS. Algorithm 232. Heapsort. *Communications of the ACM*, **7** (1964), 347–348.

Index